Studies in Systems, Decision and Control

Volume 439

The series "Studies in Systems, Decision and Control" (SSDC) covers both new developments and advances, as well as the state of the art, in the various areas of broadly perceived systems, decision making and control–quickly, up to date and with a high quality. The intent is to cover the theory, applications, and perspectives on the state of the art and future developments relevant to systems, decision making, control, complex processes and related areas, as embedded in the fields of engineering, computer science, physics, economics, social and life sciences, as well as the paradigms and methodologies behind them. The series contains monographs, textbooks, lecture notes and edited volumes in systems, decision making and control spanning the areas of Cyber-Physical Systems, Autonomous Systems, Sensor Networks, Control Systems, Energy Systems, Automotive Systems, Biological Systems, Vehicular Networking and Connected Vehicles, Aerospace Systems, Automation, Manufacturing, Smart Grids, Nonlinear Systems, Power Systems, Robotics, Social Systems, Economic Systems and other. Of particular value to both the contributors and the readership are the short publication timeframe and the world-wide distribution and exposure which enable both a wide and rapid dissemination of research output.

Indexed by SCOPUS, DBLP, WTI Frankfurt eG, zbMATH, SCImago.

All books published in the series are submitted for consideration in Web of Science.

Volodymyr Eremenko · Artur Zaporozhets
Editors

Advanced Information-Measuring Technologies and Systems I

Springer

Editors
Volodymyr Eremenko
Department of Information-Measuring Technologies, Igor Sikorsky Kyiv Polytechnic Institute
National Technical University of Ukrain
Kyiv, Ukraine

Artur Zaporozhets
General Energy Institute
National Academy of Sciences of Ukraine
Kyiv, Ukraine

Green Technology Research Center
Yuan Ze University
Taoyuan, Taiwan

ISSN 2198-4182 ISSN 2198-4190 (electronic)
Studies in Systems, Decision and Control
ISBN 978-3-031-40717-8 ISBN 978-3-031-40718-5 (eBook)
https://doi.org/10.1007/978-3-031-40718-5

This Springer imprint is published by the registered company Springer Nature Switzerland AG
The registered company address is: Gewerbestrasse 11, 6330 Cham, Switzerland

Preface

The book presents the main scientific directions and issues of research conducted at the Department of Information and Measurement Technologies of the National Technical University of Ukraine "Ihor Sikorskyi Kyiv Polytechnic Institute". The presented results cover almost all scientific directions related to information and measurement technologies—metrological support of measurement channels of information and measurement systems, methods of reproducing units of electric circuit parameters, development of specialized information and measurement systems, mathematical methods of processing measurement information, models of formation of information signals and fields, statistical diagnostic methods, information support of testing and calibration laboratories.

The book consists of 7 chapters.

Chapter "Metrological Support of Measurement Channels with Bridge Circuits" presents the design of the simulator of the sensor output voltage, which has the advantages of improved accuracy of the measuring channels calibration. Analysis of the conversion equation was carried out and the method of error correction in bridge circuits of measuring channels was investigated.

Chapter "Application of Exponential Splines in the Measurement and Control of Electric Circuit Parameters" discusses the practical aspects of using exponential splines in the tasks of measuring and controlling the parameters of electric circuits for the synthesis of test signals of a special form. The advantage of exponential splines is the ease of generation in linear electrical circuits. The formation of test signals and examples of control of the parameters of electric circuits are presented. The influence of the type of approximating functions on the quality of reproduction using the feedback of information about the parameters of the circle and, accordingly, on the error of their determination was analyzed.

Chapter "Improving of Methods of Impedance Parameters Units Reproduction and Measurement Accuracy Increasing for Ensuring Metrological Traceability" presents a structural-algorithmic method of increasing the accuracy of the calibration of measures of electrical impedance units, as well as a developed calibration method and a method of assessing the uncertainty of measurements when calibrating measures of electrical capacity and inductance of precision LCR-meters. General approaches to calibration are given, measurement equations (models) are presented, as well as examples of calculation of uncertainty budgets during calibration.

Chapter "Implementation of Information and Measurement Systems at the Base of Specialized Internet Protocols" gives the organization of distributed microcontroller systems for collecting measurement information with remote Internet access to microserver networks of virtual intelligent sensors based on specialized web protocols. The research method is theoretical and experimental based on the analysis and synthesis of hardware and software of microcontroller network systems for remote collection of experimental data. Structures of intelligent sensors for studying the composition of substances have been developed. A basic set of hardware and software tools for microcontroller modules of intelligent sensors has been developed.

Chapter "Model of Information Signals Formation in the Diagnostics of Composite Products" presents the structure of information processes for diagnosing products made of composites, which made it possible to analyze the formation of the information field, which became the basis for the development of a mathematical model of the generating information-signal field, which takes into account the main mechanisms of the formation of mechanical disturbances in the composite, in the form of a random Hilbert linear field model. The model made it possible to propose a new method of standard-free diagnosis of products made of composite materials, methods of processing measurement information, definition and construction of vectors of diagnostic signs, and decision-making rules.

Chapter "Theory and Practice of Ensuring the Validity in Testing Laboratories" shows the necessity of building a systematic model of ensuring the validity to achieve a balance between the customer's expectations regarding quality and the laboratories' quality statement and quality concept. The standards that allow applying qualitative and quantitative methods of validity assurance are considered. The algorithm of actions to improve the reliability of test results depending on the identified risks, available financial and human resources, and adopted quality goals is shown. Also, Chapter "Theory and Practice of Ensuring the Validity in Testing Laboratories" discusses the features of the organization of ensuring the validity in testing laboratories. The processes in the laboratory were analyzed in terms of the requirements of the ISO 17025 standard and the developed system model of ensuring the reliability. Examples of practical assurance measures for processes are given: equipment and personnel management facilities, deviation from the contract, externally provided products, LIMS, impartiality, sampling, risk management, managements reviews, and improvement. The role of method validation to ensure a validity is defined. The uncertainty of the result is considered as one of the options for generalizing the quantitative indicator of validity. The algorithm of actions to improve the reliability of test results depending on the identified risks, available financial and human resources, and adopted quality goals is shown.

Chapter "Methodology for Controlling Greenhouse Microclimate Parameters and Yield Forecast Using Neural Network Technologies" presents structural schema of the greenhouse control and monitoring system. The creation of a methodology using neural network technologies is intended to control the parameters of the greenhouse microclimate, which should have a positive influence on the quality of the harvest and increase yields. The optimal greenhouse microclimate conditions were also analyzed and determined: temperature, humidity, and carbon dioxide concentration, according to which it is possible to predict the yield. Based on the mathematical model, a program was designed to train the neural network. For an accurate forecast, a neural network was developed that is based on a multilayer perceptron with three hidden layers.

The authors will be grateful to all readers who will send feedback, comments, and suggestions on the material presented in the book.

Kyiv, Ukraine
March 2023

Volodymyr Eremenko
Artur Zaporozhets

Contents

Metrological Support of Measurement Channels with Bridge Circuits

Yulian Tuz, Bogdan Kokotenko, and Yuriy Samartsev

Abstract Many resistive sensors are based on the bridge circuit. It enables accurate measurements of the small resistance changes, which is essential for weight, pressure, temperature, and other types of sensors. Achieving the most accurate measurements by bridge circuits-based sensors requires calibration and certification of the measuring channels. This paper presents the design of the simulator of the sensor output voltage, which has the advantage of the improved accuracy of the measuring channels calibration. The equivalent circuit of such sensor is described. The influence of the source voltage errors is considered. The error correction of the bridge circuit-based sensor output is described. The conversion equation of the measuring channel is analyzed.

Keywords Bridge circuit · Resistive sensor simulator · Resistance measurement · Measuring channels · Error correction · Conversion equation

1 Bridge Circuit-Based Sensor Simulator

The measuring channel of a non-electrical physical quantity usually consists of the serial connection of a transducer (sensor), that transforms a non-electrical physical quantity into an electrical signal, a signal conditioning circuit [1], and an analog-to-digital converter [2].

Many sensors imply the transformation of a non-electrical physical quantity into a resistance change. As far as changes in a resistance usually are not big comparing with the nominal sensor resistance, the bridge circuits are used for the resistance change transformation [3].

The bridge circuits can operate in full or partially balanced mode.

Y. Tuz · B. Kokotenko (✉) · Y. Samartsev
National Technical University of Ukraine "Igor Sikorsky Kyiv Polytechnic Institute", Kyiv, Ukraine
e-mail: kokotenko.bogdan@gmail.com

V. Eremenko and A. Zaporozhets (eds.), *Advanced Information-Measuring Technologies and Systems I*, Studies in Systems, Decision and Control 439,
https://doi.org/10.1007/978-3-031-40718-5_1

In the case of partial balancing, the output voltage ΔU of the symmetrical bridge with equal resistors R and the resistance change of a single element in one branch is given by (1):

$$\Delta U = \frac{U\Delta R}{4R} \cdot \frac{1}{1 + \frac{\Delta R}{2R}}, \tag{1}$$

where ΔR—the resistance changes of a single element, R—the nominal resistance value, U—the bridge supply voltage.

In the case of the small relative resistance changes $\Delta R/R \ll 1$, the relative changes of the output voltage are proportional to the relative resistance change $\Delta R/R$.

The absolute resistance change value could be calculated by the relative output voltage change:

$$\Delta R = 4R\frac{\Delta U}{U} \cdot \frac{1}{1 - 2\frac{\Delta U}{U}}. \tag{2}$$

The best way of the metrological certification of the measuring channel implies calculations of the dependency between a measuring channel output voltage and a multi-valued measure of a physical quantity.

The element-by-element certification method of measuring channel parts is widely used in the metrological practice. Namely, the sensor certification by the physical quantity measure and the electrical part of the measuring channel with a further evaluation of the cumulative error.

To certify the electrical part of the measuring channel, it is necessary to create a simulator of the sensor output voltage with the closest parameters to the real sensor.

The simulator must produce the same voltage, have an adequate output impedance, a similar unbalanced or a symmetrical connection scheme, and the same nominal voltage levels relative to the common point at a symmetrical output.

If the real sensor is based on the bridge circuit, then it makes sense to create a simulator using the bridge circuit as well.

It is recommended to create a simulator of a bridge circuit in such a way that the output voltage is not created by the resistance change of a single or more arms of the bridge, but by including in a series with the resistance of a one arm of the bridge circuit an additional adjustable voltage, that will have the function of a test value, that is equivalent to a measure of a non-electrical physical quantity [4].

Since the real voltage sources have a non-zero output resistance, their presence creates an additional imbalance in the bridge. To eliminate the additional imbalance, it is necessary to include the same resistance in the adjacent bridge arm.

Using a digital-to-analog converter (DAC) is the most appropriate way to create a test voltage. In case of low levels of the output voltage, it is recommended to supply the test voltage from the DAC through a voltage divider using the maximum number of DAC bits to create it [5].

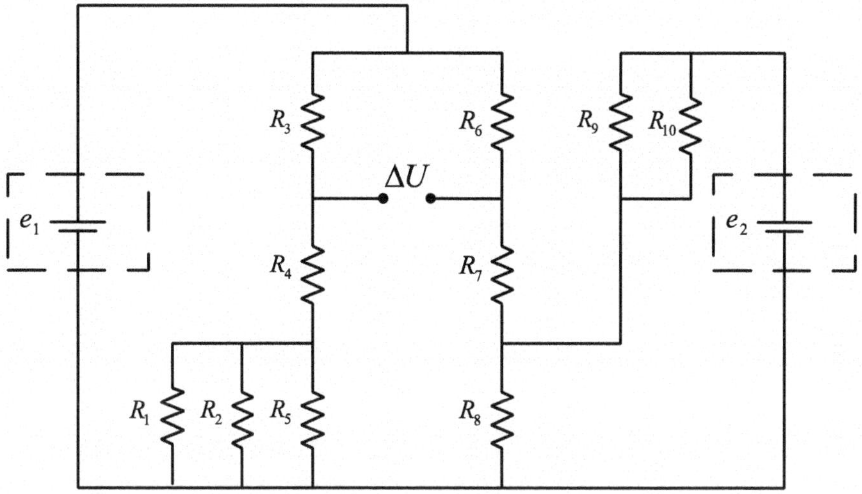

Fig. 1 The sensor simulator schematic diagram with the bridge circuit

The output voltage of the symmetrically balanced bridge is zero and does not depend on its supply voltage. And it equals half of the test voltage connected in series with the resistance of a one arm if the output resistance of the test voltage source is zero.

This property of the bridge circuit makes it possible to create a sensor simulator with the most adequate parameters, if the supply voltage of the bridge, the output resistance, and the output voltage changes are the same as in a real sensor. The output voltage meter must have a differential input.

Figure 1 offers a schematic diagram of the possible variant of the sensor simulator. The parallel connection of resistors with the same resistance is due to consideration of the unification and the greatest probability of identity of their properties.

Figure 2 shows the equivalent circuit of the simulator, which takes into account the resistances of the voltage sources e_1 and e_2.

The equivalent circuit in Fig. 2 has the following designations:

r_1, r_8—the output resistance of the voltage source e_1 and e_2 respectively;

e_1—the EMF of the power source of the bridge circuit;

r_6—the upper arm of the voltage divider;

r_7—the lower arm of the voltage divider where the voltage is isolated, creating the output voltage ΔU;

r_2,r_3,$r_4 r_5$—the four-arm bridge resistors;

z_3—the symmetrical resistance, that value equals the resistance of a parallel connection of the resistor r_7 and a serial connection of the resistors r_6 and r_8;

e_2—the EMF of the digital-to-analog converter that creates the output voltage;

ΔU—the sensor simulator output voltage.

The matrix (3) for the equivalent circuit Fig. 2 could be obtained by using the Kirchhoff's law for voltages and currents:

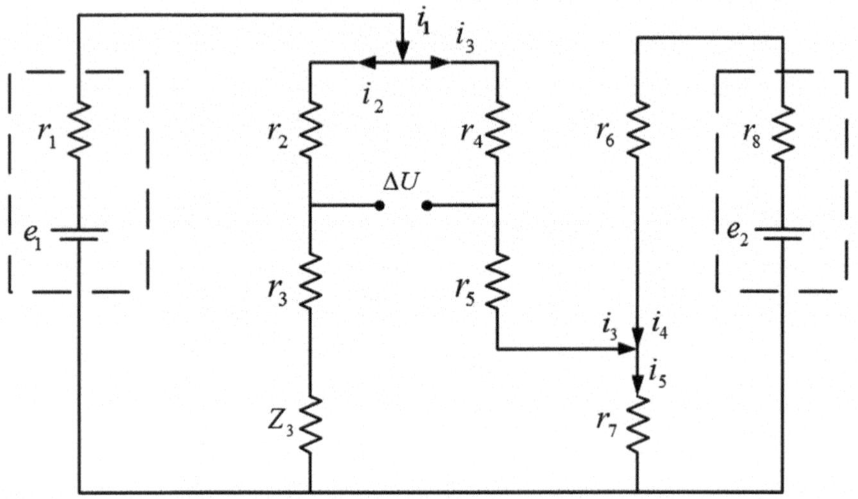

Fig. 2 The sensor simulator equivalent circuit with the bridge circuit

$$\begin{bmatrix} r_1 & z_5 & 0 & 0 & 0 \\ 0 & -z_5 & z_4 & 0 & r_7 \\ 0 & 0 & 0 & -z_6 & -r_7 \\ 1 & -1 & -1 & 0 & 0 \\ 0 & 0 & 1 & 1 & -1 \end{bmatrix} \times \begin{bmatrix} i_1 \\ i_2 \\ i_3 \\ i_4 \\ i_5 \end{bmatrix} = \begin{bmatrix} e_1 \\ 0 \\ -e_2 \\ 0 \\ 0 \end{bmatrix}, \quad (3)$$

where:

r_1—the output resistance of the voltage source e_1;
$z_5 = r_2 + r_3 + z_3$;
$z_3 = \frac{r_7 \cdot (r_6 + r_8)}{r_7 + r_6 + r_8}$;
$z_4 = r_4 + r_5$;
$z_6 = r_6 + r_8$;
r_6—the upper arm of the voltage divider;
r_7—the lower arm of the voltage divider;
$r_2 = r_3 = r_4 = r_5$—the resistance of the main bridge circuit.
The output voltage is:

$$\Delta U = i_2 r_2 - i_3 r_4. \quad (4)$$

According to the Crammer's rule [6] for (4) the current i_2 and i_3 could be determined by:

$$i_2 = \det 2 / \det C; \quad (5)$$

$$i_3 = \det 3 / \det C, \quad (6)$$

where:

det C—the determinant of the system (3);

det 2—the determinant of the system, where the second column is replaced by the right side of the system;

det 3—the determinant of the system, where the third column is replaced by the right side of the system.

$$\det 2 = \begin{vmatrix} r_1 & e_1 & 0 & 0 & 0 \\ 0 & 0 & z_4 & 0 & r_7 \\ 0 & -e_2 & 0 & -z_6 & -r_7 \\ 1 & 0 & -1 & 0 & 0 \\ 0 & 0 & 1 & 1 & -1 \end{vmatrix} = -e_2 r_1 r_7 - e_1 r_7 z_4 - e_1 r_7 z_6 - e_1 z_4 z_6; \quad (7)$$

$$\det 3 = \begin{vmatrix} r_1 & z_5 & e_1 & 0 & 0 \\ 0 & -z_5 & 0 & 0 & r_7 \\ 0 & 0 & -e_2 & -z_6 & -r_7 \\ 1 & -1 & 0 & 0 & 0 \\ 0 & 0 & 0 & 1 & -1 \end{vmatrix} = e_2 r_1 r_7 - e_1 r_7 z_5 + e_2 r_7 z_5 - e_1 z_5 z_6; \quad (8)$$

$$\det C = \begin{vmatrix} r_1 & z_5 & 0 & 0 & 0 \\ 0 & -z_5 & z_4 & 0 & r_7 \\ 0 & 0 & 0 & -z_6 & -r_7 \\ 1 & -1 & -1 & 0 & 0 \\ 0 & 0 & 1 & 1 & -1 \end{vmatrix} = -r_1 r_7 z_4 - r_1 r_7 z_5 - r_1 r_7 z_6 - r_1 z_4 z_6 - r_1 z_5 z_6$$
$$- r_7 z_4 z_5 - r_7 z_5 z_6 - z_4 z_5 z_6. \quad (9)$$

After substituting the (5)–(9) into (4), the dependency between output voltage ΔU on EMF e_1 and e_2 could be calculated by:

$$\Delta U = \frac{1}{\det C}(r_2 \cdot (-e_2 r_1 r_7 - e_1 r_7 z_4 - e_1 r_7 z_6 - e_1 z_4 z_6) - \\ -r_4 \cdot (e_2 r_1 r_7 - e_1 r_7 z_5 + e_2 r_7 z_5 - e_1 z_5 z_6)). \quad (10)$$

If a respective EMF of the output voltage ΔU is set for a sensor simulator, then a required EMF value of the digital-to-analog converter could be determined by (11):

$$e_2 = -\frac{\det C \cdot \Delta U + e_1(r_2 r_7 z_4 + r_2 r_7 z_6 - r_4 r_7 z_5 + r_2 z_4 z_6 - r_4 z_5 z_6)}{r_2 \cdot (r_1 r_2 + r_1 r_4 + r_4 z_5)}. \quad (11)$$

Extracting the multiplicative proportional ΔU and the additive proportional e_1 parts of $e_2(\Delta U)$ results in:

$$e_2 = -\frac{\det C \cdot \Delta U}{r_7(r_1r_2 + r_1r_4 + r_4z_5)} - \\ -e_1\frac{r_2r_1z_4 + r_2r_7z_6 - r_4r_7z_5 + r_2z_4z_6 - r_4z_5z_6}{r_7(r_1r_2 + r_1r_4 + r_4z_5)}. \quad (12)$$

if $r_2 = r_4$.

The output voltage is:

$$\Delta U = -\frac{r_4}{\det C} \cdot (2e_2r_1r_7 + e_1r_7z_4 - e_1r_7z_5 + e_1r_7z_6 + e_2r_7z_5 + e_1z_4z_6 - e_1z_5z_6). \quad (13)$$

The EMF of the DAC e_2 in case of the output voltage ΔU is

$$e_2 = -\frac{\det C \cdot \Delta U + e_1r_4 \cdot (r_7z_4 - r_7z_5 + r_7z_6 + z_4z_6 - z_5z_6)}{r_4r_7 \cdot (2r_1 + z_5)}. \quad (14)$$

Taking into account that DAC produces a maximum voltage in the (±2.5; ± 5; ± 10) V range, to reproduce the output voltage of the sensor simulator of the bridge circuit type in the millivolt range (1 ÷ 100) mV, it makes sense to apply the EMF e_2 to the bridge circuit via a corresponding voltage divider (on the equivalent circuit Fig. 2, the resistors r_6, r_7) to utilize the maximum number of digits of the DAC.

For further analysis, the (14) could be converted into the combination of the multiplicative and additive parts:

$$e_2 = -\frac{\det C}{r_4r_7 \cdot (2r_1 + z_5)} \cdot \Delta U - e_1 \cdot \frac{r_7z_4 - r_7z_5 + r_7z_6 + z_4z_6 - z_5z_6}{r_7 \cdot (2r_1 + z_5)}. \quad (15)$$

In the case of the values of the following resistors:

$$\left.\begin{aligned} r_7 : r_6 &= 1 : 50 \\ r_1 &= 2 \\ r_2 = r_4 &= 4700 \\ r_7 &= 47 \\ z_4 &= 9400 \\ z_5 &= 9446.07919967 \\ z_6 &= 2352 \end{aligned}\right\} \quad (16)$$

the following calculation could be done:

- the determinant of the system (3) is

$$\det C = \begin{vmatrix} r_1 & z_5 & 0 & 0 & 0 \\ 0 & -z_5 & z_4 & 0 & r_7 \\ 0 & 0 & 0 & -z_6 & -r_7 \\ 1 & -1 & -1 & 0 & 0 \\ 0 & 0 & 1 & 1 & -1 \end{vmatrix} = -2.14149605555 12663914 \cdot 10^{11}; \tag{17}$$

- the multiplicative part coefficient is

$$-\frac{\det C}{r_4 r_7 \cdot (2r_1 + z_5)} = +102.5855319149; \tag{18}$$

- the additive part coefficient is

$$-\frac{r_7 z_4 - r_7 z_5 + r_7 z_6 + z_4 z_6 - z_5 z_6}{r_7 \cdot (2r_1 + z_5)} = +1.8762590467 \cdot 10^{-11}; \tag{19}$$

- the numerical equation of e_2 depending on ΔU is:

$$e_2 = 102.5855319149 \cdot \Delta U + e_1 \cdot 1.8762590467 \cdot 10^{-11} \tag{20}$$

For resistors value r_7 ten times less than in (16):

$$\left.\begin{aligned} r_7 : r_6 &= 1 : 500 \\ r_1 &= 2 \\ r_2 = r_4 &= 4700 \\ r_7 &= 4.7 \\ z_4 &= 9400 \\ z_5 &= 9404.690626724 \\ z_6 &= 2352 \end{aligned}\right\}; \tag{21}$$

the following calculation could be done.

- the determinant of the system (3) is

$$\det C = \begin{vmatrix} r_1 & z_5 & 0 & 0 & 0 \\ 0 & -z_5 & z_4 & 0 & r_7 \\ 0 & 0 & 0 & -z_6 & -r_7 \\ 1 & -1 & -1 & 0 & 0 \\ 0 & 0 & 1 & 1 & -1 \end{vmatrix} = -2.0853454270966829621 \cdot 10^{11}; \tag{22}$$

- the multiplicative coefficient part is

$$-\frac{\det C}{r_4 r_7 \cdot (2r_1 + z_5)} = +1.0033514894 \cdot 10^{3}; \tag{23}$$

- the additive coefficient part is

$$\frac{r_7 z_4 - r_7 z_5 + r_7 z_6 + z_4 z_6 - z_5 z_6}{r_7(2r_1 + z_5)} = +1.0109142561 \cdot 10^{-11}; \tag{24}$$

- the numerical equation for e_2 calculation depending on ΔU results in

$$e_2 = 1.0033514894 \cdot 10^3 \cdot \Delta U + e_1 \cdot 1.0109142561 \cdot 10^{-11}. \tag{25}$$

As it is shown in (20) and (25), the additive part of the EMF e_2 is neglectable comparing with the multiplicative part and is a calculation error, because in the symmetrical bridge circuit (in case of no imbalance) the output voltage theoretically does not depend on the supply voltage of the bridge circuit. Minimization of the output voltage dependency on the supply voltage of the sensor simulator is one of the main advantages of the described simulator circuit comparing with the bridge circuit, where the output voltage is produced by resistance change of the single bridge resistor.

The other advantages also are:

1. The symmetrical (differential) output signal is produced by EMF sources, which have a common reference point.
2. The output voltage dependency ΔU on e_2 is linear.
3. A large number of the DAC bits are utilized to create a small ΔU value by using a voltage divider r_6, r_7. That results in very high resolution.
4. The bridge imbalance caused by unequal values of its resistors could be compensated by the EMF of the test signal source e_2.

2 Error Correction of the Sensor Simulator Output Signal with Bridge Circuit Caused by Difference Between the Real Circuit Resistance and the Calculated Values

This error could be caused either by simulator production or during its operation process due to influence of the external factors and the aging of the circuit elements.

The bridge output voltage change ΔU_r caused by a single resistor change could be calculated by (1), by changing any resistor value or EMF e_1 according to (10).

The change correction ΔU_r, namely reaching zero value ($\Delta U_r = 0$), could be done by changing one of the bridge arms or changing the EMF e_2.

The second option has an advantage, because ΔU for different sensors could be very small. Even in the case of the small values $\Delta r / r$ of the bridge arms, the change ΔU_r could be greater than the necessary value of the output voltage. And the dynamic range of the DAC could be insufficient to create a correction voltage. In that case, it is necessary to change the voltage divider coefficient e_2, or apply the previous bridge balancing by changing the circuit resistance.

To take into account these design features and the operation of the sensor simulator based on the bridge circuit let's find the equation of the output voltage errors, as a function of errors that depends on ΔU_r. Using the (10) and (4)–(9) yields the results in

$$\Delta U = \frac{r_2 \cdot \det 2 - r_4 \cdot \det 3}{\det C}. \tag{26}$$

The Eq. (26) by its structure is a fraction

$$\Delta U_r = \frac{A}{B}. \tag{27}$$

A relative error of the fraction is

$$\delta_{\Delta U} = \frac{\delta A - \delta B}{1 + \delta B}, \tag{28}$$

where δA and δB—are relative errors of the numerator and denominator of the (27), and A is a sum. A relative error of the sum is

$$\delta A = \frac{1}{A}(r_2 \cdot \det 2 \cdot \delta(r_2 \cdot \det 2) - r_4 \cdot \det 3 \cdot \delta(r_4 \cdot \det 3)). \tag{29}$$

And the error of multiplication is

$$\delta(r_2 \cdot \det 2) = \delta r_2 + \delta \det 2 + \delta r_2 \cdot \delta \det 2; \tag{30}$$

$$\delta(r_4 \cdot \det 3) = \delta r_4 + \delta \det 3 + \delta r_4 \cdot \delta \det 3. \tag{31}$$

Considering the Eqs. (28)–(31), the relative error of the determinant could be calculated by

$$\delta \det = \frac{\Delta \det}{\det}, \tag{32}$$

where $\Delta \det$—absolute error of the determinant caused by errors of its elements. The procedure of $\Delta \det$ calculation is described in [7]:

$$\Delta \det = \sum_{k=1}^{n} B^k,$$

where B^k—a sum of the determinants of the k-order error.

In the case of the small errors of the determinant elements, the determinants of the first-order errors are taken into account. Their number equals the determinant order. The absolute error of the first order equals the sum of n determinants, where column elements are replaced by columns of the determinant errors that are taken from the error matrix determinant.

The matrix error determinant of the system determinant is:

$$\det C(\Delta) = \begin{vmatrix} \Delta r_1 & \Delta z_5 & 0 & 0 & 0 \\ 0 & -\Delta z_5 & \Delta z_4 & 0 & \Delta r_7 \\ 0 & 0 & 0 & -\Delta z_6 & -\Delta r_7 \\ 0 & 0 & 0 & 0 & 0 \\ 0 & 0 & 0 & 0 & 0 \end{vmatrix}. \tag{33}$$

Thus, the absolute error of the first order of the system determinant (9) is

$$\det C = \begin{vmatrix} \Delta r_1 & z_5 & 0 & 0 & 0 \\ 0 & -z_5 & z_4 & 0 & r_7 \\ 0 & 0 & 0 & -z_6 & -r_7 \\ 0 & -1 & -1 & 0 & 0 \\ 0 & 0 & 1 & 1 & -1 \end{vmatrix} + \begin{vmatrix} r_1 & \Delta z_5 & 0 & 0 & 0 \\ 0 & -\Delta z_5 & z_4 & 0 & r_7 \\ 0 & 0 & 0 & -z_6 & -r_7 \\ 1 & 0 & -1 & 0 & 0 \\ 0 & 0 & 1 & 1 & -1 \end{vmatrix} +$$

$$+ \begin{vmatrix} r_1 & z_5 & 0 & 0 & 0 \\ 0 & -z_5 & z_4 & 0 & r_7 \\ 0 & 0 & 0 & -z_6 & -r_7 \\ 1 & -1 & -1 & 0 & 0 \\ 0 & 0 & 1 & 1 & -1 \end{vmatrix} + \begin{vmatrix} r_1 & z_5 & 0 & 0 & 0 \\ 0 & -z_5 & z_4 & 0 & r_7 \\ 0 & 0 & 0 & -\Delta z_6 & -r_7 \\ 1 & -1 & -1 & 0 & 0 \\ 0 & 0 & 1 & 0 & -1 \end{vmatrix} + \tag{34}$$

$$+ \begin{vmatrix} r_1 & z_5 & 0 & 0 & 0 \\ 0 & -z_5 & z_4 & 0 & -\Delta r_7 \\ 0 & 0 & 0 & -z_6 & -\Delta r_7 \\ 1 & -1 & -1 & 0 & 0 \\ 0 & 0 & 1 & 1 & 0 \end{vmatrix}.$$

If it is necessary to take into account the errors of all orders and not only the first, it is recommended to use the capabilities of the software for mathematical calculations, for example, Mathcad [8], presenting the elements of the determinants as a sum of the nominal values and its absolute error.

The equation of the determinants det C, det 2, det 3 with elements' nominal values and their absolute errors as a multiplication of the nominal values and their relative errors are presented in (35)–(37). Note that the relative error of the element e_2 in (36) and (37) are not taken into account since it does not influence the output voltage, which is obvious from (25).

The result of the calculation in Mathcad for each of the determinants is an ordered sum, where the first group of terms in the form of products of the nominal values of the non-repeating elements of the determinant is its nominal value (represented in black). The second group of terms in the form of products of the nominal values of non-repeating elements by their relative errors represents the absolute error of the determinant of the first order (represented in blue). The third group of terms in the form of products of the nominal values of non-repeating elements by two non-repeating relative errors of the two elements represents the absolute error of the determinant of the second order (represented in purple). The fourth group of terms in the form of products of non-repeating elements by three non-repeating relative errors of the three elements is the absolute error of the determinant of the third order (represented in green).

Expressions of nominal values and absolute errors of various orders obtained in an analytical form are very productive for further analysis and optimization.

$$\det C = \begin{vmatrix} r_1 + r_1 \cdot \delta r_1 & z_5 + z_5 \cdot \delta z_5 & 0 & 0 & 0 \\ 0 & -z_5 - z_5 \cdot \delta z_5 & z_4 + z_4 \cdot \delta z_4 & 0 & r_7 + r_7 \cdot \delta r_7 \\ 0 & 0 & 0 & -z_6 - z_6 \cdot \delta z_6 & -r_7 - r_7 \cdot \delta r_7 \\ 1 & -1 & -1 & 0 & 0 \\ 0 & 0 & 1 & 1 & -1 \end{vmatrix} =$$

$$\begin{aligned}
&= -r_1 r_7 z_4 - r_1 r_7 z_5 - r_1 r_7 z_6 - r_1 z_4 z_6 - r_1 z_5 z_6 - r_7 z_4 z_5 - r_7 z_5 z_6 - z_4 z_5 z_6 - \\
&-r_1 r_7 \cdot \delta r_1 \cdot z_4 - r_1 r_7 \cdot \delta r_1 \cdot z_5 - r_1 r_7 \cdot \delta r_1 \cdot z_6 - \\
&-r_1 r_7 \cdot \delta r_7 \cdot z_4 - r_1 r_7 \cdot \delta r_7 \cdot z_5 - r_1 r_7 \cdot \delta r_7 \cdot z_6 - r_1 r_7 \cdot \delta z_4 \cdot z_4 - \\
&-r_1 r_7 \cdot \delta z_5 \cdot z_5 - r_1 r_7 \cdot \delta z_6 \cdot z_6 - r_1 \cdot \delta r_1 \cdot z_4 z_6 - r_1 \cdot \delta r_1 \cdot z_5 z_6 - \\
&-r_7 \cdot \delta r_7 \cdot z_4 z_5 - r_7 \cdot \delta r_7 \cdot z_5 z_6 - r_1 \cdot \delta z_4 \cdot z_4 z_6 - r_1 \cdot \delta z_5 \cdot z_5 z_6 - \\
&-r_1 \cdot \delta z_6 \cdot z_4 z_6 - r_1 \cdot \delta z_6 \cdot z_5 z_6 - r_7 \cdot \delta z_4 \cdot z_4 z_5 - r_7 \cdot \delta z_5 \cdot z_4 z_5 - \\
&-r_7 \cdot \delta z_5 \cdot z_5 z_6 - r_7 \cdot \delta z_6 \cdot z_5 z_6 - \delta z_4 \cdot z_4 z_5 z_6 - \delta z_5 \cdot z_4 z_5 z_6 - \\
&-\delta z_6 \cdot z_4 z_5 z_6 - \\
&\quad -r_1 r_7 \cdot \delta r_1 \cdot \delta r_7 \cdot z_4 - r_1 r_7 \cdot \delta r_1 \cdot \delta r_7 \cdot z_5 - \\
&\quad -r_1 r_7 \cdot \delta r_1 \cdot \delta r_7 \cdot z_6 - r_1 r_7 \cdot \delta r_1 \cdot \delta z_4 \cdot z_4 - r_1 r_7 \cdot \delta r_1 \cdot \delta z_5 \cdot z_5 - \\
&\quad -r_1 r_7 \cdot \delta r_1 \cdot \delta z_6 \cdot z_6 - \\
&\quad -r_1 r_7 \cdot \delta r_7 \cdot \delta z_4 \cdot z_4 - r_1 r_7 \cdot \delta r_7 \cdot \delta z_5 \cdot z_5 - r_1 r_7 \cdot \delta r_7 \cdot \delta z_6 \cdot z_6 - \\
&\quad -r_1 \cdot \delta r_1 \cdot \delta z_4 \cdot z_4 z_6 - r_1 \cdot \delta r_1 \cdot \delta z_5 \cdot z_5 z_6 - r_1 \cdot \delta r_1 \cdot \delta z_6 \cdot z_4 z_6 - \\
&\quad -r_1 \cdot \delta r_1 \cdot \delta z_6 \cdot z_5 z_6 - r_7 \cdot \delta r_7 \cdot \delta z_4 \cdot z_4 z_5 - r_7 \cdot \delta r_7 \cdot \delta z_5 \cdot z_4 z_5 - \\
&\quad -r_7 \cdot \delta r_7 \cdot \delta z_5 \cdot z_5 z_6 - r_7 \cdot \delta r_7 \cdot \delta z_6 \cdot z_5 z_6 - r_1 \cdot \delta z_4 \cdot \delta z_6 \cdot z_4 z_6 - \\
&\quad -r_1 \cdot \delta z_5 \cdot \delta z_6 \cdot z_5 z_6 - r_7 \cdot \delta z_4 \cdot \delta z_5 \cdot z_4 z_5 - r_7 \cdot \delta z_5 \cdot \delta z_6 \cdot z_5 z_6 - \\
&\quad -\delta z_4 \cdot \delta z_5 \cdot z_4 z_5 z_6 - \delta z_4 \cdot \delta z_6 \cdot z_4 z_5 z_6 - \delta z_5 \cdot \delta z_6 \cdot z_4 z_5 z_6 - \\
&\quad -r_1 r_7 \cdot \delta r_1 \cdot \delta r_7 \cdot \delta z_4 \cdot z_4 - r_1 r_7 \cdot \delta r_1 \cdot \delta r_7 \cdot \delta z_5 \cdot z_5 - \\
&\quad -r_1 r_7 \cdot \delta r_1 \cdot \delta r_7 \cdot \delta z_6 \cdot z_6 - r_1 \cdot \delta r_1 \cdot \delta z_4 \cdot \delta z_6 \cdot z_4 z_6 - \\
&\quad -r_1 \cdot \delta r_1 \cdot \delta z_5 \cdot \delta z_6 \cdot z_5 z_6 - r_7 \cdot \delta r_7 \cdot \delta z_4 \cdot \delta z_5 \cdot z_4 z_5 - \\
&\quad -r_7 \cdot \delta r_7 \cdot \delta z_5 \cdot \delta z_6 \cdot z_5 z_6 - \delta z_4 \cdot \delta z_5 \cdot \delta z_6 \cdot z_4 z_5 z_6;
\end{aligned} \tag{35}$$

$$\det 2 = \begin{vmatrix} r_1 + r_1 \cdot \delta r_1 & e_1 + e_1 \cdot \delta e_1 & 0 & 0 & 0 \\ 0 & 0 & z_4 + z_4 \cdot \delta z_4 & 0 & r_7 + r_7 \cdot \delta r_7 \\ 0 & -e_2 - e_2 \cdot \delta e_2 & 0 & -z_6 - z_6 \cdot \delta z_6 & -r_7 - r_7 \cdot \delta r_7 \\ 1 & 0 & -1 & 0 & 0 \\ 0 & 0 & 1 & 1 & -1 \end{vmatrix} =$$

$$= -e_2 r_1 r_7 - e_1 r_7 z_4 - e_1 r_7 z_6 - e_1 z_4 z_6 -$$

$$-e_2 r_1 r_7 \cdot \delta r_1 - e_2 r_1 r_7 \cdot \delta e_2 - e_2 r_1 r_7 \cdot \delta r_7 - e_1 r_7 \cdot \delta e_1 \cdot z_4 -$$
$$-e_1 r_7 \cdot \delta e_1 \cdot z_6 - e_1 r_7 \cdot \delta r_7 \cdot z_4 - e_1 r_7 \cdot \delta r_7 \cdot z_6 - e_1 r_7 \cdot \delta z_4 \cdot z_4 -$$
$$-e_1 r_7 \cdot \delta z_6 \cdot z_6 - e_1 \cdot \delta e_1 \cdot z_4 z_6 - e_1 \cdot \delta z_4 \cdot z_4 z_6 - e_1 \cdot \delta z_6 \cdot z_4 z_6 -$$
$$-e_2 r_1 r_7 \cdot \delta r_1 \cdot \delta e_2 - e_2 r_1 r_7 \cdot \delta r_1 \cdot \delta r_7 - e_2 r_1 r_7 \cdot \delta e_2 \cdot \delta r_7 -$$
$$-e_1 r_7 \cdot \delta e_1 \cdot \delta r_7 \cdot z_4 - e_1 r_7 \cdot \delta e_1 \cdot \delta r_7 \cdot z_6 - e_1 r_7 \cdot \delta e_1 \cdot \delta z_4 \cdot z_4 -$$
$$-e_1 r_7 \cdot \delta e_1 \cdot \delta z_6 \cdot z_6 - e_1 r_7 \cdot \delta r_7 \cdot \delta z_4 \cdot z_4 - e_1 r_7 \cdot \delta r_7 \cdot \delta z_6 \cdot z_6 -$$
$$-e_1 \cdot \delta e_1 \cdot \delta z_4 \cdot z_4 z_6 - e_1 \cdot \delta e_1 \cdot \delta z_6 \cdot z_4 z_6 - e_1 \cdot \delta z_4 \cdot \delta z_6 \cdot z_4 z_6 -$$

$$-e_2 r_1 r_7 \cdot \delta r_1 \cdot \delta e_2 \cdot \delta r_7 - e_1 r_7 \cdot \delta e_1 \cdot \delta r_7 \cdot \delta z_4 \cdot z_4 -$$
$$-e_1 r_7 \cdot \delta e_1 \cdot \delta r_7 \cdot \delta z_6 \cdot z_6 - e_1 \cdot \delta e_1 \cdot \delta z_4 \cdot \delta z_6 \cdot z_4 z_6; \quad (36)$$

$$\det 3 = \begin{vmatrix} r_1 + r_1 \cdot \delta r_1 & z_5 + z_5 \cdot \delta z_5 & e_1 + e_1 \cdot \delta e_1 & 0 & 0 \\ 0 & -z_5 - z_5 \cdot \delta z_5 & 0 & 0 & r_7 + r_7 \cdot \delta r_7 \\ 0 & 0 & -e_2 - e_2 \cdot \delta e_2 & -z_6 - z_6 \cdot \delta z_6 & -r_7 - r_7 \cdot \delta r_7 \\ 1 & -1 & 0 & 0 & 0 \\ 0 & 0 & 0 & 1 & -1 \end{vmatrix} =$$

$$= e_2 r_1 r_7 - e_1 r_7 z_5 + e_2 r_7 z_5 - e_1 z_5 z_6 +$$

$$+e_2 r_1 r_7 \cdot \delta r_1 + e_2 r_1 r_7 \cdot \delta e_2 + e_2 r_1 r_7 \cdot \delta r_7 - e_1 r_7 \cdot \delta e_1 \cdot z_5 +$$
$$+e_2 r_7 \cdot \delta e_2 \cdot z_5 - e_1 r_7 \cdot \delta r_7 \cdot z_5 + e_2 r_7 \cdot \delta r_7 \cdot z_5 - e_1 r_7 \cdot \delta z_5 \cdot z_5 +$$
$$+e_2 r_7 \cdot \delta z_5 \cdot z_5 - e_1 \cdot \delta e_1 \cdot z_5 z_6 - e_1 \cdot \delta z_5 \cdot z_5 z_6 - e_1 \cdot \delta z_6 \cdot z_5 z_6 +$$
$$+e_2 r_1 r_7 \cdot \delta r_1 \cdot \delta e_2 + e_2 r_1 r_7 \cdot \delta r_1 \cdot \delta r_7 + e_2 r_1 r_7 \cdot \delta e_2 \cdot \delta r_7 -$$
$$-e_1 r_7 \cdot \delta e_1 \cdot \delta r_7 \cdot z_5 + e_2 r_7 \cdot \delta e_2 \cdot \delta r_7 \cdot z_5 - e_1 r_7 \cdot \delta e_1 \cdot \delta z_5 \cdot z_5 +$$
$$+e_2 r_7 \cdot \delta e_2 \cdot \delta z_5 \cdot z_5 - e_1 r_7 \cdot \delta r_7 \cdot \delta z_5 \cdot z_5 + e_2 r_7 \cdot \delta r_7 \cdot \delta z_5 \cdot z_5 -$$
$$-e_1 \cdot \delta e_1 \cdot \delta z_5 \cdot z_5 z_6 - e_1 \cdot \delta e_1 \cdot \delta z_6 \cdot z_5 z_6 - e_1 \cdot \delta z_5 \cdot \delta z_6 \cdot z_5 z_6 +$$

$$+e_2 r_1 r_7 \cdot \delta r_1 \cdot \delta e_2 \cdot \delta r_7 - e_1 r_7 \cdot \delta e_1 \cdot \delta r_7 \cdot \delta z_5 \cdot z_5 +$$
$$+e_2 r_7 \cdot \delta e_2 \cdot \delta r_7 \cdot \delta z_5 \cdot z_5 - e_1 \cdot \delta e_1 \cdot \delta z_5 \cdot \delta z_6 \cdot z_5 z_6. \quad (37)$$

3 Conversion Equation of Measuring Channels

The conversion equation of measuring channels of measuring systems are created using control and verification equipment (CVE). CVE creates electrical signals according to sensor ranges. Thus it is possible to create test values, that are necessary to create additional equations [9].

A priori, within the declared accuracy, the electronic component of the measuring channels is linear, with additive and multiplicative errors, respectively.

$$y = kx + b = k_0(1 + \gamma) + b_1, \tag{38}$$

where k—the conversion coefficient; k_0—the nominal value of the conversion coefficient; γ—the multiplicative error; b—the additive error in the output value range.

Due to the usage of the low accuracy active and passive components in the schematic diagrams, it is necessary to determine the value of the real coefficient k_p and the additive component in (38), then substitute them in the formula for the measured value calculation and, having the obtained initial value y^*, find the measured value x^* with (39).

$$y^* = k_p x^* + b_p, \tag{39}$$

where y^*—the output value of the channel; x^*– the unknown measured value.

In the case of the known x and y it is enough to create two following equations to find k_p and b_p:

$$y_1 = k_p x_1 + b_p; \tag{40}$$

$$y_2 = k_p x_2 + b_p. \tag{41}$$

Using them the real values k_p and b_p could be found by:

$$k_p = \frac{y_1 - y_2}{x_1 - x_2}; \tag{42}$$

$$b_p = y_1 - \frac{y_1 - y_2}{x_1 - x_2} x_1; \tag{43}$$

or

$$b_p = y_2 - \frac{y_1 - y_2}{x_1 - x_2} x_2. \tag{44}$$

Using (39), (42) and (43) the measured value x^* could be calculated

$$x^* = \frac{y^* - b_p}{k_p} = \frac{y^* - \left[y_1 - \frac{y_1 - y_2}{x_1 - x_2} x_1\right]}{\frac{y_1 - y_2}{x_1 - x_2}}. \tag{45}$$

Making a conversion in (45) yields the results:

$$x^* = \frac{y^*(x_1 - x_2) + y_1 x_2 - y_2 x_1}{y_1 - y_2}. \tag{46}$$

If the test value x_2 is set to zero,

$$x_2 = 0, \tag{47}$$

then:

$$x^* = \frac{y^* x_1 - y_2 x_2}{y_1 - y_2}. \tag{48}$$

In case of rapid changes of k_p and b_p it makes sense to calculate their values before each measurement, in case of slow changes—calculate them before a series of measurements using (49):

$$x^* = \frac{y^* - b_p}{k_p}, \tag{49}$$

using calculated values k_p and b_p by (42) and (43), saved in the device memory, during measuring values processing.

The k_p and b_p could be calculated either before the output value measurement x^*, or after its measurement using (49).

The location of the test values x_1 and x_2 relative to the measured value x^* can be more optimal after finding its location, based on the calculation error minimization using (46).

References

1. Pallas-Areny, R., Webster, J.G.: Sensors and Signal Conditioning. Wiley, New York (1991)
2. Sheingold, D.: Analog-Digital Conversion Handbook, 3rd edn. Prentice-Hall, Norwood (1986)
3. Lizon, B.: A Basic Guide to Bridge Measurements (B. Lizon, J. Wu). Application Note. Texas Instruments, Inc. (2022)
4. Tuz, Y.M.: Analysis Automation of the Measuring Instruments: Tutorial (Y.M. Tuz, Y.S. Shumkov, O.V. Kosir), 312p. Publishing house "Helvetika", Odessa (2022). ISBN 978-966-992-770-5
5. Tsumbalen, H.: Linear Circuits. Design Manual. (H. Tsumbalen), 1128p. Analog Devices, Inc., Newnes (2008). ISBN 978-0-7506-8703-4
6. Strizhak, T.H.: Mathematic. Numerical methods. Polynomials. Determinants. Tutorial, 459 p. T.H. Strizhak, Y.M. Tuz, H.H. Baranovska, I.V. Veklich, Kyiv (1991)

7. Tuz, Y.M.: Measuring Instruments Analysis Automation: Monograph (Y.M. Tuz, Y.S. Shumkov, O.V. Kosir), 172p. K: "Korniichuk" (2014)
8. Dyakonov, V.P.: Mathcad 11/12/13 in Mathematics: Manual (V.P. Dyakonov), 958p. Telecom (2007). ISBN 5-93517-332-8
9. Tuz, Y.M.: Structural Methods of Measuring Instruments Accuracy Improvement (Y.M. Tuz), 256p. K: High School Main Publishing House (1976)

Application of Exponential Splines in the Measurement and Control of Electric Circuit Parameters

Yulian Tuz, Yurii Shumkov, and Oleh Kozyr

Abstract Some practical aspects of the use of exponential splines for measuring and controlling parameters of electric circuits are considered. The parameter control technique of complex two-pole electric circuits based on the zeros and poles method is presented. The special shape test signal model reproduces the immittance inverse function of a multielement two-pole electric circuit under control. The model is used to achieve the step response of a two-pole circuit with nominal parameters. The usage of exponential spline models to synthesize special-form test signals is described. It is shown that exponential splines are the optimal basis for the discrete synthesis of exponential test signals. The advantage of exponential splines usage is the ease of test signal generation in linear electric circuits. Test signals creation and examples of electric circuit parameters control are presented. The effect of the type of approximating functions on the quality of the immittance inverse function reproduction and the accuracy of circuit parameters estimation using circuit response is studied. It is shown that such a basis can increase the accuracy of determining parameters of electric circuits with a limited number of approximation sections.

Keywords Electric circuit · Special shape test signals · Discrete synthesis · Exponential spline model · Basis exponential spline

1 Complex Electric Circuit Parameters Control

In terms of production, an important task is the internal circuit control of complex electric circuit element parameters [1–3]. At the same time, the task is traditionally reduced to sequential control of *R*,*L*,*C*-parameters of selected sections of a circuit in the form of electric two-pole elements [4, 5]. In the general case, the model of selected sections of a circuit will be a *multielement two-pole electric circuit* (MTPEC).

Y. Tuz · Y. Shumkov · O. Kozyr (✉)
National Technical University of Ukraine "Igor Sikorsky Kyiv Polytechnic Institute", Kyiv, Ukraine
e-mail: kozyr.oleh@lll.kpi.ua

V. Eremenko and A. Zaporozhets (eds.), *Advanced Information-Measuring Technologies and Systems I*, Studies in Systems, Decision and Control 439,
https://doi.org/10.1007/978-3-031-40718-5_2

To increase control productivity a direct assessment of circuit parameters in the transient process is used based on the circuit transient response to test signals [6–8]. However, the circuit response to standard test signals (TS) with simple shapes generally has a shape that is difficult to analyze. This leads to the use of complicated analysis equipment and a significant loss of control time when evaluating individual values of MTPEC *R*, *L*, *C*-parameters. In this regard, it is relevant to use methods that can simplify the control procedure and reduce control time.

The main goal of parameter control is to find deviations in the real model from the nominal one. Therefore, it is crucial to assess the deviation of the circuit response shape from the nominal one. The circuit response signal shape can be used as a "generalized" informative parameter to control MTPEC circuit parameters.

In this regard, control methods based on the use of special shape TS, which immediately provide easily analyzed circuit response signals during the transient process, are preferred [9, 10]. For example, if the parameters of circuit elements are equal to the nominal ones, then the circuit response will have a predetermined shape of some unified response (UR).

To synthesize the above-mentioned TS (in the form of voltage or current) the information about the nominal circuit model in the form of a rational immittance function with universal zeros and poles will be used. TS are described by functions in a form of a sum of exponents, when $t \geq 0$, and reproduce the inverse model of the circuit immittance function in the Laplace domain. The zeros and poles of the TS model are chosen to compensate for the poles and zeros of the MTPEC immittance function. By carefully selecting the TS model it is possible to get the nominal circuit response in the form of a single-step function $1(t)$. Uncompensated zeros and poles determine the deviation of the circuit response from the nominal one, that is, from some constant level of voltage or current at $t \geq 0$, which is easy to assess and, accordingly, to control. Therefore, it is possible to invariantly determine individual R, L, C—parameters of MTPEC using the deviation of circuit response from the nominal one at specific points of the transition process.

The parameters of MTPEC elements are determined during the steady-state period of a transient process in an electrical circuit. It requires the synthesis of special shape TS for a small number of approximation sections. Using such signals during the measurement and control of parameters of complex electrical circuits solves an important task of increasing the productivity of automated control systems.

A complex multi-pole electric circuit to which nodes the measuring equipment is connected can be reduced to a three-pole circuit due to external switching. A controlled two-pole element is a branch of such a triangle that is either a simple single-element or multi-element R, L, C-two-pole circuit [5]. The second and third branches of the triangle consist of other elements of the complex circuit. Such a triangle is included in the measuring circuit of the "immittance-voltage" converter, which provides a special mode at the poles of the triangle, which allows for electrical isolation of the controlled two-pole (TP) circuit [11–14].

The delta circuit provides:

1. The voltage source at the poles of controlled TP in a parallel circuit. The current in TP is converted into a voltage at the output of the “immittance-voltage” converter.
2. Or the current source in TP in the sequential circuit. The voltage on the poles of the TP circuit is converted into the output value of the “immittance-voltage” converter.

The current in the controlled TP (or the voltage on the TP) is determined only by the parameters of the TP regardless of the other branches of the triangle. An invariant transformation of the immittance of the controlled TP into an active value, which is a voltage generated at the output of the converter, is taken place.

According to the zeros and poles method (ZPM), TS of a special form is used in the form of a sum of exponents (at $t \geq 0$) [9, 10]. The TS model reproduces the inverse nominal immittance function of MTPEC $H_{x0}(\boldsymbol{\alpha}^{z}_{x0}, \boldsymbol{\beta}^{p}_{x0}, p)$ in the Laplace domain (this is either the operator resistance $Z_{x0}(p)$ of the serial TP or the operator conductivity $Y_{x0}(p)$ of the parallel TP). The nominal response model in the time domain is chosen as a unit step function $1(t)$.

A condition for the synthesis of a model of an ideal TS $S_0(\boldsymbol{\alpha}^{z}_{0}, \boldsymbol{\beta}^{p}_{0}, p)$ is:

$$S_0(\boldsymbol{\alpha}^{z}_{0}, \boldsymbol{\beta}^{p}_{0}, p) \cdot H_{x0}(\boldsymbol{\alpha}^{z}_{x0}, \boldsymbol{\beta}^{p}_{x0}, p) = \frac{1}{p} \cdot \mathrm{A}_0, \quad \boldsymbol{\alpha}^{z}_{0} = \boldsymbol{\beta}^{p}_{x0}; \quad \boldsymbol{\beta}^{p}_{0} = \boldsymbol{\alpha}^{z}_{x0},$$

where $\boldsymbol{\alpha}^{z}_{0}, \boldsymbol{\beta}^{p}_{0}$ are the nominal values of the zeros and poles of TS; $\boldsymbol{\alpha}^{z}_{x0}, \boldsymbol{\beta}^{p}_{x0}$ are the nominal values of the zeros and poles of the immittance function of the multi-element TP. An example is

$$I_{S0}(p) Z_{x0}(p) = \frac{1}{p} \cdot \mathrm{U}_0 \text{ or } U_{S0}(p) \cdot Y_{x0}(p) = \frac{1}{p} \cdot \mathrm{I}_0.$$

That is, the synthesis condition is the compensation of the zeros and poles of the immittance function of the TP by the zeros and poles of the TS in the Laplace domain.

The model of the synthesized TS is defined as

$$S_0(\boldsymbol{\alpha}^{z}_{0}, \boldsymbol{\beta}^{p}_{0}, p) = \frac{1}{p} H^{-1}_{x0}(\boldsymbol{\alpha}^{z}_{x0}, \boldsymbol{\beta}^{p}_{x0}, p) \cdot \mathrm{A}_0.$$

The response model normalized by the level when TS is ideal is equal to

$$U_x(p) = S_0(\boldsymbol{\alpha}^{z}_{0}, \boldsymbol{\beta}^{p}_{0}, p) \cdot H_x(\boldsymbol{\alpha}^{z}_{x}, \boldsymbol{\beta}^{p}_{x}, p) =$$
$$= \frac{1}{p} H^{-1}_{x0}(\boldsymbol{\alpha}^{z}_{x0}, \boldsymbol{\beta}^{p}_{x0}, p) \cdot H_x(\boldsymbol{\alpha}^{z}_{x0} + \Delta\boldsymbol{\alpha}^{z}_{x}, \boldsymbol{\beta}^{p}_{x0} + \Delta\boldsymbol{\beta}^{p}_{x}, p) \cdot \mathrm{A}_0.$$

If $\boldsymbol{\alpha}^{z}_{x} = \boldsymbol{\alpha}^{p}_{x0}; \quad \boldsymbol{\beta}^{p}_{x} = \boldsymbol{\beta}^{p}_{x0}$, then the normalized response will have the form of a unit step function with a transform operator such as $U_x(p) = (1/p) \cdot \mathrm{A}_0$. If the zeros and poles are not compensated then the deviation of the response in the time domain from step function (at $t \geq 0$) reproduces the deviation of individual parameters of the circuit from the nominal ones.

$$U_x(p) = \frac{1}{p} \cdot A_0 + \Delta(p),$$

$$\Delta(p) \cong \sum_{i=1}^{m} \frac{A_0}{p} H_{x0}^{-1}\left(\boldsymbol{\alpha}_{x0}^{z}, \boldsymbol{\beta}_{x0}^{p}, p\right)\left(\frac{dH_x\left(\boldsymbol{\alpha}_x^z, \boldsymbol{\beta}_x^p, p\right)}{d\alpha_i}\right)_{/\alpha_x = \alpha_{x0},\ \beta_x = \beta_{x0}} \cdot \Delta\alpha_i +$$

$$+ \sum_{j=1}^{n} \frac{A_0}{p} H_{x0}^{-1}\left(\boldsymbol{\alpha}_{x0}^{z}, \boldsymbol{\beta}_{x0}^{p}, p\right)\left(\frac{dH_x\left(\boldsymbol{\alpha}_x^z, \boldsymbol{\beta}_x^p, p\right)}{d\alpha_i}\right)_{/\alpha_x = \alpha_{x0},\ \beta_x = \beta_{x0}} \cdot \Delta\beta_j,$$

where $\boldsymbol{\alpha}_{x0} = \{\alpha_{i0}\}_{i=1}^{m}$; $\boldsymbol{\beta}_{x0} = \left\{\beta_{j0}\right\}_{j=1}^{n}$. Such a response can be considered as a "generalized parameter» of a multi-element TP that is convenient for the control.

Let is simulate in Micro-CAP software the controlling of the parameters of the sequential (Fig. 1) and the parallel (Fig. 2) RC-two-pole circuit using the method of the zeros and poles.

An example of TS $i_{S0}(t)$ to determine the parameters of the sequential R, C-two-pole circuit is presented in Table 1, item 1 (Fig. 3).

The model of the ideal TS reproduces the inverse nominal function of the impedance of sequential R_x, C_x-two-pole elements and is described by the equation

$$I_{S0}(p) = \frac{1}{Z_{x0}(p)} \cdot \frac{1}{p} U_0 = \frac{p}{R_{x0}(p + 1/R_{x0}C_{x0})} \cdot \frac{1}{p} U_0 = \frac{1}{R_{x0}(p + 1/R_{x0}C_{x0})} \cdot U_0;$$

The unified response is

$$U_{out}(p) = I_{S0}(p) Z_x(p) = \frac{R_x(p + 1/R_x C_x)}{p \cdot R_{x0}(p + 1/R_{x0}C_{x0})} \cdot U_0 =$$

$$= \frac{R_x(p + \alpha_{x_T})}{R_{x0} \cdot p\,(p + \beta_{0_T})} U_0; \quad \alpha_{x_T} = 1/R_x C_x; \quad \beta_{0_T} = 1/R_{x0}C_{x0};$$

$$U_{out}(p) = \frac{R_x}{R_{x0}} \cdot \frac{1}{(p + \beta_{0_T})} \cdot U_0 + \frac{R_x}{R_{x0}} \cdot \frac{\alpha_{x_T}}{p\,(p + \beta_{0_T})} \cdot U_0;$$

$$u_{out}(t) = \frac{R_x}{R_{x_0}} e^{-\beta_{0_T} t} \cdot U_0 + \frac{C_{x_0}}{C_x}\left(1 - e^{-\beta_{0_T} t}\right) \cdot U_0 = u_{R_x}(t) + u_{C_x}(t), \quad t \geq 0,$$

where $u_{R_x}(t) = \frac{R_x}{R_{x_0}} e^{-\beta_{0_T} t} U_0$; $u_{C_x}(t) = \frac{C_{x_0}}{C_x}\left(1 - e^{-\beta_{0_T} t}\right) U_0$, $t \geq 0$.

The parameters are determined by the equations

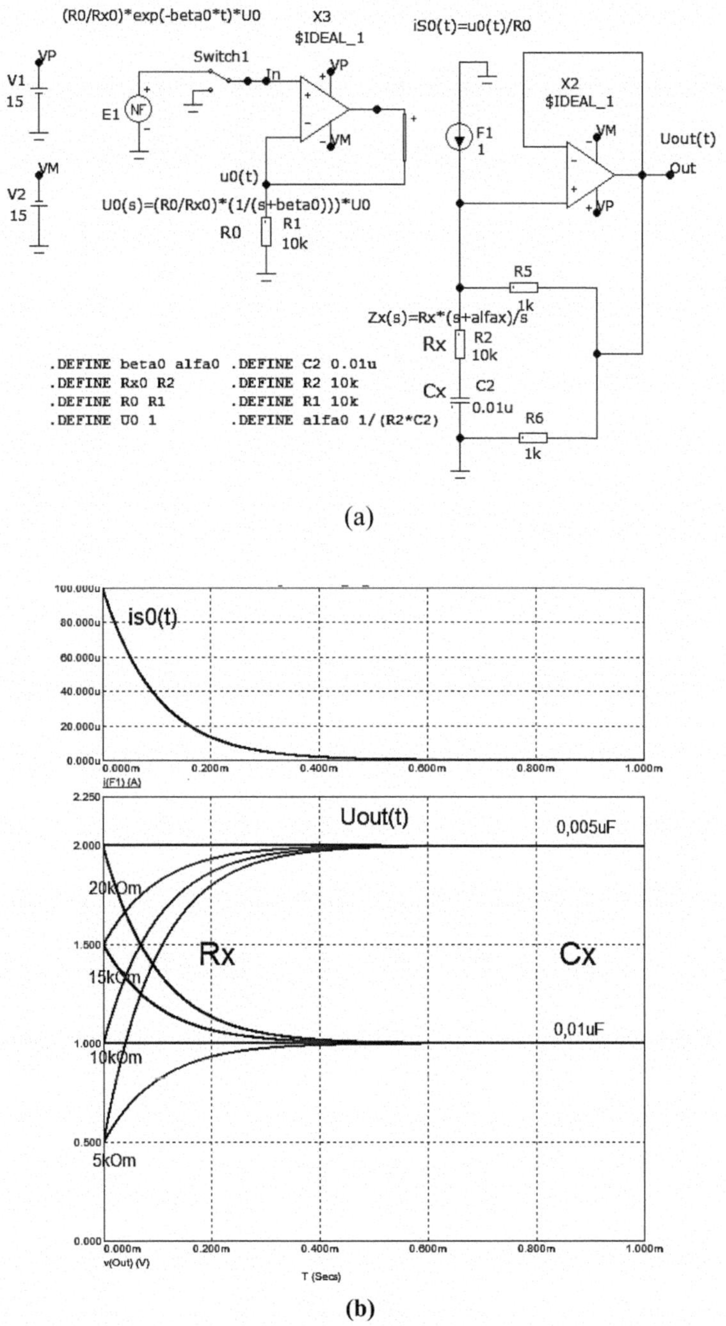

Fig. 1 The definition of parameters of sequential R_x, C_x-two-pole circuit: **a** is the circuit of the "immittance-voltage" converter; **b** is the response $u_{out}(t)$, if $C_x = 0.01\mu F$, $C_x = 0.005\mu F$; $R_x \in$ [5kOm; 20kOm] is changing with step ΔR_x = 5kOm

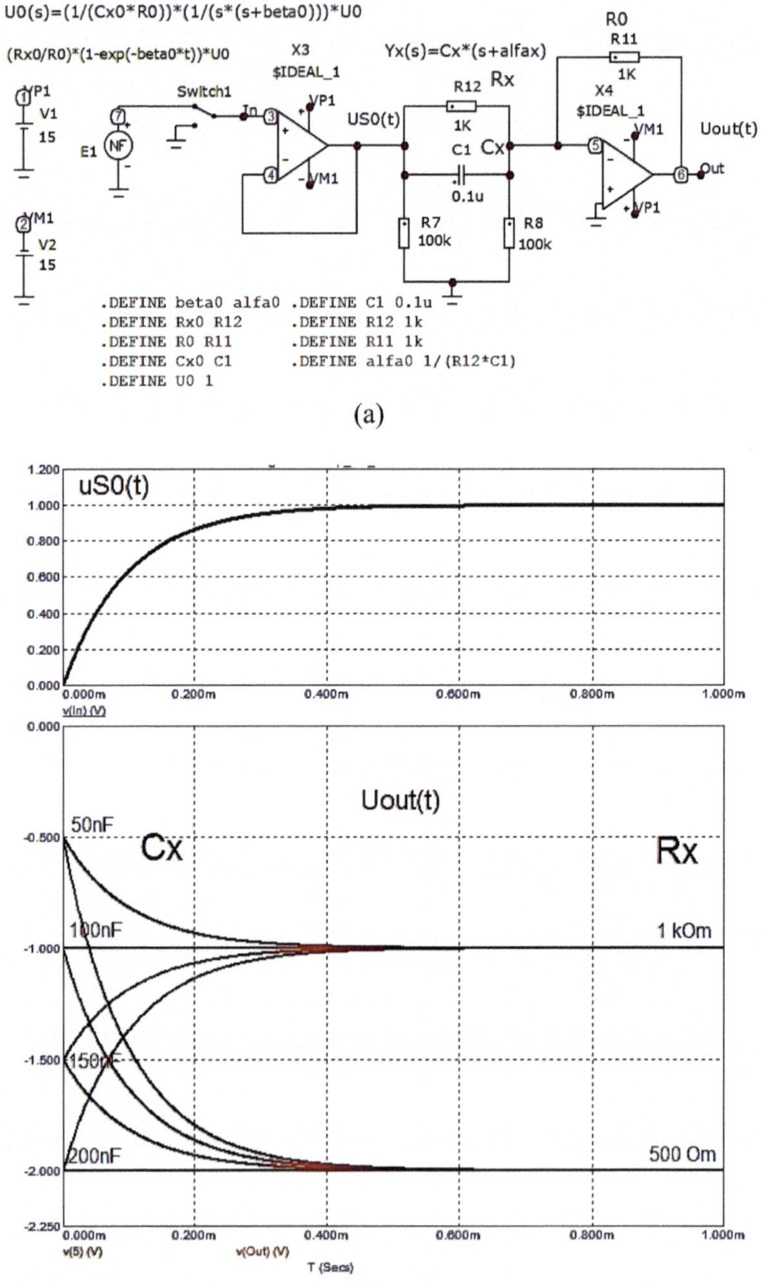

Fig. 2 The parameter definition of the parallel R_x, C_x—two-pole circuit: **a** is the circuit of the "immittance-voltage" converter; **b** is the circuit response $u_{out}(t)$ if $R_x = 1\,\text{kOm}$, $R_x = 500\,\text{Om}$, $C_x \in [50\text{nF};\ 200\text{nF}]$ is changing with step $\Delta C_x = 50\text{nF}$

Table 1 Exponential test signals

TP model	Test signal
Item 1 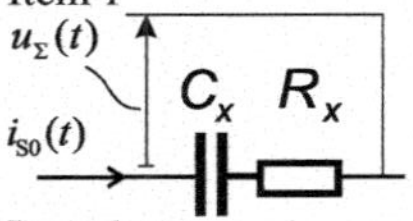 Impedance $Z(p) = R_x \cdot \frac{p+\frac{1}{R_x C_x}}{p}$ $R_x = R_{x0} + \Delta R_x$, $C_x = C_{x0} + \Delta C_x$	The model of ideal TS for the current is $i_{S0}(t) = \begin{cases} \frac{U_0}{R_{x0}} \cdot e^{-\frac{1}{R_{x0}C_{x0}} \cdot t}, & t \geq 0; \\ 0, & t < 0 \end{cases}$ The synthesis condition is $I_{S0}(p) \cdot Z_{x0}(p) = \frac{1}{p} \cdot U_0$ The Laplace transform is $I_{S0}(p) = \frac{1}{R_{x0}} \cdot \frac{U_0}{(p+\frac{1}{R_{x0}C_{x0}})}$ The nominal response (the voltage on TP) is $u_\Sigma(t) = 1(t)$, when $C_x = C_{x0},\ R_x = R_{x0},\ U_0 = 1$

(continued)

Table 1 (continued)

TP model	Test signal
Item 2 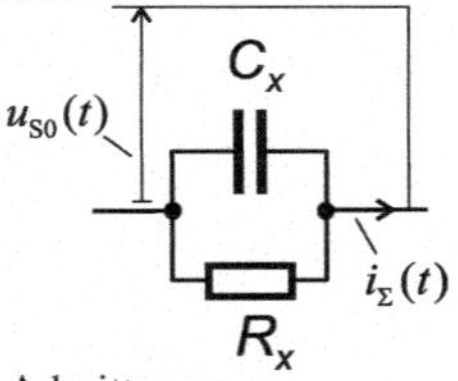 Admittance $Y_x(p) = C_x(p + \frac{1}{R_x C_x})$ $R_x = R_{x0} + \Delta R_x, C_x = C_{x0} + \Delta C_x$	The model of the ideal TS for the voltage is $u_{S0}(t) = \begin{cases} I_0 R_{x0} \cdot (1 - e^{-\frac{1}{R_{x0}C_{x0}} \cdot t}), & t \geq 0; \\ 0, & t < 0 \end{cases}$ The synthesis condition is $U_{S0}(p) \cdot Y_{x0}(p) = \frac{1}{p} \cdot \mathrm{I}_0$ The Laplace transform is $U_{S0}(p) = \frac{1}{C_{x0}} \cdot \frac{1}{p(p+\frac{1}{R_{x0}C_{x0}})} \cdot \mathrm{I}_0$ The nominal response (the current in TP) $i_\Sigma(t) = 1(t)$, when $C_x = C_{x0},\ R_x = R_{x0},\ I_0 = 1$
Item 3 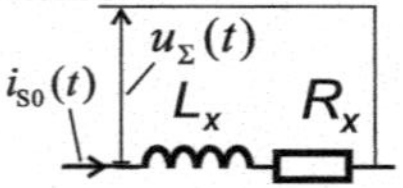 Impedance $Z_x(p) = L_x(p + \frac{R_x}{L_x})$ $R_x = R_{x0} + \Delta R_x, C_x = C_{x0} + \Delta C_x$	The model of ideal TS for the current is $i_{S0}(t) = \begin{cases} \frac{U_0}{R_{x0}} \cdot (1 - e^{-\frac{R_{x0}}{L_{x0}} \cdot t}), & t \geq 0; \\ 0, & t < 0 \end{cases}$ The synthesis condition is $I_{S0}(p) Z_{x0}(p) = \frac{1}{p} \cdot \mathrm{U}_0$ The Laplace transform is $I_{S0}(p) = \frac{U_0}{L_{x0}} \cdot \frac{1}{p \cdot (p+\frac{R_{x0}}{L_{x0}})}$ The nominal response (the voltage on TP) is $u_\Sigma(t) = 1(t)$, when $L_x = L_{x0},\ R_x = R_{x0},\ U_0 = 1$

(continued)

Table 1 (continued)

TP model	Test signal
Item 4 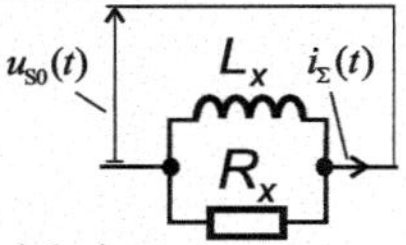 Admittance $Y_x(p) = \frac{1}{R_x} \cdot \frac{p+\frac{R_x}{L_x}}{p}$ $R_x = R_{x0} + \Delta R_x, C_x = C_{x0} + \Delta C_x$	The model of the ideal TS for the current is $u_{S0}(t) = \begin{cases} R_{x0} I_0 e^{-\frac{R_{x0}}{L_{x0}} \cdot t}, & t \geq 0; \\ 0\,, & t < 0 \end{cases}$ The synthesis condition is $U_{S0}(p) \cdot Y_{x0}(p) = \frac{1}{p} \cdot \mathrm{I}_0$ The Laplace transform is $U_{S0}(p) = R_{x0} \cdot \frac{1}{(p+\frac{R_{x0}}{L_{x0}})} \cdot \mathrm{I}_0$ The nominal response (the current in TP) $i_\Sigma(t) = 1(t)$, when $L_x = L_{x0}, \; R_x = R_{x0}, \; I_0 = 1$

(continued)

Table 1 (continued)

TP model	Test signal
Item 5 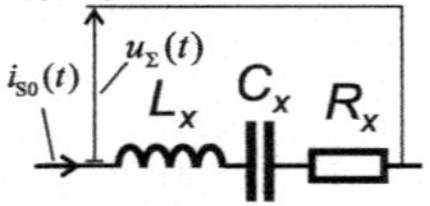 Impedance $Z_x(p) = \frac{L_x(p^2 + \frac{R_x}{L_x}p + \frac{1}{L_x C_x})}{p}$ The synthesis condition is $I_{S0}(p) \cdot Z_{x0}(p) = \frac{1}{p} \cdot U_0$ The Laplace transform is $I_{S0}(p) = \frac{U_0}{L_{x0}(p^2 + \frac{R_{x0}}{L_{x0}}p + \frac{1}{L_{x0}C_{x0}})}$	The model of ideal TS for the current is $R_{x0}^2 C_{x0} < 4L_{x0}$: $i_{S0}(t) = \frac{U_0}{L_{x0}\lambda_{x0}} \cdot e^{-\gamma_{x0}\cdot t} \sin\lambda_{x0}t, \quad t \geq 0;$ $\lambda_{x0} = \frac{1}{2L_{x0}}\sqrt{\frac{4L_{x0} - C_{x0}R_{x0}^2}{C_{x0}}}; \quad \gamma_{x0} = \frac{R_{x0}}{2L_{x0}}$ $R_{x_0}^2 C_{x_0} > 4L_{x_0}$: $i_{S0}(t) = \frac{U_0}{L_{x0}(\alpha_{10} - \alpha_{20})}(e^{-\alpha_{20}t} - e^{-\alpha_{10}t}), \ t \geq 0;$ $\alpha_{10,20} = \frac{R_{x_0}}{2L_{x_0}} \pm \frac{1}{2L_{x_0}}\sqrt{\frac{C_{x_0}R_{x_0}^2 - 4L_{x_0}}{C_{x_0}}}$ $R_{x0}^2 C_{x0} = 4L_{x0}$: $i_{S0}(t) = \frac{U_0}{L_{x0}} \cdot t \cdot e^{-\frac{R_{x0}}{2L_{x0}}\cdot t}, \quad t \geq 0$ The nominal response (the voltage on TP) is $u_\Sigma(t) = 1(t)$, when $R_x = R_{x0}, \ L_x = L_{x0}, \ C_x = C_{x0}, \ U_0 = 1$

(continued)

Table 1 (continued)

TP model	Test signal
Item 6 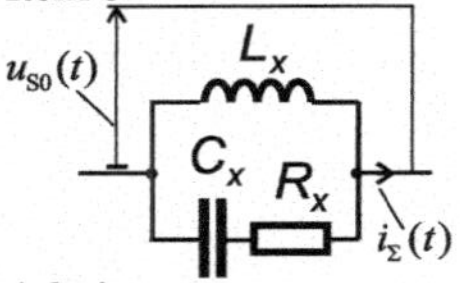 Admittance $Y_x(p) = \frac{p^2+\frac{R_x}{L_x}p+\frac{1}{L_xC_x}}{R_x\cdot p\cdot(p+\frac{1}{R_xC_x})}$ The synthesis condition is $U_{S0}(p)\cdot Y_{x0}(p) = \frac{1}{p}\times I_0$ The Laplace transform is $U_{S0}(p) = \frac{R_{x0}(p+\frac{1}{R_{x0}C_{x0}})\cdot I_0}{p^2+\frac{R_{x0}}{L_{x0}}p+\frac{1}{L_{x0}C_{x0}}}$	The model of the ideal TS for the voltage is $R_{x0}^2C_{x0} < 4L_{x0}$: $u_{S0}(t) = -R_{x0}M_0e^{-\gamma_0 t}\sin(\lambda_0 t-\Theta)\cdot I_0,\ t\geq 0;$ $M_0 = \frac{1}{\lambda_0}\sqrt{\lambda_0^2+(\gamma_0-b_0)^2};$ $\lambda_0 = \frac{1}{2L_{x0}}\sqrt{\frac{4L_{x0}-C_{x0}R_{x0}^2}{C_{x0}}};$ $\gamma_0 = \frac{R_{x0}}{2L_{x0}};\ \Theta = arctg\frac{\lambda_0}{\gamma_0-b_0};\ b_0 = \frac{1}{C_{x0}R_{x0}}$ $R_{x0}^2C_{x0} > 4L_{x0}$: $u_{S0}(t) = \frac{R_{x0}}{(\alpha_{10}-\alpha_{20})}\cdot\left[(\alpha_{10}-b_0)\cdot e^{-\alpha_{10}t}-\right.$ $\left.-(\alpha_{20}-b_0)\cdot e^{-\alpha_{20}t}\right]\cdot I_0,\quad t\geq 0;$ $\alpha_{10,20} = \frac{R_{x0}}{2L_{x0}}\pm\frac{1}{2L_{x0}}\sqrt{\frac{C_{x0}R_{x0}^2-4L_{x0}}{C_{x0}}};\quad b_0 = \frac{1}{C_{x0}R_{x0}}$ The nominal response (the current in TP) $i_\Sigma(t) = 1(t),$ when $R_x = R_{x0},\ L_x = L_{x0},\ C_x = C_{x0},\ I_0 = 1$

(continued)

Table 1 (continued)

TP model	Test signal
Item 7 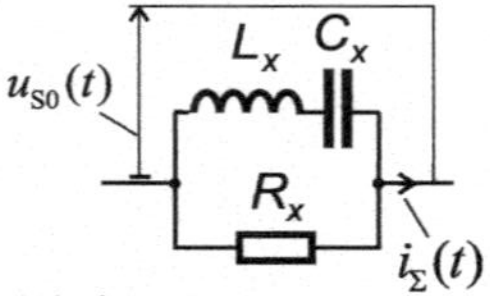 Admittance $Y_x(p) = \frac{p^2+\frac{R_x}{L_x}p+\frac{1}{L_xC_x}}{R_x(p^2+\frac{1}{L_xC_x})}$ The synthesis condition is $U_{S0}(p)\ Y_x(p) = \frac{1}{p}\cdot I_0$ The Laplace transform is $U_{S0}(p) = \frac{R_{x0}(p^2+\frac{1}{L_{x0}C_{x0}})\cdot I_0}{p\left[p^2+\frac{R_{x0}}{L_{x0}}p+\frac{1}{L_{x0}C_{x0}}\right]}$	The model of the ideal TS for the voltage is $R_{x0}^2C_{x0} < 4L_{x0}$: $u_{S0}(t) = R_{x0}\left[\frac{b_0}{\rho_0} - M_0e^{-\gamma_0 t}\sin(\lambda_0 t - \Theta)\right]\cdot I_0,\quad t \geq 0;$ $M_0 = \frac{\sqrt{\upsilon_0^2+u_0^2}}{\lambda_0\sqrt{\rho_0}};\quad \rho_0 = \gamma_0^2+\lambda_0^2;$ $\lambda_0 = \frac{1}{2L_{x0}}\sqrt{\frac{4L_{x0}-C_{x0}R_{x0}^2}{C_{x0}}};\quad \gamma_0 = \frac{R_{x0}}{2L_{x0}};$ $b_0 = \frac{1}{L_{x0}C_{x0}};\ \Theta = arctg\frac{\upsilon_0}{u_0} + arctg\frac{\lambda_0}{\gamma_0};$ $\upsilon_0 = -2\gamma_0\lambda_0;\quad u_0 = \gamma_0^2 - \lambda_0^2 + b_0$ $R_{x0}^2C_{x0} > 4L_{x0}$: $u_{S0}(t) = \frac{R_{x0}b_0^2}{\alpha_{10}\alpha_{20}}\cdot\left[1+\frac{\alpha_{20}(\alpha_{10}+b_0)}{b_0(\alpha_{10}-\alpha_{20})}\cdot e^{-\alpha_{10}t} - \right.$ $\left. - \frac{\alpha_{10}(\alpha_{20}+b_0)}{b_0(\alpha_{10}-\alpha_{20})}\cdot e^{-\alpha_{20}t}\right]\cdot I_0,\quad t \geq 0;$ $\alpha_{10,20} = \frac{R_{x0}}{2L_{x0}} \pm \frac{1}{2L_{x0}}\sqrt{\frac{C_{x0}R_{x0}^2-4L_{x0}}{C_{x0}}};\quad b_0 = \frac{1}{C_{x0}R_{x0}}$ The nominal response (the current in TP) $i_\Sigma(t) = 1(t)$, when $R_x = R_{x0},\ L_x = L_{x0},\ C_x = C_{x0},\ I_0 = 1$

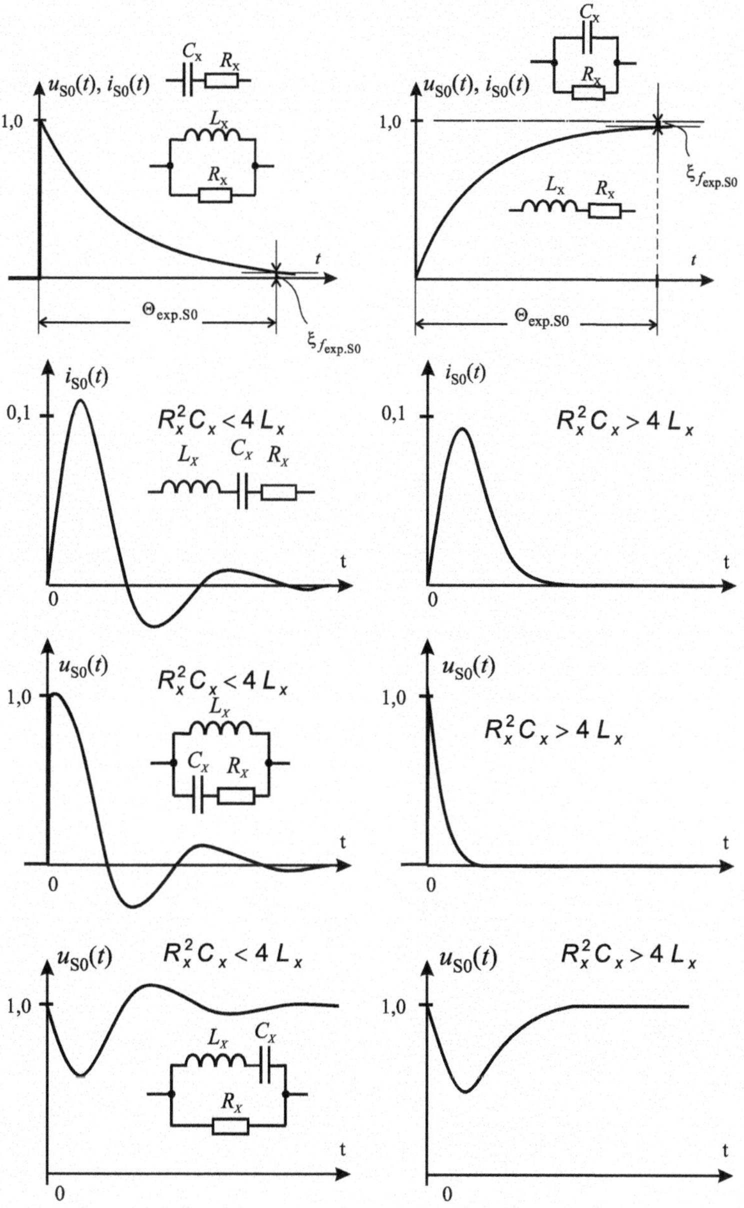

Fig. 3 The exponential TS to control R, L, C-parameters of TP element

$$\lim_{t\to 0} u_{out}(t) = u_{out}(0) = \frac{R_x}{R_{x0}} U_0; \qquad R_x = \frac{u_{out}(0)}{U_0} R_{x0}.$$

$$\lim_{t\to\infty} u_{out}(t) = u_{out}(\infty) = \frac{C_{x0}}{C_x} U_0; \qquad C_x = \frac{U_0}{u_{out}(\infty)} C_{x0}.$$

The parameters are controlled by the equations

$$u_{out}(t) \cong (1 + \frac{\Delta R_x}{R_{x0}}) e^{-\beta_{0_T} t} U_0 + (1 - \frac{\Delta C_x}{C_{x0}})(1 - e^{-\beta_{0_T} t})\, U_0 =$$

$$= 1 \cdot U_0 + \frac{\Delta R_x}{R_{x0}} e^{-\beta_{0_T} t} U_0 - \frac{\Delta C_x}{C_{x0}} (1 - e^{-\beta_{0_T} t}) U_0, \quad t \ge 0.$$

The invariance conditions are

$$u_{R_x} \cong \lim_{t\to 0} u_{out}(t) = \left(1 + \frac{\Delta R_x}{R_{x0}}\right) U_0 = [1 + \delta R_x] U_0; \text{ when } t \to 0\ u_{C_x}(t) \to 0.$$

$$u_{C_x} \cong \lim_{t\to\infty} u_{out}(t) = \left(1 - \frac{\Delta C_x}{C_{x0}}\right) U_0 = [1 - \delta C_x] U_0; \text{ when } t \to \infty\ u_{Rx}(t) \to 0.$$

An example of TS $u_{S0}(t)$ to define parameters of the parallel R_x, C_x-two-pole circuit is presented in Table 1, item 2.

The model of the ideal TS reproduces the inverse admittance function of the parallel R_x, C_x—two-pole elements and is described by the equation

$$U_{S0}(p) = \frac{1}{Y_{x0}(p)} \cdot \frac{1}{p} \cdot I_0 = \frac{1}{C_{x0}(p + 1/R_{x0}C_{x0})} \cdot \frac{1}{p} \cdot I_0 = \frac{1}{C_{x0}\ p(p + 1/R_{x0}C_{x0})} \cdot I_0;$$

The unified response is

$$U_{out}(p) = I_x(p) \cdot R_0 = U_{S0}(p) Y_x(p) \cdot R_0 = \frac{C_x(p + 1/R_x C_x)}{C_{x0} \cdot p\ (p + 1/R_{x0} C_{x0})} \cdot I_0 R_0 =$$

$$= \frac{C_x(p + \alpha_{x_T})}{C_{x0} \cdot p\ (p + \beta_{0_T})} U_0; \quad U_0 = I_0 R_0; \quad \alpha_{x_T} = 1/R_x C_x; \quad \beta_{0_T} = 1/R_{x0} C_{x0}.$$

$$U_{out}(p) = \frac{C_x}{C_{x0}} \cdot \frac{1}{(p + \beta_{0_T})} U_0 + \frac{C_x}{C_{x0}} \cdot \frac{\alpha_{x_T}}{p\,(p + \beta_{0_T})} U_0;$$

$$u_{out}(t) = \frac{C_x}{C_{x0}} e^{-\beta_{0_T} t} U_0 + \frac{R_{x0}}{R_x} (1 - e^{-\beta_{0_T} t}) U_0 = u_{C_x}(t) + u_{R_x}(t), \quad t \ge 0.$$

where $u_{C_x}(t) = \frac{C_x}{C_{x0}} e^{-\beta_{0_T} t} U_0; \quad u_{R_x}(t) = \frac{R_{x0}}{R_x}(1 - e^{-\beta_{0_T} t}) U_0, \quad t \ge 0.$

The parameters are determined by the equations

$$\lim_{t\to 0} u_{\text{out}}(t) = u_{\text{out}}(0) = \frac{C_x}{C_{x0}} U_0; \qquad C_x = \frac{u_{\text{out}}(0)}{U_0} C_{x0}.$$

$$\lim_{t\to\infty} u_{\text{out}}(t) = u_{\text{out}}(\infty) = \frac{R_{x0}}{R_x} U_0; \qquad R_x = \frac{U_0}{u_{\text{out}}(\infty)} R_{x0}.$$

The parameters are controlled by the equations

$$u_{\text{out}}(t) \cong (1 + \frac{\Delta C_x}{C_{x0}}) e^{-\beta_{0_T} t} U_0 + (1 - \frac{\Delta R_x}{R_{x0}})(1 - e^{-\beta_{0_T} t}) U_0 =$$
$$= 1 \cdot U_0 + \frac{\Delta C_x}{C_{x0}} e^{-\beta_{0_T} t} U_0 + \frac{\Delta R_x}{R_{x0}} (1 - e^{-\beta_{0_T} t}) U_0, \quad t \geq 0.$$

The invariance conditions are

$$u_{C_x} \cong \lim_{t\to 0} u_{\text{out}}(t) = \left(1 + \frac{\Delta C_x}{C_{x0}}\right) U_0 = [1 + \delta C_x] U_0; \text{ at } t \to 0 \; u_{R_x}(t) \to 0.$$

$$u_{R_x} \cong \lim_{t\to\infty} u_{\text{out}}(t) = \left(1 - \frac{\Delta R_x}{R_{x0}}\right) U_0 = [1 - \delta R_x] U_0; \text{ at } t \to \infty \; u_{C_x}(t) \to 0.$$

The duration of the TS is definitely related to the settling time of the circuit response which is determined by the largest time constant. Let's estimate the TS duration time on the example of a two-element R_x, C_x-two-pole circuit.

The relative error of the settling time of the exponential TS $\xi_{f_{\text{exp.S0}}}$ is equal to

$$\xi_{f_{\text{exp.S0}}} \approx \xi_{f_x},$$

where $\xi_{f_x} = e^{-\beta_{0_T} \Theta_x}$ is the relative error of the settling time of the circuit response. That is, the minimum duration of the exponential TS is equal to:

$$\Theta_{\text{exp.S0}} = \Theta_x = (1/\beta_{0_T}) \cdot \left|\ln \xi_{f_x}\right|.$$

where Θ_x is the response observation interval. According to the synthesis conditions $\beta_{0_T} = 1/\tau_{x0}$, τ_{x0} is the nominal value of the time constant of the investigated circuit.

At $\xi_{f_x} = 0, 01$ the minimum duration of the exponential TS is

$$\Theta_{\text{exp.S0}} = \Theta_x = 4.6 \; \tau_{x0}.$$

So, for determining the parameters of circuits from the time constant τ_{x0} in the range of 0.87 μs to 100 ms minimum duration of TS is in the range of 4 μs–460 ms (Table 2).

In turn, the duration of the TS is related to the number of approximation sections N_a and discretization step h by the formula

$$\Theta_{\text{exp.S0}} = N_a h.$$

Table 2 The parameter estimation ranges of the two-element two-pole circuit using MNP

Duration of TS	Parameters, range				
$\Theta_{exp.S0}$	τ_x	R_x	$C_{x\,(min)}$	R_x	$L_{x\,(min)}$
4 μs ($\gamma_{non\text{-}invariance} \leq 1\,\%$)	0,87 μs	100 Ω	8.7 nF	100 Ω	87 μH
		100 kOhm	8.7 pF	100 kOhm	87 mH
$\Theta_{exp.S0}$	τ_x	R_x	$C_{x\,(max)}$	R_x	$L_{x\,(max)}$
460 ms ($\gamma_{non\text{-}invariance} \leq 1\,\%$)	100 ms	100 kOhm	1 μF	1 kOhm	100 H
		100 Ω	100 μF	100 Ω	10 H

Given that $\xi_{f_x} = 0, 01$ then we have

$$\Theta_{exp.S0} = \Theta_x = 4.6\ \tau_{x0} = N_a h.$$

During the control of two-element TP with $\tau_{x0} = 0.87$ μs (when $\xi_{f_x} = 0.01$) the minimum duration is $\Theta_{exp.S0} = 4.6\ \tau_{x0} = N_a h = 4$ μs. When $\xi_{f_x} = 0, 01$ the DAC sample frequency will be $f_T = h^{-1} = N_a/(4.6 \cdot \tau_{x0})$. Given that $N_a = 4$ the DAC sample frequency will be $f_T = 1$ MHz (or $h = 1$ μs). For a small amount N_a the approximation error is decisive. When the creation of TS is based on piecewise step interpolation (PSI) or piecewise linear interpolation (PLI) the error of determining the parameters of the circuit by MNP will be 47% and 14% respectively. This error is caused by the method of approximation of the shape to the ideal TS.

To ensure an error of less than 1% to create TS based on PSI it is necessary to set $N_a = 256$ which requires the DAC sample frequency to be $f_T = 64$ MHz. To ensure an error of less than 0.2%, you need to choose a number $N_a = 1024$ that requires $f_T = 256$ MHz. Therefore, it is required a high-precision DAC with very high speed to create TS. They should operate with the sample rate of the order of 256 MHz and higher. At the same time, they have a very short settling time of the output signal $t_{set} \cong 4 \div 5$ ns. It is required to use high-performance digital signal processors that work on large data sets.

In the case of PLI to ensure an error of less than 1% ones should select $N_a = 16$ which requires sample frequency $f_T = 4$ MHz when $\tau_{x0} = 0.87$ μs. At the same time, the real model during the creation of TS will have some piecewise exponential shape and requires clarification.

2 The Discrete Synthesis of Test Signals Based on Exponential Splines

To generate test signals an analog filter is used to obtain smooth dependencies, for example, from a piecewise-step signal generated by a DAC. If there are a limited number of approximation sections of the reproduced continuous signal, in other terms of the conditions of the Whittaker-Kotelnikov-Shannon sampling theorem

[15] are not met, then the spectra of the continuous signal shifted due to discretization are superimposed. At the same time, accurate reproduction of a continuous signal by filtering high-frequency spectral components through discretization is not possible. With a limited number of approximation sections, the filtering of a piecewise step signal leads to some piecewise exponential functions. The problem of discrete synthesis of the TS becomes an approximation task and can be solved in the time domain in the class of exponential spline functions.

Such splines can be constructed from functions that correspond to real signals in linear electrical circuits that respond to standard impulse signals, that is, they are creation friendly. At the same time, exponential splines are optimal for the synthesis of TS which belong to the class of exponential signals and are described by the same functions as the approximation ones [16].

The theory of splines has found practical use in various problems of computational mathematics [17–25]. Many works are devoted to the development of the theory of exponential splines, among which we note only [22–25]. The results can be used in the problems of approximation of the investigated processes to obtain their mathematical models, i.e., an "approximant" in an analytical form, and to process measurement information. The theory of exponential splines in problems of filtering and analog–digital signal processing is devoted in [26, 27].

In the work, we will consider some practical aspects of the use of exponential splines (ES) in the tasks of measuring and controlling the parameters of electric circuits:

- a creation of the special shape electric TS using finite basic ES, where the finite exponential B-spline represents the response of some spline-approximating filter to an instantaneous impulse. That is, obtaining an "approximant" in the form of an electrical signal;
- using an "approximant" in the form of an electrical signal to obtain the "immittance-voltage" measuring transformation. As a result, at the output of the converter, the information about the parameters of the investigated circuit is determined as the response to the special shape TS.

That is, by the ES method in the tasks of measuring and controlling the parameters of electric circuits, we understand the acquisition and transformation of the "approximant" in the form of an electrical signal [28]. At the same time, the properties of the approximating functions affect the quality of the information about the parameters of the circuit from its response and, accordingly, the parameter estimation errors.

2.1 The Definition

Let's consider the type of ES that can be obtained in the form of an electrical signal in linear electrical circuits with lumped parameters [29]. Suppose that on a segment $[a, b]$ is the system of nodes (a grid) $\Delta_N : a = t_0 < t_1 < \cdots < t_N = b$. Denote by $C^{m-\nu}[a, b]$ the set of $(m - \nu)$-times continuously differentiable functions on the segment $[a, b]$.

Definition 1. *An exponential spline of the order m of class* $C^{m-\nu}$ $(1 \leq \nu \leq m)$ *with nodes on the grid* Δ_N *is a piecewise function* $sf_{Gm}(t)$, *which* ***(a)*** *on each segment* $[t_i, t_{i+1}]$, $i = 0, 1, \ldots, N-1$ *is the solution of some linear inhomogeneous differential equation with constant coefficients.*

$$\sum_{j=0}^{l} a_j \cdot D^{l-j} s_i(t) = \mu_i \cdot 1(t); \quad D^j = \left(\frac{d}{dt}\right)^j,$$

(b) *while is* $sf_{Gm}(t) \in C^{m-\nu}[a, b]$.

Exponential splines can be represented by a system of nonzero finite basis splines on a finite interval. Basic exponential splines (BES) will be called G-splines, which can be formed in linear electric circuits by their impulse transient characteristic [16].

Definition 2. *Basic exponential splines or G-splines of the order m of class* $C^{m-\nu}[a, b]$ *are such functions* $G_{m,i}(t) \in sf_{Gm}(t)$, *that* $G_{m,i}(t) \neq 0$ *are on the interval* $t \in [t_i, t_{i+m}]$ *and* $G_{m,i}(t) \equiv 0$ *out of the interval* $t \in [t_i, t_{i+m}]$.

The order of splines $m = l + 1$ is the "order" of reproducing spline functions, which is determined by the number of roots of the characteristic polynomial $Q_m(s)$ taking into account the right-hand side of the linear inhomogeneous differential equation (LIDE), which describes the transient process in a linear electric circuit. The interval on which the G-spline is different from zero is minimal and uniquely determined by the order of the spline. Functions $G_{m,i}(t)$, $i = \overline{0, N-1}$ are linearly independent and form a basis in the spline space $sf_{Gm}(t)$.

2.2 The Discrete Synthesis

The essence of discrete synthesis is in signal creation based on the sum of time-shifted finite base exponential splines with their weight coefficients [29]. Consider the relative scale where the variable $\overline{t} = n + \varepsilon$ is a relative time that links to actual time by $\overline{t} = t/h$, where h is a uniform sampling interval and $n = 0, 1, 2, \ldots;$ $0 \leq \varepsilon \leq 1$.

Finite basic exponential spline or G-order spline m is

$$G_{\mathrm{m}}(\boldsymbol{a}, \overline{t}) = \begin{cases} A \cdot (\boldsymbol{a}) \cdot b_0(\varepsilon), & \overline{t} \in [0,\ 1]; \\ A \cdot (\boldsymbol{a}) \cdot b_1(\varepsilon), & \overline{t} \in [1,\ 2]; \\ \quad \ldots\ldots\ldots\ldots & \\ A \cdot (\boldsymbol{a}) \cdot b_{\mathrm{m}-1}(\varepsilon), & \overline{t} \in [m-1,\ m]; \\ 0, \quad \overline{t} < 0, \quad \overline{t} > m, & \end{cases}$$

where $A\cdot(\boldsymbol{a}) = 1 / |b_0(\varepsilon)|_{/\varepsilon=1}$ (or $A\cdot(\boldsymbol{a}) = 1 / \left|\sum_{k=0}^{m-1} b_{\mathrm{k}}(\varepsilon)\right|_{/\varepsilon=1}$) is the normalizing factor; m is the order of the characteristic polynomial of the LIDE, taking into account the right-hand side; $\bar{t} \in [0, m]$ is a binding to a time interval; $\boldsymbol{a}$ is a parameter determined by the roots of the characteristic polynomial.

Functions $b_{\mathrm{k}}(\varepsilon)$ are determined by the sum $b_{\mathrm{k}}(\varepsilon) = \sum_{i=0}^{k} \mu_i \phi(k - i + \varepsilon)_+$, of $k = \overline{0, m-1}$; $0 \le \varepsilon \le 1$ truncated functions

$$\phi(\bar{t})_+ = \begin{cases} \phi(\bar{t}), & \bar{t} \ge 0; \\ 0, & \bar{t} < 0, \end{cases}$$

where $\phi(\bar{t})$ is a solution of LIDE under zero initial conditions for which $\phi(0)_+ = \phi^{/}(0)_+ = \cdots = \phi^{(m-v)}(0)_+ = 0$.

The values of the coefficients μ_i with elementary functions $\phi(k - i + \varepsilon)_+$ in expressions $b_{\mathrm{k}}(\varepsilon)$ are determined under the condition $G_{\mathrm{m}}(\bar{t}) \equiv 0$ out of the range $\bar{t} \in [0,\ m]$.

The exponential spline function (ESF) $sf_{\mathrm{G}m}(\bar{t})$ of the order m is

$$sf_{\mathrm{G}m}(\bar{t}) = sf_{\mathrm{Gm}}(n + \varepsilon) = \sum_{i=-\infty}^{\infty} f[i+1] \cdot G_{\mathrm{m}}(n + \varepsilon - i) =$$
$$= A \cdot (\boldsymbol{a}) \left[0 + f[1] \cdot b_{\mathrm{m}-1}(\varepsilon) + f[2] \cdot b_{\mathrm{m}-2}(\varepsilon) + \cdots + f[n] \cdot b_1(\varepsilon) + f[n+1] \cdot b_0(\varepsilon) + 0\right].$$

where $\{f_i\}_{i=-\infty}^{\infty}$ are ESF coefficients which represent, for example, discrete samples (instantaneous values) in moments $\bar{t} = n$, $n = 0,\ 1,\ 2,\ \ldots$ of some continuous dependency $f(\bar{t})$ reproduced.

The ESF expression is a discrete convolution of the lattice function $f[n]$ with an impulse function $G_{\mathrm{m}}(\bar{t}) = G_{\mathrm{m}}(n + \varepsilon)$. In practice, another form of discrete convolution is more useful

$$sf_{\mathrm{G}m}(\bar{t}) = sf_{\mathrm{G}m}(n + \varepsilon) = \sum_{i=-\infty}^{\infty} f[n+1-i] \cdot G_{\mathrm{m}}(i + \varepsilon) =$$
$$= A \cdot (\boldsymbol{a})\left[0 + f[n+1] \cdot b_0(\varepsilon) + f[n] \cdot b_1(\varepsilon) + f[n-1] \cdot b_2(\varepsilon) + \cdots + f[1] \cdot b_{\mathrm{m}-1}(\varepsilon) + 0\right].$$

2.3 *Signal-Forming Circuit Model*

The signal-forming circuit model in the form of an open structure (Fig. 4) includes a combined continuous part (CP) with a transfer function (TF) $W_{\mathrm{G}}(q) = A(\boldsymbol{a}_{\mathrm{T}}, h) \cdot W_{\mathrm{reduced}}(q)$ and the combined discrete part (DP) with the TF $H(e^{-q})$, where $q = ph$ is a complex variable on a relative scale, $q = \overline{\sigma} + j\overline{\omega}$, $\overline{\omega} = \omega h$ is the relative frequency, h is a uniform sampling interval, $A(\mathbf{a}_{\mathrm{T}}, h)$ is a normalizing factor; $\perp$ is an ideal impulse element with a sample step h, which reflects the discretization process

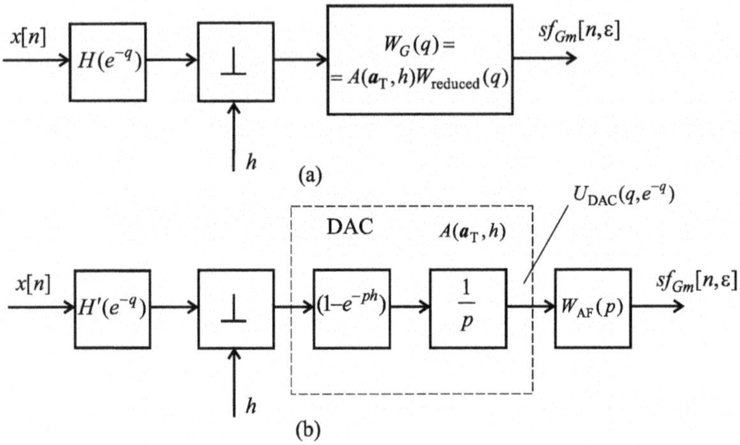

Fig. 4 The mathematical model of the signal formation by G-splines: **a** a generalized model; **b** the model that takes into account DAC with memory registers

or an already existing discrete representation of the input signal $X^*(\bar{t}) = x[n]$, where $\bar{t} = n;\ n = 0,\ 1,\ 2, \ldots$.

Type of TF of the combined CP $W_G(q) = A(\mathbf{a}_T, h) \cdot W_{reduced}(q)$ (Fig. 4a) determines the type of piecewise functions that form the spline and ensures the continuity of splines and their derivatives at nodes. Thus, to exclude discontinuities of the first kind at nodes of a piecewise function the following is a necessary condition: the degree of the polynomial in the denominator of the TF CP should be $m \geq 2 + l$, *where l is the degree of the polynomial in the numerator. To ensure the continuity of the first derivative at the nodes a condition must be met $m \geq 3 + l$ for polynomials and so on.*

The selection of the type of TF $H(e^{-q})$ of the composite discrete part (a digital filter that can be implemented, for example, based on a signal processor) ensures the finite duration (finiteness) of the reaction of the impulse system, that is, of the basic functions $G_m(\bar{t})$, for each input sample $x[n]$.

The dynamic model of DAC with memory registers (Fig. 4b) can be represented by a non-minimum-phase link with a TF $\frac{1-e^{-ph}}{p} = \frac{1-e^{-q}}{q} \cdot h$ with continuous and discrete parts. The multiplier $(1 - e^{-q}) = 1 - e^{-ph}$ can be taken into account in the TF of the combined DP $H(e^{-q}) = (1 - e^{-q})H'(e^{-q})$, where $H'(e^{-q}) = H_{DF}(z^{-1})$ is the transfer function of the digital filter, $e^{-q} = z^{-1}$. The multiplier $1/p$ can be taken into account in the TF of the combined CP. Then we will have

$$W_G(q) = A(\boldsymbol{a}_T, h) \cdot W_{reduced}(q) = A(\boldsymbol{a}_T, h) \cdot K_\Phi(q)\,\tilde{W}_{reduced}(q),$$

$$\tilde{W}_{reduced}(q) = \left[\frac{1}{p} W_{AF}(p)\right]_{p=q/h},$$

where $K_\Phi(q)$ is the transfer function of the pulse element, which takes into account the shape of the pulses at its output and the period of their repetition h; $W_{AF}(p)$ is TF of some linear continuous electric circuits (for example, an analog filter); index "t" means that the function or parameter is defined in the real time scale. The multiplier $1/p$ takes into account the step shape of the signal formed at the output of the DAC. TF of the ideal impulse element ⊥ is equal to:

$$K_\Phi(q) = \frac{1}{h} S_T(p)_{/p=\frac{q}{h}} = \frac{1}{h},$$

where $S_T(p)$ takes into account the shape of the pulses at element output. If the impulse element is ideal, that is, at its output perfect instantaneous pulses that are modulated in the area (intensity) by the input value at discrete moments of time $\bar{t} = n$ are formed, than $S_T(p) = 1$, $K_\Phi(q) = \frac{1}{h}$.

Therefore, the TF of the combined LF, taking into account the ideal pulse element in the relative time scale, is equal to

$$W_{\text{reduced}}(q) = K_\Phi(q)\, \tilde{W}_{\text{reduced}}(q) = \frac{1}{h}\left[\frac{1}{p} W_{AF}(p)\right]_{p=q/h}.$$

2.4 The TF of the Forming Circuit

In the presence of a discretization process, the output signal of the CP is described by the *shifted lattice functions* [30]. At the same time, the continuous part corresponds to the TF for the discretized signal, which can be obtained based on $\mathfrak{D}$-transform of the discrete Laplace transform (DLT):

$$W_{\text{reduced}}(q) = \frac{R_l(q)}{Q_m(q)} \xrightarrow{\mathfrak{D}} W^*_{\text{reduced}}(q, \varepsilon) = K_h \cdot \frac{\sum_{k=0}^{m-1} b_k(\varepsilon) \cdot e^{q(m-k)}}{\sum_{j=0}^{m} a_j \cdot e^{q \cdot j}}, \qquad m > l,$$

where $q = ph$, $0 \le \varepsilon \le 1$. The index (*) is used to indicate the TF for discretized signals and $\mathfrak{D}$ is the operator that establishes a connection between the Laplace transform of a continuous function and the corresponding shifted lattice function [30].

TF $W^*_{\text{reduced}}(q, \varepsilon)$ can be expanded into an infinite power series with respect to the variable e^{-q}. That is, as for continuous systems, functions $W^*_{\text{reduced}}(q, \varepsilon)$ in the time domain correspond to an impulse function, as a response to an instantaneous unit impulse, which is different from zero on an infinite time interval.

The general TF of the forming circuit in the **D**—domain (DLT-domain) of shifted lattice functions will look like:

$$K^*_{\Sigma_{\text{Gm}}}(q, \varepsilon) = H(e^{-q}) \cdot A(\mathbf{a}_T, h) \cdot W^*_{\text{reduced}}(q, \varepsilon).$$

A necessary and sufficient condition for obtaining a finite response to each single pulse input (impulse transient characteristic of the forming circuit) is the coincidence of the poles of the analog part with the zeros of the discrete part.

So, if the TF of the discrete part will have the form $H(e^{-q}) = \sum_{j=0}^{m} a_j \cdot e^{q(j-m)}$, then the total TF of such a system can be represented by a finite number of power terms e^{-q}, i.e. contains only zeros:

$$K^*_{\Sigma\mathrm{Gm}}(q,\varepsilon) =$$

$$= A(\mathbf{a}_\mathrm{T},h)H(e^{-q})K_\mathrm{h}\frac{\sum_{k=0}^{m-1} b_k(\varepsilon)\,e^{q(m-k)}}{\sum_{j=0}^{m} a_j\,e^{q\cdot j}} = A(\mathbf{a}_\mathrm{T},h)K_\mathrm{h}e^{-mq}\sum_{k=0}^{m-1} b_k(\varepsilon)\,e^{q(m-k)}.$$

So, for **D**-transform of the *G*-spline we will have:

$$G^*_\mathrm{m}(q,\varepsilon) = K^*_{\Sigma\mathrm{Gm}}(q,\varepsilon) = A(\mathbf{a}_\mathrm{T},h)\cdot K_\mathrm{h}\cdot e^{-mq}\sum_{k=0}^{m-1} b_k(\varepsilon)\,e^{q(m-k)}\;.$$

2.5 Exponential Splines. Time Domain, Relative Scale

The impulse transient characteristic of the forming circuit (or finite basic spline) has the form:

$$G_\mathrm{m}(\mathbf{a},\bar{t}) = A\cdot(\mathbf{a})\cdot\mathbf{D}^{-1}\left\{e^{-mq}\cdot\sum_{k=0}^{m-1} b_k(\varepsilon)\cdot e^{q(m-k)}\right\} =$$

$$= A\cdot(\mathbf{a})\cdot\mathbf{D}^{-1}\left\{0 + b_0(\varepsilon) + b_1(\varepsilon)e^{-q} + \cdots + b_{\mathrm{m}-1}(\varepsilon)e^{-q(m-1)} + 0\right\},$$

where $A\cdot(\mathbf{a}) = A(\mathbf{a}_\mathrm{T},h)\cdot K_\mathrm{h}$; $\mathbf{D}^{-1}$ is an operator of the inverse DLT [30]; **a** are parameters of the TF CP (roots of the characteristic polynomial in the relative time scale), $\mathbf{a} = \mathbf{a}_\mathrm{T}h$. The variable ε acts as a real parameter that causes a simple transition in the time domain to the expression:

$$G_\mathrm{m}(\mathbf{a},\bar{t}) = \begin{cases} A\cdot(\mathbf{a})\cdot b_0(\varepsilon), & \bar{t}\in[0,\ 1];\\ A\cdot(\mathbf{a})\cdot b_1(\varepsilon), & \bar{t}\in[1,\ 2];\\ \quad\cdots\cdots\cdots\cdots & \\ A\cdot(\mathbf{a})\cdot b_{\mathrm{m}-1}(\varepsilon), & \bar{t}\in[m-1,\ m];\\ 0,\ \ \bar{t}<0,\ \ \bar{t}>m, & \end{cases}$$

When applied to the input of an impulse system, the transfer function of which $G^*_{\mathrm{m}}(q,\varepsilon)$, sequences of discrete samples $x[n] = f[n+1]$, the **D**-transform (DLT-domain) of the output signal will look like

$$sf^*_{\mathrm{Gm}}(q,\varepsilon) = G^*_{\mathrm{m}}(q,\varepsilon)\cdot F^*(q,0)\cdot e^q =$$
$$= F^*(q,0)\cdot A\cdot(\mathbf{a})\big[0 + b_0(\varepsilon)\cdot e^q + b_1(\varepsilon)\cdot e^0 + \cdots + b_{\mathrm{m}-1}(\varepsilon)\cdot e^{-q(m-2)} + 0\big],$$

where $F^*(q,0) = \mathbf{D}\{f[n]\}$, **D** is the operator of the direct DLT. The expression $sf^*_{\mathrm{Gm}}(q,\varepsilon)$ gives a spline in the time domain $sf_{\mathrm{Gm}}(\bar{t}) = sf_{\mathrm{Gm}}[n,\varepsilon]$, where $\bar{t} = n+\varepsilon$. Shift by one sample in the direction of advance in the values of the input discrete signal in relation to $f[n]$, i.e. $x[n] = f[n+1]$ or $X^*(q,0) = e^q\cdot F^*(q,0)$, serves to compensate for the time delay during the reproduction of a continuous dependence $f(\bar{t})$. The function $H'(e^{-q})$ can be taken into account in the input discrete signal in the form $X^*(q,0) = H'(e^{-q})\cdot F^*(q,0)\cdot e^q$.

Spline function $sf_{\mathrm{Gm}}(\bar{t})$ is the sum of responses to each input discrete sample (i.e., to the input sequence) $x[n] = f[n+1]$, $n = 0, 1, 2, \ldots$:

$$sf_{\mathrm{Gm}}(\bar{t}) = A\cdot(\mathbf{a})\big\{\,0 + f[n+1]\cdot b_0(\varepsilon) + f[n]\cdot b_1(\varepsilon) + \cdots + f[1]\cdot b_{\mathrm{m}-1}(\varepsilon) + 0\,\big\};$$

$$\bar{t} = n+\varepsilon;\ n = 0, 1, 2, \ldots;\ 0 \le \varepsilon \le 1.$$

Spline function $sf_{\mathrm{Gm}}(\bar{t})$ can be written as:

$$sf_{\mathrm{G}m}(\bar{t}) = \sum_{i=-\infty}^{\infty} f[n+1-i]\cdot G_m(i+\varepsilon),$$

where $\bar{t} = t/h$ is a relative time, $\bar{t} = n+\varepsilon$; h is a uniform sampling interval.

In terms of shifted lattice functions, the spline function has the form:

$$sf_{\mathrm{G}m}[n,\varepsilon] = \sum_{i=-\infty}^{\infty} f[n+1-i]\cdot G_{\mathrm{m}}[i,\varepsilon].$$

2.6 Examples of Synthesized Splines and Signal Generation

Exponential splines of the second order. Consider an example of the formation of an exponential spline of the second order (Figs. 5 and 6). The TF model of the combined continuous part in the real time scale, taking into account the multiplier $1/p$ of TF of DAC and normalizing factor $A(\alpha_{\mathrm{T}}, h)$, is:

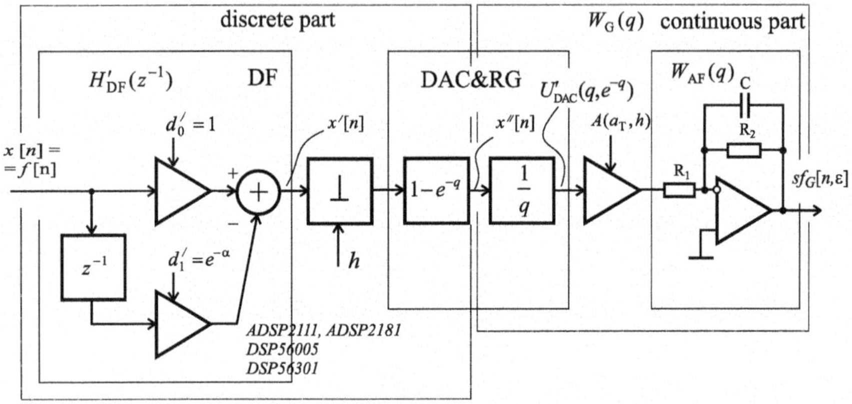

Fig. 5 Mathematical model of the forming circuit

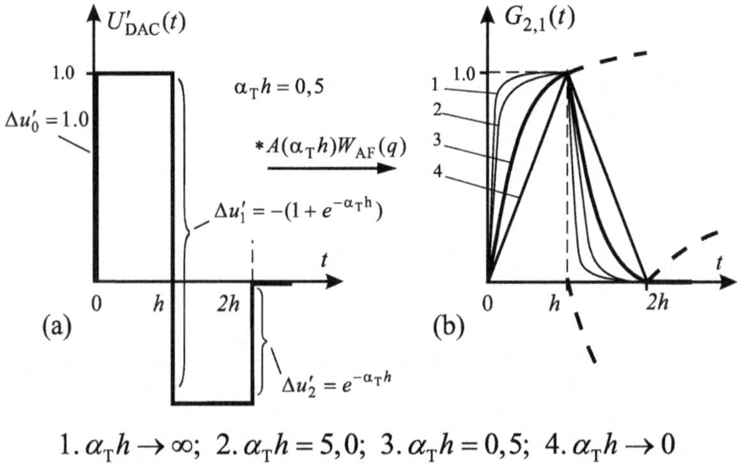

Fig. 6 Impulse weighting functions: **a** the impulse response at the DAC output; **b** the response at the output of the continuous part that is a finite spline $G_{2,1}(t)$ of the second order

$$W_{G2,1}(\alpha_T, h, p) = A(\alpha_T, h) W_{reduced}(p) = A(\alpha_T, h) \cdot \frac{1}{p(p + \alpha_T)}.$$

In the case of hardware implementation (see example in Fig. 5) the normalizing factor $A(\alpha_T, h) = 1/K_{10}(1 - e^{-\alpha_T h})$, $K_{10} = R_2/R_1$; $\alpha_T = 1/\tau = 1/R_2 C$.

The finite basis spline and spline function are shown below

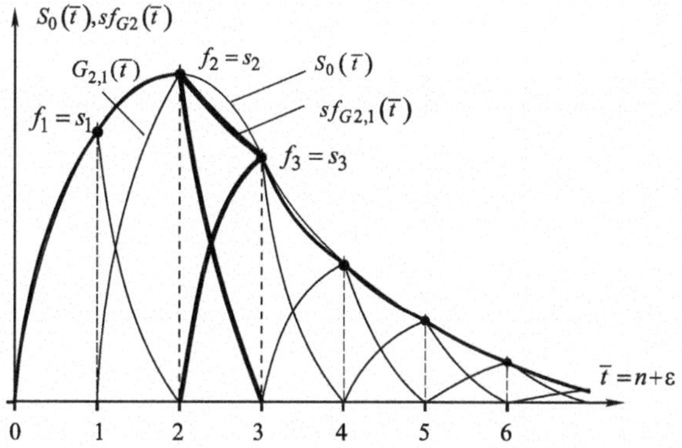

Fig. 7 Formation of signals by splines $G_{2,1}(t)$

$$G_{2,1}(\bar{t}) = \begin{cases} \frac{1}{(1-e^{-\alpha})} \cdot (1 - e^{-\alpha\varepsilon}), & \bar{t} \in [0,\ 1]; \\ \frac{1}{(1-e^{-\alpha})} \cdot (e^{-\alpha\varepsilon} - e^{-\alpha}), & \bar{t} \in [1,\ 2]; \\ 0, \ \bar{t} < 0, \ \bar{t} > 2. \end{cases}$$

$$sf_{G2,1}(\bar{t}) = \frac{1}{(1-e^{-\alpha})}\left\{f[n+1]\cdot(1-e^{-\alpha\varepsilon}) + f[n]\cdot(e^{-\alpha\varepsilon} - e^{-\alpha})\right\}, \quad \alpha = \alpha_T h;$$

$$\bar{t} = t/h = n + \varepsilon; \quad n = 0, 1, 2, \ldots; \quad 0 \le \varepsilon \le 1.$$

In Fig. 7 shows an example of signal formation $S_0(\bar{t})$, where $f[n]$ are discrete readings applied to the input of the forming circuit.

The values of the spline function $sf_{G2,1}(\bar{t})$ at the nodes when $\varepsilon = 0$ and $\varepsilon = 1$ and shaped signal $S_0(\bar{t})$, which is given by discrete samples, $f[n] = s[n]$ at the moments of discretization are coincided. That is, a spline of the second order $sf_{G2,1}(\bar{t})$ is used for interpolation. It is continuous at the nodes, but the continuity of the first derivative is not ensured.

Other models of second-order splines can be obtained by choosing another model of the continuous part of the forming circuit.

Exponential splines of the third order. Let's consider the synthesis of exponential splines of the 3rd order in more detail. Let TF model of the combined continuous part in the real time scale with $1/p$ DAC and normalizing factor $A(\alpha_T, h)$ is:

$$W_{G3,1}(\alpha_T, h, p) = A(\alpha_T, h)W_{\text{reduced}}(p) = A(\alpha_T, h)\frac{1}{p^2(p+\alpha_T)}.$$

Let's consider a relative time scale. The transfer function of the combined CP with consideration of the impulse element ⊥ is:

$$W_{G3,1}(\alpha_T, h, q) = A(\alpha_T, h)K_\Phi(q)\tilde{W}_{reduced}(q) = A(\alpha_T, h)\frac{1}{h}\left[\frac{1}{p^2(p+\alpha_T)}\right]_{/p=\frac{q}{h}} =$$

$$= A(\alpha_T, h)\frac{1}{h}\frac{h^3}{q^2(q+\alpha)} = A(\alpha_T, h)\frac{h^2}{q^2(q+\alpha)}, \quad \alpha = \alpha_T h,$$

where $K_\Phi(q) = \frac{1}{h}$ is the transfer coefficient of an ideal impulse element ⊥; $\tilde{W}_{reduced}(q) = \left[\frac{1}{p}W_{AF}(p)\right]_{p=q/h}$; $W_{AF}(p)$ is an ordinary TF of the continuous analog part.

The transfer function of the combined TF for the discretized signal is equal to

$$K_i^*(q, \varepsilon) = \mathfrak{D}\{W_{G3,1}(\alpha_T, h, q)\}.$$

But the tables [30] with $\mathfrak{D}$-transformations are not complete. Therefore, to receive $K_i^*(q, \varepsilon)$ use another way, namely use **D**-transformation of impulse transient characteristics (ITC), presented in the form of shifted lattice functions $g(\bar{t}) = g[n, \varepsilon]$, where $\bar{t} = n + \varepsilon$, or tables from **Z**-transformations [31], where ε—is a valid parameter.

The impulse transient characteristic of the combined CP (ITC by definition is a response to a single instantaneous impulse) is found from the expression for the TF of the combined CP according to tables of the continuous $\mathscr{L}$-Laplace transform as for a regular variable p [31]:

$$\mathscr{L}^{-1}\left\{\frac{h^2}{q^2(q+\alpha)}\right\} = \frac{h^2}{\alpha^2}[-1 + \alpha\bar{t} + e^{-\alpha\bar{t}}].$$

That is, the ITC of the combined CP taking into account the normalizing factor $A(\alpha_T, h)$ will look like:

$$g(\bar{t}) = A(\alpha_T, h)\frac{h^2}{\alpha^2}[-1 + \alpha\bar{t} + e^{-\alpha\bar{t}}],\ \bar{t} = n + \varepsilon,$$

or in the form of shifted lattice function:

$$g[n, \varepsilon] = A(\alpha_T, h)\frac{h^2}{\alpha^2}[-1 + \alpha(n+\varepsilon) + e^{-\alpha(n+\varepsilon)}].$$

Then the transfer function of the combined CP for the discretized signal is equal to:

$$K_i^*(q, \varepsilon) = \mathfrak{D}\{W_{G3,1}(\alpha, h, q)\} =$$

$$= \mathbf{D}\{g[n, \varepsilon]\} = A(\alpha_T, h)\left[\frac{h}{\alpha}\right]^2 \mathbf{D}\{-1 + \alpha(n+\varepsilon) + e^{-\alpha(n+\varepsilon)}\} =$$

$$= A(\alpha_T, h)\left[\frac{h}{\alpha}\right]^2\left[-\frac{e^q}{(e^q-1)} + \alpha\left(\frac{e^q}{(e^q-1)^2} + \frac{e^q}{(e^q-1)}\varepsilon\right) + \frac{e^q}{(e^q-e^{-\alpha})}e^{-\alpha\varepsilon}\right],$$

where ε is the real parameter.

After conversion $K_i^*(q,\varepsilon)$ to a simpler expression, taking into account the TF $H(e^{-q})$ of the combined discrete part, we get the general transfer function $K^*_{\Sigma_{G3}}(q,\varepsilon) = H(e^{-q}) \cdot K_i^*(q,\varepsilon)$ of the forming circuit for a discretized signal:

$$K^*_{\Sigma_{G3}}(q,\varepsilon) =$$
$$= H(e^{-q})A(\alpha_{\mathrm{T}},h)\left[\frac{h}{\alpha}\right]^2\left[\frac{e^q}{(e^q-1)^2(e^q-e^{-\alpha})}\right]\{[-1+\alpha\varepsilon+e^{-\alpha\varepsilon}]e^{2q}+$$
$$+[1+\alpha+e^{-\alpha}-(1+e^{-\alpha})\alpha\varepsilon-2e^{-\alpha\varepsilon}]e^q+e^{-\alpha}[-1+\alpha(\varepsilon-1)+e^{-\alpha(\varepsilon-1)}]\}.$$

Given

$$H(e^{-q}) = e^{-3q}(e^q-1)^2(e^q-e^{-\alpha}),$$

we will get **D**-transform finite $G_{3,1}(\overline{t})$-spline:

$$G_3^*(q,\varepsilon) =$$
$$= K^*_{\Sigma_{G3}}(q,\varepsilon) = A(\alpha_{\mathrm{T}},h)e^{-2q}\left(\frac{h}{\alpha}\right)^2\{[-1+\alpha\varepsilon+e^{-\alpha\varepsilon}]e^{2q}+$$
$$+[1+\alpha+e^{-\alpha}-(1+e^{-\alpha})\alpha\varepsilon-2e^{-\alpha\varepsilon}]e^q+e^{-\alpha}[-1+\alpha(\varepsilon-1)+e^{-\alpha(\varepsilon-1)}]\} =$$
$$= A\cdot(\alpha)e^{-q}\{[-1+\alpha\varepsilon+e^{-\alpha\varepsilon}]e^q+[1+\alpha+e^{-\alpha}-(1+e^{-\alpha})\alpha\varepsilon-2e^{-\alpha\varepsilon}]+$$
$$+e^{-\alpha}[-1+\alpha(\varepsilon-1)+e^{-\alpha(\varepsilon-1)}]e^{-q}\} = A\cdot(\alpha)e^{-q}[b_0(\varepsilon)e^q+b_1(\varepsilon)+b_2(\varepsilon)e^{-q}],$$

where $A\cdot(\alpha) = A(\alpha_{\mathrm{T}},h)\left(\frac{h}{\alpha}\right)^2$.

Piecewise functions that form a spline are:

$$b_0(\varepsilon) = -1+\alpha\varepsilon+e^{-\alpha\varepsilon},\ 0\le\varepsilon\le1;$$

$$b_1(\varepsilon) = 1+\alpha+e^{-\alpha}-(1+e^{-\alpha})\alpha\varepsilon-2e^{-\alpha\varepsilon},\ 0\le\varepsilon\le1;$$

$$b_2(\varepsilon) = e^{-\alpha}[-1+\alpha(\varepsilon-1)+e^{-\alpha(\varepsilon-1)}], 0\le\varepsilon\le1.$$

The multiplier e^{-q} gives a time delay that can be compensated for by shifting one count forward in the values of the input discrete signal with respect to $f[n]$, i.e. $x[n] = f[n+1]$ or $X^*(q,0) = e^q F^*(q,0)$.

Then **D**-transform spline function will be:

$$sf^*_{G3}(q,\varepsilon) =$$
$$= K^*_{\Sigma_{G3}}(q,\varepsilon)\cdot X^*(q,0) = K^*_{\Sigma_{G3}}(q,\varepsilon)\cdot e^q\cdot F^*(q,0) =$$

$$= A \cdot (\alpha)\left\{F^*(q,0)\left[-1+\alpha\varepsilon+e^{-\alpha\varepsilon}\right]e^{q} + F^*(q,0)\left[1+\alpha+e^{-\alpha}-\right.\right.$$
$$\left.\left.-(1+e^{-\alpha})\alpha\varepsilon - 2e^{-\alpha\varepsilon}\right]+F^*(q,0)e^{-\alpha}\left[-1+\alpha(\varepsilon-1)+e^{-\alpha(\varepsilon-1)}\right]e^{-q}\right\}.$$

The normalizing factor is chosen from the condition:

$$A\cdot(\alpha) = \left(\frac{h}{\alpha}\right)^2 \cdot A(\alpha_{\mathrm{T}}, h) = [b_0(\varepsilon)]^{-1}_{/\varepsilon=1} = \frac{1}{(-1+\alpha+e^{-\alpha})}.$$

From here we have:

$$A(\alpha_{\mathrm{T}}, h) = \left(\frac{\alpha}{h}\right)^2 \cdot \frac{1}{(-1+\alpha_{\mathrm{T}}h+e^{-\alpha_{\mathrm{T}}h})}.$$

The base ES $G_{3,1}(\bar{t}) = \mathbf{D}^{-1}\left\{G_3^*(q,\varepsilon)\right\}$ is equal to

$$G_{3,1}(\bar{t}) = \begin{cases} \dfrac{1}{-1+\alpha+e^{-\alpha}}\left[-1+\alpha\varepsilon+e^{-\alpha\varepsilon}\right], & \bar{t}\in[0,1]; \\ \dfrac{1}{-1+\alpha+e^{-\alpha}}\left[1+\alpha+e^{-\alpha}-(1+e^{-\alpha})\cdot\alpha\varepsilon-2e^{-\alpha\varepsilon}\right], & \bar{t}\in[1,2]; \\ \dfrac{1}{-1+\alpha+e^{-\alpha}}e^{-\alpha}\left[-1+\alpha(\varepsilon-1)+e^{-\alpha(\varepsilon-1)}\right], & \bar{t}\in[2,3]; \\ 0, \quad \bar{t}<0, \quad \bar{t}>3. \end{cases}$$

The spline function $sf_{G3,1}(\bar{t}) = \mathbf{D}^{-1}\left\{sf^*_{G3}(q,\varepsilon)\right\}$ is equal to

$$sf_{G3,1}(\bar{t}) = \frac{1}{-1+\alpha+e^{-\alpha}}\left\{f[n+1]\cdot(-1+\alpha\varepsilon+e^{-\alpha\varepsilon})+\right.$$
$$\left.+f[n]\cdot\left[1+\alpha+e^{-\alpha}-(1+e^{-\alpha})\cdot\alpha\varepsilon-2e^{-\alpha\varepsilon}\right]+f[n-1]e^{-\alpha}\left[\alpha(\varepsilon-1)-1+e^{-\alpha(\varepsilon-1)}\right]\right\}.$$

It can be shown that for a piecewise polynomial function $sf_{\mathrm{G3,1}}(\bar{t}) = sf_{\mathrm{G3,1}}[n,\varepsilon]$, where $\bar{t} = n+\varepsilon$; $\varepsilon \in [0,1][0,1]$, the continuity condition is met at the nodes both for the function itself and for its first derivative.

So indeed, the value of the function at the nodal points on n-th segment is determined by $\varepsilon = 0$ and $\varepsilon = 1$. The value of the spline function at the end of the previous segment $sf_{\mathrm{G3,1}}[n-1,\varepsilon]$ is determined by $\varepsilon = 1$. That is, for everyone n we have:

$$sf_{\mathrm{G3,1}}[n-1,1] = sf_{\mathrm{G3,1}}[n,0].$$

Therefore, the condition of continuity of the spline function at the nodes is met. The expression for the first derivative of the spline function has the form:

$$sf^{/}_{G3}[n,\varepsilon] = \alpha\frac{f[n+1](1-e^{-\alpha\varepsilon})+f[n](-1-e^{-\alpha}+2e^{-\alpha\varepsilon})+f[n-1]e^{-\alpha}[1-e^{-\alpha(\varepsilon-1)}]}{(-1+\alpha+e^{-\alpha})}.$$

Similarly, at $\varepsilon = 0$ and $\varepsilon = 1$ we obtain the value of the derivative spline function at the endpoints of the n-th segment. The value of the derivative at the end of the previous segment $sf^{/}_{G3,1}[n-1, \varepsilon]$ is determined by $\varepsilon = 1$. At the same time:

$$sf^{/}_{G3,1}[n-1, 1] = sf^{/}_{G3,1}[n, 0]$$

That is, the condition of continuity of the first derivative at the nodes is met.

TF of the combined discrete part is:

$$H(e^{-q}) = e^{-3q}(e^{q} - 1)^2(e^{q} - e^{-\alpha}).$$

where $e^{-q}(e^{q} - 1) = 1 - e^{-q}$ DAC with memory registers is implemented.

Then we have:

$$H(e^{-q}) = H^{/}_{DF}(e^{-q})[1 - e^{-q}] = H^{/}_{DF}(e^{-q}) \cdot e^{-q}(e^{q} - 1).$$

From here TF of the digital filter:

$$H^{/}_{DF}(e^{-q}) = e^{-2q}(e^{q} - 1)(e^{q} - e^{-\alpha});$$
$$H^{/}_{DF}(z^{-1}) = d^{/}_{0} + d^{/}_{1}z^{-1} + d^{/}_{2}z^{-2},$$

where

$$z^{-1} = e^{-q}, \quad d^{/}_{0} = 1, \quad d^{/}_{1} = -(1 + e^{-\alpha}), \quad d^{/}_{2} = e^{-\alpha}.$$

Models of the combined continuous and discrete part (digital filter structure) of the forming circuit are shown in Fig. 8.

Figure 9 is shown the pulse weight functions (as a reaction to a single pulse at the input of the forming circuit): in Fig. 9a—weighted piecewise step function formed by DAC; in Fig. 9b—a response of the CP to this weighting function. The reaction at the output of the CP is a finite spline $G_{3,1}(t)$.

Parameter α (Fig. 9b) allows you to change the shape of the finite spline $G_{3,1}(t)$, which makes it possible to adapt to the dependency model that is reproduced. Other models of splines of the third order can be obtained in the same way by choosing another model of the continuous part of the forming circuit.

Figure 10 is shown an example of the formation of some signal $S_0(\overline{t})$ by finite splines $G_{3,1}(t)$, where $f[n]$ are discrete samples that are used as inputs of the forming system. The values of the spline function $sf_{G3,1}(\overline{t})$ at the nodes when $\varepsilon = 0$ and $\varepsilon = 1$ and formed signal $S_0(\overline{t})$, if the spline is given by discrete samples $f[n] = s[n]$, at the moments of discretization $\overline{t} = n$, do not match. Hence, a finite exponential spline of the third order $G_{3,1}(\overline{t})$ should be considered as approximating, which complicates discrete synthesis. Namely, it requires preliminary determination of the values of discrete samples $f[n]$, which will be applied to the input of the forming pulse system, which should be considered as a spline approximating filter.

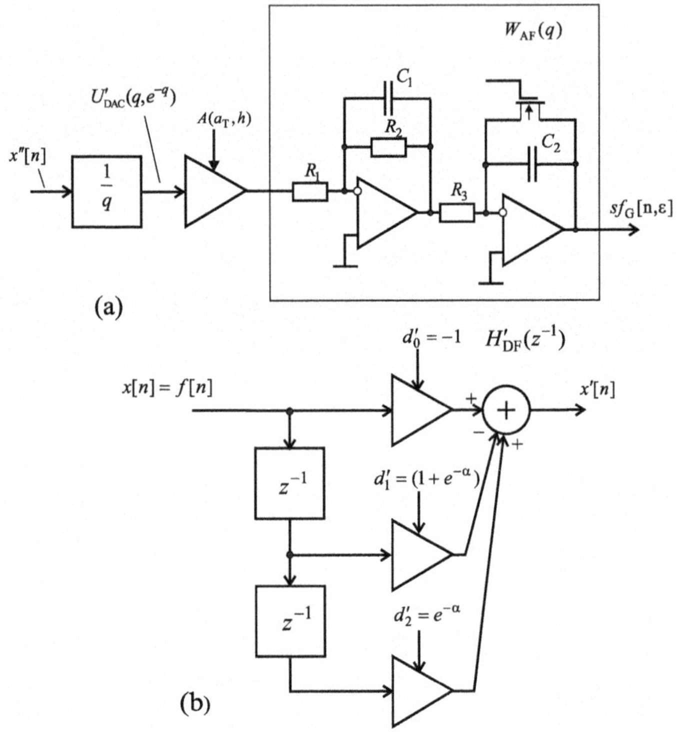

Fig. 8 Models of combined continuous and discrete parts of the forming circuit: **a** the model of the continuous part; **b** the structure of a digital filter

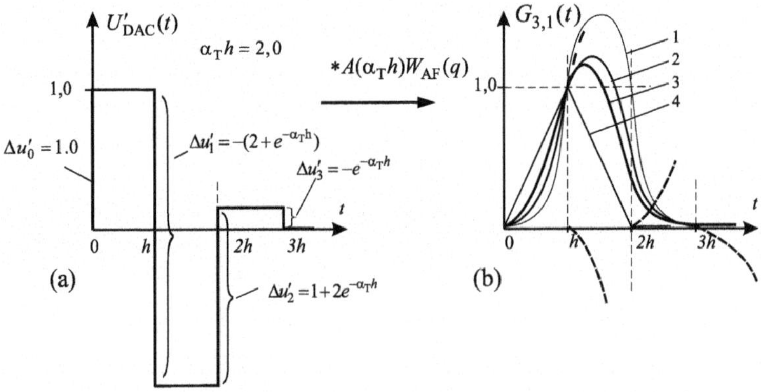

$1.\ \alpha_T h \rightarrow 0;\ \ 2.\ \alpha_T h = 1.0;\ \ 3.\ \alpha_T h = 2.0;\ \ 4.\ \alpha_T h \rightarrow \infty$

Fig. 9 Impulse weighting functions: **a** impulse weighting function at the DAC output; **b** the view of the finite spline $G_{3,1}(t)$

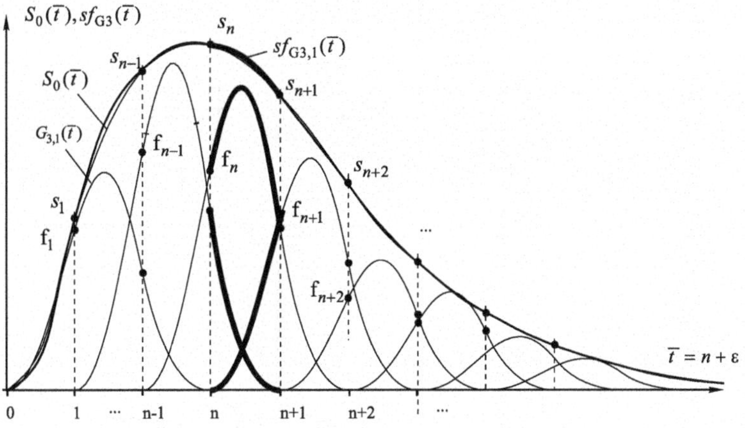

Fig. 10 Formation of signals by splines $G_{3,1}(t)$

Consider the case of interpolation by finite splines. So, for nth segment at $\varepsilon = 1$ we have:

$$s[n+1] = sf_{G3,1}(\bar{t})_{/\varepsilon=1},\ n = 0, 1, 2, \ldots$$

Given that (see the expression $sf_{G3,1}(\bar{t})$)

$$f[n-1]e^{-\alpha}\left[\alpha(\varepsilon-1) - 1 + e^{-\alpha(\varepsilon-1)}\right]_{/\varepsilon=1} = 0,$$

we obtain the recurrence relation:

$$s[n+1] = f[n+1] - \left[\frac{1 - e^{-\alpha} - \alpha e^{-\alpha}}{1 - \alpha - e^{-\alpha}}\right] \cdot f[n]\ = f[n+1] - A \cdot f[n],$$

where $A = \left[\frac{1-e^{-\alpha}-\alpha e^{-\alpha}}{1-\alpha-e^{-\alpha}}\right]$. That is, if one set the initial values $f[0] = 0$, $f[1] = s[1]$, then other values $f[n]$ are calculated by formulas $f[2] = s[2] + A \cdot f[1]$, $f[3] = s[3] + A \cdot f[2]$, …., $f[n+1] = s[n+1] + A \cdot f[n]$.

Implementation of spline interpolation can be achieved by preliminary including a digital filter (Fig. 11), which implements this algorithm.

Other models of splines of the third order can be obtained by choosing another model of the continuous part of the forming circuit. An additional parameter α in contrast to power splines allows you to change the shape of the exponential spline and adapt it to the model of the dependence that is reproduced.

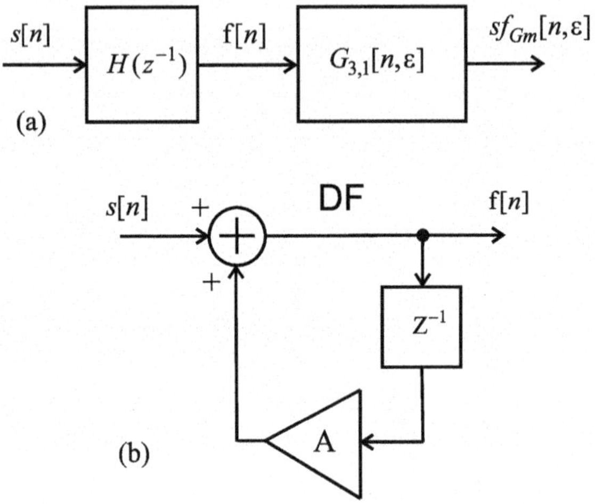

Fig. 11 Signal interpolation $s(\bar{t})$ splines $G_{3,1}(\bar{t})$: **a** interpolation scheme; **b** digital filter structure

3 Analysis of the Errors in Determining the Parameters of Electric Circuits Due to the Approximation of Test Signals by Splines

Let's consider the error of parameter determination by ZPM of the sequential R_x, C_x-two-pole circuit (Fig. 1) with the use of TS that formed based on exponential spline models (ESM) of signals, that is caused by the imperfection of the TS model, that is the methodological component. With a limited number of approximation sections N_a the indicated component of the error in determining the parameters of electric circuits is the main one.

Ideally shaped test signal. The impedance function of the sequential R_x, C_x—two-pole circuit (Table 1, item 1) is:

$$Z_x(p, \alpha_{x_T}) = R_x \cdot \frac{p + \alpha_{x_T}}{p}, \ \alpha_{x_T} = \frac{1}{R_x C_x}.$$

$$R_x = R_{x0} + \Delta R_x; \ C_x = C_{x0} + \Delta C_x.$$

Model of the ideal TS (current source is in sequential R_x, C_x-two-pole circuit):

$$I_{S0}(p) = \frac{1}{Z_{x0}(p)} \cdot \frac{1}{p} U_0 = \frac{1}{R_{x0}} \cdot \frac{1}{(p + \beta_{0_T})} U_0, \quad \beta_{0_T} = \alpha_{x0_T} = \frac{1}{R_{x0} C_{x0}}.$$

The model of the ideal TS (time domain) is:

$$i_{S0}(t) = \begin{cases} \frac{U_0}{R_{x0}} \cdot e^{-\beta_{x0_T} \cdot t}, & t \geq 0; \\ 0, & t < 0. \end{cases}$$

The unified response of the ideal shape TS is:

$$u_{out}(t) = \frac{R_x}{R_{x0}} e^{-\beta_{0_T} t} U_0 + \frac{R_x}{R_{x0}} \frac{\alpha_{x_T}}{\beta_{0_T}} (1 - e^{-\beta_{0_T} t}) U_0 =$$

$$= \frac{R_x}{R_{x0}} e^{-\beta_{0_T} t} U_0 + \frac{C_{x0}}{C_x} (1 - e^{-\beta_{0_T} t}) U_0 = u_{Rx}(t) + u_{Cx}(t), \quad t \geq 0.$$

Exponential spline model test signal. The non-ideality of the TS model is determined by the type of ES and by the method of building the model. Let's consider the case of using splines of the type $G_{2,1}(\overline{t})$ and $sf_{G2,1}(\overline{t})$ that is mathematically "similar" to the model of an ideal TS $S_0(\overline{t})$ (at each segment the analytical expression of the piecewise function coincides to $S_0(\overline{t})$).

The basic exponential spline is:

$$G_{2,1}(\overline{t}) = \begin{cases} \frac{1}{(1 - e^{-\alpha_G})} \cdot (1 - e^{-\alpha_G \varepsilon}), & \overline{t} \in [0, 1]; \\ \frac{1}{(1 - e^{-\alpha_G})} \cdot (e^{-\alpha_G \varepsilon} - e^{-\alpha_G}), & \overline{t} \in [1, 2]; \\ 0, \ \overline{t} < 0, \ \overline{t} > 2. \end{cases}$$

The piecewise exponential spline function is:

$$sf_{G2,1}(\overline{t}) = \sum_{i=-\infty}^{\infty} f[i+1] \cdot G_{2,1}(n + \varepsilon - i),$$

or

$$sf_{G2,1}(\overline{t}) = \frac{1}{(1 - e^{-\alpha_G})} \{ f[n+1](1 - e^{-\alpha_G \varepsilon}) + f[n](e^{-\alpha_G \varepsilon} - e^{-\alpha_G}) \}, \quad \alpha_G = \alpha_{G_T} h.$$

$$\overline{t} = t/h = n + \varepsilon; \quad n = 0, 1, 2, \ldots; \quad 0 \leq \varepsilon \leq 1.$$

where $\{f_i\}_{i=-\infty}^{\infty}$ are coefficients of the spline function (the sequence of discrete samples is $f[n], \ n = 0, 1, 2, \ldots$).

The model of TS formed by splines $G_{2,1}(\overline{t})$ is:

$$\frac{U_0}{R_{x0}} e^{-\beta_0 \overline{t}} \approx \frac{U_0}{R_{x0}} \left[1(\overline{t}) - sf_{G2,1}(\overline{t}) \right], \quad t \geq 0,$$

where $\bar{t} = n + \varepsilon$ is a relative time, $\bar{t} = t/h$, where h is a uniform sampling interval; $n = 0,\ 1,\ 2,\ \ldots; 0\ \le \varepsilon \le\ 1$; The unit step function $1(\bar{t})$ reproduces the initial step of TS.

3.1 Methods of Building a Spline Model Test Signal

Let's consider the following methods of constructing an exponential spline model (ESM) of the test signal.

1. Interpolation of the ideal TS.
2. "Uniform" approximation, close to optimal, on a uniform grid of nodes (with minimum absolute error at each approximation segment).
3. "Interpolation" of ideal response.

For comparison, let's use the error estimates that correspond to the application of TS formed based on piecewise step (PS) functions and piecewise linear (PL) functions with the same number of approximation segments N_a.

So, when $\alpha_{\mathrm{G}} \to 0$ and $\alpha_{\mathrm{G}} \to \infty$ ESM of TS is transformed into PL- and PS-dependences, respectively. As a result, according to the expressions of UR $U_{\mathrm{ZG}}[n, \varepsilon]$ the corresponding estimates of parameter determination error can be obtained for the case of piecewise step interpolation (PSI) and piecewise linear interpolation (CLI).

Figure 12 shows: 1—interpolation of an ideal TS; 2—approximation in the form of an ideal TS with a minimum absolute error at each segment; 3—construction of a spline model of TS by "interpolation" of the ideal response; 4—error by interpolation of an ideal TS; 5—error by "uniform" approximation (with minimal absolute error at each segment); 6—error by "interpolation" of an ideal response.

3.2 Interpolation of the Ideal Test Signal

$$i_{\mathrm{SO}}(\bar{t}) = \frac{U_0}{R_{\mathrm{x}0}} e^{-\beta_0 \bar{t}},\ i_{\mathrm{SO}}[n, \varepsilon] = \frac{U_0}{R_{\mathrm{x}0}} \mathrm{e}^{-\beta_0 \cdot n} \mathrm{e}^{-\beta_0 \cdot \varepsilon}.$$

Interpolation conditions are:

$$\frac{U_0}{R_{\mathrm{x}0}} \mathrm{e}^{-\beta_0 \cdot n} = \frac{U_0}{R_{\mathrm{x}0}} \left[1 - s f_{\mathrm{G}21}[n, \varepsilon]_{/\varepsilon=0}\right],$$

so $f_{\mathrm{G}21}[n, \varepsilon]_{/\varepsilon=0} = 1 - \mathrm{e}^{-\beta_0 \cdot n},\ \ n = 0,\ 1,\ 2, \ldots\ N_a - 1.$

Spline $f_{\mathrm{G}21}[n, \varepsilon]$ is interpolating. From here we have:

$$f[n] = 1 - \mathrm{e}^{-\beta_0 \cdot n},\ \ n = 0,\ 1,\ 2, \ldots\ N_a - 1.$$

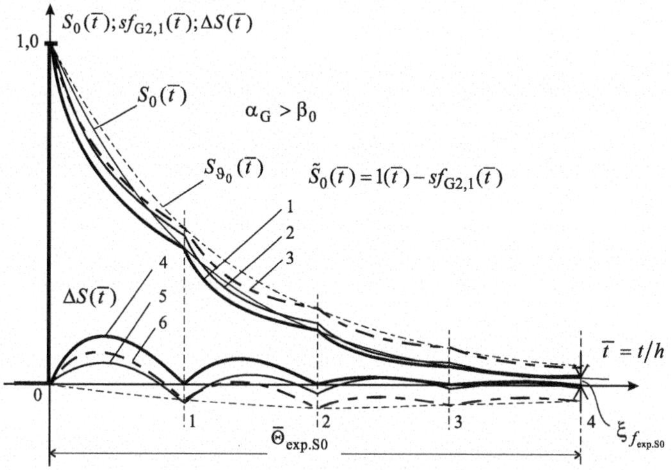

Fig. 12 Ways of building a spline model of the TS ($S_0(\bar{t}) = A_0 e^{-\beta_0 \bar{t}}, \bar{t} \geq 0,\ A_0 = 1,0$—the normalized function is considered)

The response at the output of the "impedance-voltage" converter $u_{\text{out}}[n, \varepsilon]$, which reproduces the impedance $Z_{\text{x}}(p, \alpha_{\text{x}_\text{T}})$ of a sequential R, C-two-pole circuit, in the case that TS is formed by $G_{2,1}(\bar{t})$-splines, will be denoted as $u_{\text{out}}[n, \varepsilon] = u_{\text{ZG}}[n, \varepsilon]$:

$$
\begin{gathered}
u_{\text{ZG}}[n, \varepsilon] = \frac{R_{\text{x}} U_0}{R_{\text{x}0}} \left\{ \frac{\alpha_{\text{x}}}{\alpha_{\text{G}}} + \frac{\alpha_{\text{x}}(e^{-\beta_0} - e^{-\alpha_{\text{G}}})}{(1 - e^{-\alpha_{\text{G}}})(1 - e^{-\beta_0})} + \right. \\
+ \left[(1 - \frac{\alpha_{\text{x}}}{\alpha_{\text{G}}}) \cdot \frac{e^{-\beta_0} - e^{-\alpha_{\text{G}}}}{(1 - e^{-\alpha_{\text{G}}})} - \frac{\alpha_{\text{x}}(e^{-\beta_0} - e^{-\alpha_{\text{G}}})}{(1 - e^{-\alpha_{\text{G}}})(1 - e^{-\beta_0})} \right] \cdot e^{-\beta_0 \cdot n} + \\
\left. + \frac{\alpha_{\text{x}}(e^{-\beta_0} - e^{-\alpha_{\text{G}}})}{(1 - e^{-\alpha_{\text{G}}})} e^{-\beta_0 \cdot n} \cdot \varepsilon + (1 - \frac{\alpha_{\text{x}}}{\alpha_{\text{G}}}) \cdot \frac{1 - e^{-\beta_0}}{(1 - e^{-\alpha_{\text{G}}})} e^{-\beta_0 \cdot n} e^{-\alpha_{\text{G}} \cdot \varepsilon} \right\},
\end{gathered} \tag{1}
$$

where $\alpha_{\text{x}} = \alpha_{\text{x}_\text{T}} h = \frac{1}{R_{\text{x}} C_{\text{x}}} h$, $\alpha_{\text{x}} = \alpha_{\text{x}0} + \Delta\alpha_{\text{x}}$—is the relative time zero of the impedance function of the sequential R, C—two-pole circuit, $R_{\text{x}}, C_{\text{x}}$ is the real circuit parameters; by the condition of synthesis $\beta_0 = \alpha_{\text{x}0}$, $R_{\text{x}0}$ is the parameters of an ideal shape TS; $\beta_0 = \beta_{0_\text{T}} h$—is a parameter of TS on a relative time scale; $\alpha_{\text{G}} = \alpha_{\text{G}_\text{T}} h$—is the time-relative parameter of the spline model. Here and then the index G is used to indicate response parameters $u_{\text{ZG}}[n, \varepsilon]$ in the case when the TS is formed based on G-splines.

The unified response (normalized) is shown in Fig. 13. For clarity, a small number of approximation sections is selected ($N_a = 4$).

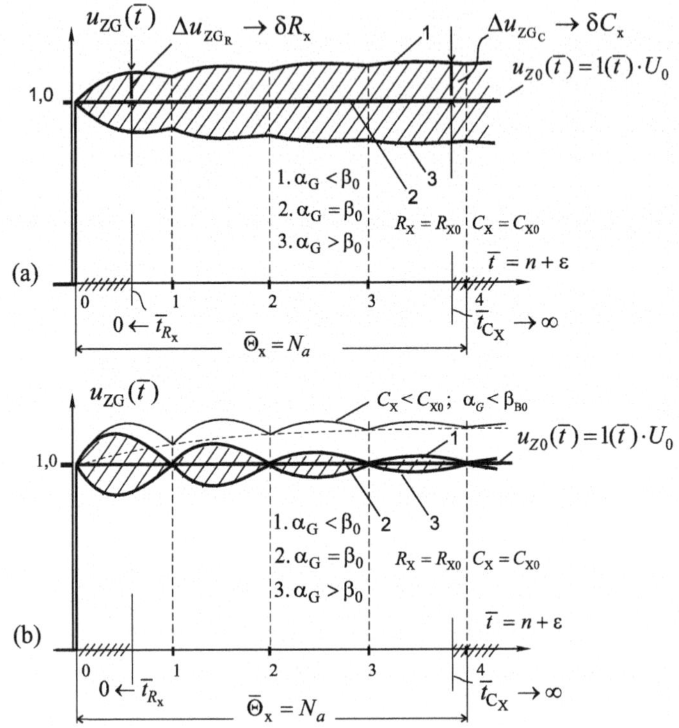

Fig. 13 Unified response (normalized): **a** interpolation of an ideal TS; **b** construction of the exponential spline model of the TS by "interpolation" according to the nominal response

3.3 Methodical Error Component Due to Interpolation

At $\alpha_G = \beta_0$—the spline and the ideal TS models at local segments are completely coincided. If $R_x \neq R_{x0}$ and $C_x \neq C_{x0}$, the condition is met $u_{ZG}[n, \varepsilon] = u_Z[n, \varepsilon]$, as in the case of an ideal shape TS, regardless of the number of approximation sections N_a:

$$u_Z[n, \varepsilon] = \frac{R_x}{R_{x0}} U_0 \, e^{-\beta_0 (n+\varepsilon)} + \frac{R_x}{R_{x0}} \frac{\alpha_x}{\beta_0} (1 - e^{-\beta_0 (n+\varepsilon)}) \, U_0.$$

It can also be shown that at $\alpha_G \neq \beta_0 \alpha_G \neq \beta_0$

$$\lim_{N_a \to \infty} u_{ZG}[n, \varepsilon] = u_Z[n, \varepsilon].$$

Similarly for the incoming VS $\lim_{N_a \to \infty} \frac{U_0}{R_{x0}} (1 - sf_{G21}[n, \varepsilon]) = i_{S0}[n, \varepsilon]$.

If $\alpha_G = \beta_0$ and $R_x = R_{x0}$ and $C_x = C_{x0}$, then we get $u_{ZG}[n, \varepsilon] = 1\,[n, \varepsilon] \cdot U_0$—the response is equal to the nominal, i.e. $u_{Z0}(\bar{t}) = 1(\bar{t}) \cdot U_0$ (line 2, Fig. 13a), as in the case of an ideal shape TS. There is no methodical component of the parameter determination error due to interpolation.

If $\alpha_G \neq \beta_0$ then when $R_x = R_{x0}$ and $C_x = C_{x0}$—the deviation appears $\Delta u_{ZG}(\bar{t})$ of the response from the nominal $u_{Z0}(\bar{t}) = 1(\bar{t}) \cdot U_0$ (the lines 1–3, Fig. 13a). At the same time deviation $\Delta u_{ZG}(\bar{t})$ displays only the methodical component of the deviation error due to the approximation method.

3.4 Approximation with Minimum Absolute Error on Each Approximation interval

According to ZPM, the information about the circuit parameters is obtained in the time domain from the response instantaneous values which reproduces the immittance. The response shape is a "generalized parameter" of the circuit.

The error of the TS formation

$$\Delta S(\bar{t}) = S_0(\bar{t}) - sf_{G2,1}(\bar{t})$$

can be estimated (the numerical characteristic of time dependence $\Delta S(\bar{t})$) as

$$\delta S_m = \max_{\bar{t}\in[0,\overline{\Theta}_{exp.S0}]} \left|\Delta S(\bar{t})\right| / S_m,$$

where $S_m = \max_{\bar{t}\in[0,\overline{\Theta}_{exp.S0}]} \{S(\bar{t})\}$, $\overline{\Theta}_{exp.S0} = N_a$. That is, the approximation problem must be solved according to the minimax criterion. But in technical applications, the problem of uniform approximation is often solved very well.

For each segment, the functions are strictly convex (or concave). In this case, at each segment, the error inside the interval changes its sign once and has an opposite sign and equal magnitude values at the ends of the segment. That is, an approximation qualitatively similar to Chebyshev's (the best uniform approximation) is possible.

So, on each segment $S_0[n, \varepsilon]$ the ESM values $sf_{G2,1}[n, \varepsilon]$ are looked for at nodes $\{f_n\}_{n=0}^{N_a-1}$ where the error modulus is equal to the error within the interval. Provided $\alpha_G > \beta_0$ the set of values for the spline model at the nodes $\{f_n\}_{n=0}^{N_a}$ can be obtained as a result of solving the system

$$S_0[n, \varepsilon]_{/\varepsilon=0} - sf_G[n, \varepsilon]_{/\varepsilon=0} = -\Delta_n;$$

$$S_0[n, \varepsilon]_{/\varepsilon=\varepsilon_m} - sf_G[n, \varepsilon]_{/\varepsilon=\varepsilon_m} = +\Delta_{n+1};$$

$$S_0[n, \varepsilon]_{/\varepsilon=1} - sf_G[n, \varepsilon]_{/\varepsilon=1} = -\Delta_{n+1};$$

for every $n = \overline{0, N_a - 1}$, where $\frac{d}{d\varepsilon}\{S_0[n, \varepsilon] - sf_G[n, \varepsilon]\}_{/\varepsilon=\varepsilon_m} = 0$.

The substitution of the expressions $sf_{G2,1}(\overline{t})$ and models of ideal TS $S_0(\overline{t}) = e^{-\beta_0 \overline{t}}$ into the system turns it into a new system

$$e^{-\beta_0 n} - f[n] = -\Delta_{\text{n}};$$

$$e^{-\beta_0[n+\varepsilon_{\text{m}}]} - \frac{1}{(1-e^{-\alpha_{\text{G}}})}\left\{f[n+1](1-e^{-\alpha_{\text{G}}\varepsilon_{\text{m}}}) + f[n](e^{-\alpha_{\text{G}}\varepsilon_{\text{m}}} - e^{-\alpha_{\text{G}}})\right\} = \Delta_{\text{n}+1};$$

$$e^{-\beta_0(n+1)} - f[n+1] = -\Delta_{\text{n}+1};$$

where $\varepsilon_m = -\frac{1}{(\beta_0-\alpha_{\text{G}})}\ln\left[\frac{\alpha_{\text{G}}}{\beta_0(1-e^{-\alpha_{\text{G}}})}\{f[n]-f[n+1]\}\right]$, which is obtained from $\frac{d}{d\varepsilon}\{S_0[n,\varepsilon] - sf_G[n,\varepsilon]\}_{/\varepsilon=\varepsilon_m} = 0$.

That leads to

$$e^{-\beta_0(n+\varepsilon_{\text{m}})} - \left[e^{-\beta_0(n+1)} + \Delta_{n+1}\right]\frac{1-e^{-\alpha_{\text{G}}\varepsilon_{\text{m}}}}{(1-e^{-\alpha_{\text{G}}})} - (e^{-\beta_0\cdot n} + \Delta_n)\frac{e^{-\alpha_{\text{G}}\varepsilon_{\text{m}}} - e^{-\alpha_{\text{G}}}}{(1-e^{-\alpha_{\text{G}}})} = \Delta_{n+1};$$

$$\varepsilon_{\text{m}} = -\frac{1}{\beta_0-\alpha_{\text{G}}}\ln\left[\frac{\alpha_{\text{G}}}{\beta_0(1-e^{-\alpha_{\text{G}}})}\left[e^{-\beta_0\cdot n}(1-e^{-\beta_0}) + \Delta_n - \Delta_{n+1}\right]\right]; \quad n = \overline{0, N_a-1},$$

where Δ_n is an error in the n-th segment. If $\alpha_{\text{G}} < \beta_0$, then it is necessary to change the sign for the error in each section of the approximation.

At the starting point we accept $\Delta_0 = 0$, $f[0] = 0$ and calculate Δ_1. Then we will get $f[1] = 1 - \text{e}^{-\beta_0} - \Delta_1$. By the error value Δ_{n} on the previous segment of the approximation we find the value $\Delta_{\text{n}+1}$ on the next one. Then we will get $f[n+1] = 1 - \text{e}^{-\beta_0\cdot n} - \Delta_{n+1}$ and we do this for every $n = \overline{0, N_a-1}$. As a result, we find all values $sf_{G2,1}[n,\varepsilon]$ in nodes $\{f_n\}_{n=0}^{N_a-1}$ (we will also find errors Δ_{n}, $n = \overline{0, N_a-1}$). These values define the input discrete signal $f[n]$, $n = 0, 1, 2, \ldots$ of the forming system.

But such a lattice function $f[n]$ cannot be provided in closed form. So, it does not have an expression in closed analytical form for the **D**-transform by discrete Laplace transform [30]. Therefore, to determine feedback $U_{\text{ZG}}[n,\varepsilon]$ at the output of the "immittance-voltage" converter, the general expression should be used for arbitrary values of the spline function at the nodes, when the input discrete signal of the forming system is simply a set of discrete readings $\{f_n\}_{n=0}^{N_a-1}$.

$$\begin{aligned} u_{\text{ZG}}[n,\varepsilon] = {} & \frac{(-)R_{\text{x}}U_0}{R_{\text{x0}}\alpha_{\text{G}}(1-e^{-\alpha_{\text{G}}})}\cdot\left\{\alpha_{\text{G}}\alpha_{\text{x}}(1-e^{-\alpha_{\text{G}}})\sum_{m=0}^{n-1} f[m] + \right. \\ & + f[n]\cdot\left[(\alpha_{\text{G}}-\alpha_{\text{x}})(e^{-\alpha_{\text{G}}\varepsilon} - e^{-\alpha_{\text{G}}}) + \alpha_{\text{G}}\alpha_{\text{x}} - \alpha_{\text{G}}\alpha_{\text{x}}e^{-\alpha_{\text{G}}}\cdot\varepsilon\right] + \\ & \left. + f[n+1]\cdot\left[(\alpha_{\text{G}}-\alpha_{\text{x}})(1-e^{-\alpha_{\text{G}}\varepsilon}) + \alpha_{\text{G}}\alpha_{\text{x}}\cdot\varepsilon\right]\right\} + \\ & + \frac{R_{\text{x}}U_0}{R_{\text{x0}}}\cdot[1+\alpha_{\text{x}}(n+\varepsilon)], \quad n = \overline{0, N_a-1}; \quad 0 \le \varepsilon \le 1. \end{aligned} \tag{2}$$

If you choose a value of $f[n]$ that is subjected to interpolation $f_{G21}[n, \varepsilon]_{/\varepsilon=0} = 1 - e^{-\beta_0 \cdot n}$, $n = 0, 1, 2, \ldots N_a - 1$, that is, the model of the TS is

$$\frac{U_0}{R_{X0}} e^{-\beta_0 \cdot n} e^{-\beta_0 \cdot \varepsilon}{}_{/\varepsilon=0} = \frac{U_0}{R_{X0}} [1 - sf_{G21}[n, \varepsilon]]_{/\varepsilon=0},$$

we will have $f[m] = 1 - e^{-\beta_0 \cdot m}$. In this case we have $\sum_{m=0}^{n-1} f[m] = n - \frac{1-e^{-\beta_0 n}}{1-e^{-\beta_0}}$ and from (2) we obtain the expression for $U_{ZG}[n, \varepsilon]$ the same as by interpolation of an ideal TS (1).

3.5 "Interpolation" of an Ideal Response

«Interpolation» of the ideal response $u_{Z0}(\bar{t}) = 1(\bar{t}) \cdot U_0$ allows us to significantly reduce the methodical component of the measurement error caused by the non-ideality of the TS model when $\alpha_G \neq \beta_0$. The essence of the method is to determine the coefficients of ESM of TS $\{f_i\}_{i=0}^{N_a-1}$ using interpolation of the ideal response $u_{Z0}(\bar{t}) = 1(\bar{t}) \cdot U_0$ by the real one $u_{ZG}(\bar{t})$. The ideal response can be obtained at the output of the "immittance-voltage" converter when there are an ideal shape TS and the parameters $C_x = C_{x0}$, $R_x = R_{x0}$ of the circuit under investigation have nominal values. That is, when forming TS by splines $G_{2,1}(\bar{t})$, and $\alpha_G \neq \beta_0$, $C_x = C_{x0}$, $R_x = R_{x0}$, the response reproduction error at the nodes is zero. The problem of the synthesis of TS, which is essentially an approximation, is solved by minimizing the error of determining the parameters C_x and R_x only at nodes.

The interpolation conditions are:

$$u_{ZG}[n, \varepsilon]_{/\varepsilon=0} = u_{Z0}[n, \varepsilon]_{/\varepsilon=0},$$

where $u_{Z0}[n, 0] = 1[n] \cdot U_0$, $n = 0, 1, 2, \ldots, N_a$; $\varepsilon = 0$.

The inverse solution of the problem of "input–output" transformation with respect to instantaneous samples at the nodes of the TS will be:

$$X_{20}[n] - \frac{U_0}{R_{x0}} e^{-\vartheta_0 \cdot n} = \frac{U_0}{R_{x0}} (1 - sf_{21}[n, \varepsilon]_{/\varepsilon=0});$$

where $\vartheta_0 = -\ln d_0$; $d_0 = \frac{\alpha_G \alpha_{x0} e^{-\alpha_G} + (\alpha_G - \alpha_{x0})(1 - e^{-\alpha_G})}{\alpha_G \alpha_{x0} + (\alpha_G - \alpha_{x0})(1 - e^{-\alpha_G})}$;

$$sf_{21}[n, \varepsilon]_{/\varepsilon=0} = \frac{U_0}{R_{x0}} (1 - e^{-\vartheta_0 \cdot n}).$$

That is, the model of the input TS will be:

$$i_0(\bar{t}) = \frac{U_0}{R_{x0}} e^{-\vartheta_0 \bar{t}}.$$

The unified feedback is:

$$
\begin{aligned}
u_{ZG}[n,\varepsilon] = {} & \frac{R_x U_0}{R_{x0}} \cdot \left\{ \frac{\alpha_x}{\alpha_G} + \alpha_x \cdot \frac{d_0 - e^{-\alpha_G}}{(1 - e^{-\alpha_G})(1 - d_0)} + \right. \\
& + \left[(1 - \frac{\alpha_x}{\alpha_G}) \cdot \frac{d_0 - e^{-\alpha_G}}{(1 - e^{-\alpha_G})} - \alpha_x \cdot \frac{d_0 - e^{-\alpha_G}}{(1 - e^{-\alpha_G})(1 - d_{B0})} \right] \cdot e^{-\vartheta_0 \cdot n} + \\
& \left. + \alpha_x \cdot \frac{d_0 - e^{-\alpha_G}}{(1 - e^{-\alpha_G})} \cdot e^{-\vartheta_0 \cdot n} \cdot \varepsilon + (1 - \frac{\alpha_x}{\alpha_G}) \cdot \frac{1 - d_0}{(1 - e^{-\alpha_G})} \cdot e^{-\vartheta_0 \cdot n} e^{-\alpha_G \varepsilon} \right\}.
\end{aligned} \tag{3}
$$

So, for this example (the determination of the parameters of the sequential R, C-two-pole circuit) the TS model will also be an exponent, but will have a different damping coefficient. This corresponds to the use of a correction in the model of the input TS, which minimizes the error due to the shift of the ordinates of the response nodes $U_{ZG}[n,\varepsilon]$ because of $\alpha_G \neq \beta_0$. The view of the URis shown in Fig. 13, b. Feedback reproduction error in nodes on condition $C_x = C_{x0}$, $R_x = R_{x0}$ is zero if $\alpha_G \neq \beta_{B0}$.

3.6 *Estimation of the Methodical Error Component Caused by the Approximation Method*

If $\alpha_G \neq \beta_{B0}$ and if $R_x = R_{x0}$ and $C_x = C_{x0}$ the deviation displays only the *methodical* component due to the approximation (Fig. 13). Let is consider the normalized uniform response, when $U_0 = 1$.

The duration of the exponential TS $\Theta_{\text{exp.S0}} = (1/\beta_{0_T}) \cdot |\ln \xi_{f_x}|$, where $\beta_0 = \beta_{0_T} h$, $\overline{\Theta}_{\text{exp.S0}} = N_a$ is determined at the relative level of significance $\xi_{f_x} = e^{-\beta_{0_T}\Theta_x}$. For example, when $\xi_f = 0{,}01$ the minimum duration of the exponential TS is $\Theta_{\text{exp.S0}} = \Theta_x = 4{,}6\ \tau_{x0}$. Moreover, $\Theta_{\text{exp.S0}} = \Theta_x = N_a h$. From where, if we set N_a, we obtain the following relations for h, β_0 and α_G:

$$
h = \frac{\Theta_x}{N_a} = \frac{|\ln \xi_{f_{\text{exp.S0}}}|}{\beta_{0_T} N_a}; \ \beta_0 = \beta_{0_T} h = \beta_{0_T} \frac{|\ln \xi_{f_{\text{exp.S0}}}|}{\beta_{0_T} N_a} = \frac{|\ln \xi_{f_{\text{exp.S0}}}|}{N_a};
$$

$$
\alpha_G = \alpha_{G_T} h = \alpha_{G_T} \frac{|\ln \xi_{f_{\text{exp.S0}}}|}{\beta_{0_T} N_a} = \chi_G \frac{|\ln \xi_{f_{\text{exp.S0}}}|}{N_a}, \ \chi_G = \frac{\alpha_{G_T}}{\beta_{0_T}} = \frac{\alpha_G}{\beta_0}.
$$

The specified ratios when substituting them into expressions for $i_{S0}(\bar{t})$ and $sf_{G2,1}(\bar{t}) = f_{G2,1}[n,\varepsilon]$, $u_{ZG}[n,\varepsilon]$ or $u_Z[n,\varepsilon]$ allow during calculations to establish a connection between the methodical component of the error in determining the parameters R, C (or reproducing in the form of the nominal model the TS during its generation) and the number of discrete readings or approximation sections used for this purpose N_a.

Evaluation of R_x by URwhen reading information about the parameter in the interval $\bar{t}_{R_x} \in [0,\ 1]$ is determined by the value $u_{ZG_R} = \max_{\varepsilon \in [0,1]} |u_{ZG}[0, \varepsilon]|$. If $\alpha_G \neq \beta_{B0}$, $R_x = R_{x0}$, $C_x = C_{x0}$ the methodical component of the definition error $R_x = R_{x0}$ is equal to:

$$\delta R_x \cong \Delta u_{ZG_R}/u_{R0}, \text{ where} \Delta u_{ZG_{Rx}} = \max_{\varepsilon \in [0,1]} |u_{ZG}[0,\ \varepsilon] - 1\,|,\ u_{R0} = 1, 0$$

When $\alpha_G = \beta_0$, $R_x \neq R_{x0}$, $C_x \neq C_{x0}$, as it follows from (1), the estimate of the parameter is $u_{ZG_{Rx}} = \frac{R_x}{R_{x0}} U_0 \cong 1 + \Delta R_x / R_{x0}$, as in the case of an ideal shape TS.

The methodological component of the definition error $C_x = C_{x0}$ is equal to:

$$\delta C_x \cong \Delta u_{ZG_{Cx}} / u_{C0},\ \text{where } \Delta u_{ZG_{Cx}} = \max_{\varepsilon \in [0;1]} \left| u_{ZG}[N_a - 1, \varepsilon] - 1 \right|,\ u_{C0} = 1, 0$$

As follows from expression (1), deviation $u_{ZG}(\bar{t})$ from nominal $u_{Z0}(\bar{t}) = 1(\bar{t}) \cdot U_0$ at $\alpha_G \neq \beta_0$ ($R_x = R_{x0}$, $C_x = C_{x0}$) and reading information about the parameter on the interval $\bar{t}_{C_x} \in [N_a - 1,\ N_a]$ is:

$$\Delta u_{ZG_{Cx}} \cong \frac{\alpha_{x0}}{\alpha_G} - 1 + \frac{\alpha_{x0}\,(e^{-\beta_0} - e^{-\alpha_G})}{(1 - e^{-\alpha_G})\,(1 - e^{-\beta_0})}.$$

$u_{C0} = 1, 0$—is the estimation of the parameter C_x according to the normalized UR, as with the ideal TS, if $C_x = C_{x0}$ ($u_{C_x} = \frac{C_{x0}}{C_x} \cdot U_0 \cong 1$).

Note that when $\alpha_G = \beta_0$; $R_x \neq R_{x0}$, $C_x \neq C_{x0}$ and $\bar{t}_{C_X} \to \infty$, as follows from (1), the estimate of the parameter C_x for normalized response will be $u_{ZG_{Cx}} = \frac{C_{x0}}{C_x} U_0 \cong 1 - \frac{\Delta C_x}{C_{x0}}$ and coincides with the estimate for an ideal TS.

As follows from the expression (3) deviation of $u_{ZG}(\bar{t})$ from nominal $u_{Z0}(\bar{t}) = 1(\bar{t}) \cdot U_0$ for the case of interpolation of an ideal shape response is

$$\Delta u_{ZG_{Cx}} \cong \frac{\alpha_{x0}}{\alpha_G} - 1 + \frac{\alpha_{x0}\,(e^{-\vartheta_0} - e^{-\alpha_G})}{(1 - e^{-\alpha_G})\,(1 - e^{-\vartheta_0})},$$

where $\vartheta_0 = -\ln\left[\frac{\alpha_G\,\alpha_{x0}\,e^{-\alpha_G} + (\alpha_G - \alpha_{x0})\,(1 - e^{-\alpha_G})}{\alpha_G\,\alpha_{x0} + (\alpha_G - \alpha_{x0})\,(1 - e^{-\alpha_G})}\right]$.

The estimates of the methodical component of the determination error of the C_x parameter for the case of interpolation of the ideal TS by splines are given in Table 3, where a comparison with the errors in the case of interpolation by piecewise-step and piecewise-linear functions is also given.

The use of basic splines, which shape of piecewise functions, that form spline in local segments, are similar to the formed test signals, provides a significantly smaller error in determining the parameters of electric circuits. So, for example, for the $G_{2,1}(\bar{t})$—splines the error in the entire range $\chi_G \in [0;\infty]$, where $\chi_G = \alpha_G/\beta_0$, is significantly less in comparison to the synthesis based on PS functions, and less in the range $\chi_G \in [0;2.6]$ in comparison to the synthesis based on PL functions.

Table 3 The error estimation of the C_{x} determination (interpolation of the ideal TS)

Clock frequency of DAC $f_{\mathrm{T}} = 1/h$ [MHz]	Duration of the TS Θ_{S0} (or the measurement time Θ_{x}) [4 μs]		The error estimates of the C_{x} determination due to interpolation with splines, if the model spline parameter α_G is equal to		The error estimates of the C_{x} determination due to interpolation with splines PSI and PLI	
	Θ_{S0} $N_a h$	Θ_{x} $\tau_{\mathrm{x0}} \cdot \lvert \ln \xi_{f_{\mathrm{x}}} \rvert$	$\alpha_G \in [0;\ 2.6\alpha_{\mathrm{x0}}]$ [%]	$\alpha_G \in [0.8\alpha_{\mathrm{x0}};\ 1.2\alpha_{\mathrm{x0}}]$ [%]	PSI [%]	PLI [%]
1	4 h	$4.6 \cdot \tau_{\mathrm{x0}} = = 4\ \mu\mathrm{s}$, where $\tau_{\mathrm{x0}} = = 0.87\ \mu\mathrm{s}$ $\xi_{f_{\mathrm{x}}} = 0.01$	≤14	≤1	47	14
2	8 h		≤3	≤0.21	26	3
4	16 h		≤0.91	≤0.07	13.6	0.91
8	32 h		≤0.18	–	6.9	0.18
16	64 h		≤0.053	–	3.45	0.053
32	128 h		–	–	1.7	0.018
64	256 h		–	–	0.83	–
128	512 h		–	–	0.4	–
256	1024 h		–	–	0.2	–
512	2048 h		–	–	0.1	–

Note PSI is the piecewise-step interpolation; PLI is the piecewise linear interpolation. The duration of TS—$\Theta_{\mathrm{S0}} = N_a h = \tau_{\mathrm{x0}} \lvert \ln \xi_{f_{\mathrm{x}}} \rvert$, i.e., the DAC clock frequency is defined as $f_{\mathrm{T}} = N_a/[4.6\tau_{\mathrm{x0}}]$

The comparison of the ESM parameters and the TS model $\chi_{\mathrm{G}} = \alpha_{\mathrm{G}}/\beta_0$, is close to $\chi_{\mathrm{G}} = 1.0$, the methodical error component regardless of N_a and the method of approximation goes to zero. On the other hand, real piecewise step and piecewise linear functions during their formation are close to piecewise exponential ones.

In the case of interpolation by the $G_{2,1}(\bar{t})$—splines with the ratio $\chi_{\mathrm{G}} = \alpha_{\mathrm{G}}/\beta_0$ in the range $\chi_{\mathrm{G}} \in [0.34;\ 1.7]$ (α_{G} is within ±70% relative to the optimal value $\alpha_{\mathrm{G0}} = \beta_0$) the determination error of R_{x} and C_{x} will be, at least, 2 times less than in the case of formation of TS based on PL-functions, and, at least, 6 times less compared to PS-functions. To get the same result, for example, when using PS-functions (with given Θ_{x} and $\xi_{f_{\mathrm{x}}}$) the N_a should be increased (the DAC sampling frequency) more than 8 times.

Thus, during the formation of the TS of 4 μs duration (to determine the parameters of circuits with $\tau_{\mathrm{x0}} = 0.87\ \mu\mathrm{s}$) based on the PS function to ensure a measurement error of less than 1% the DAC sample rate should be 64 MHz ($N_a = 256$). To ensure an error of less than 0.1% the DAC sample rate should be 512 MHz ($N_a = 2048$).

On the other hand, when forming the TS based on the ESM, to ensure an error of less than 1% without optimizing the ESM model parameters ($\chi_{\mathrm{G}} \in [\,0;\ 2.6]$),$N_a = 16$ is enough (the DAC sample rate is 4 MHz). When using the optimization ($\chi_{\mathrm{G}} \in [0.78;\ 1.2]$, the parameter α_{G} of the TS spline model is within ±20% of the optimal value $\alpha_{\mathrm{G0}} = \beta_0$), to ensure an error of less than 1% $N_a = 4$ is sufficient (the DAC sample rate is 1 MHz). To ensure an error of less than 0.1% when $\chi_{\mathrm{G}} \in [0.78;\ 1.2]$ it should be $N_a = 16$ and $f_{\mathrm{T}} = 4$ MHz.

Figures 14 and 15 dashed lines 3 indicate the estimates of the parameter determination error for the case of interpolation in the form of an ideal response. Thus, the error in determining the parameters of the MTPEC can be reduced by a factor of two

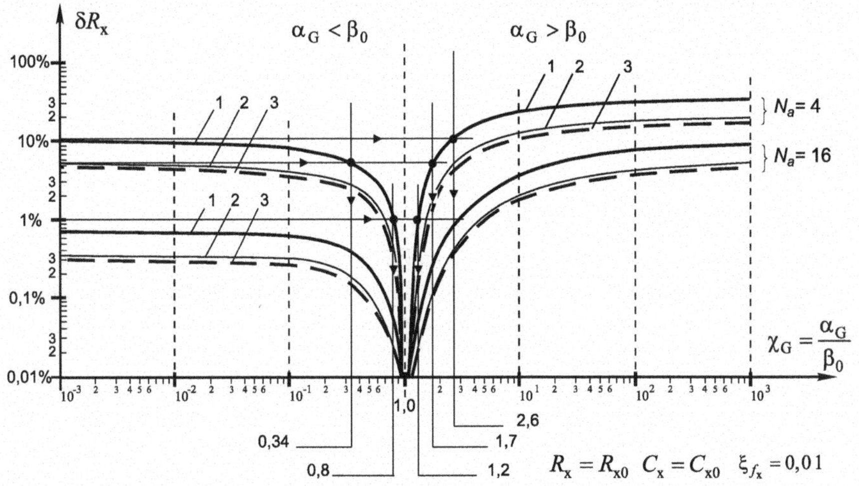

Fig. 14 The methodological error component δR_x due to the approximation method: 1—interpolation of the ideal TS; 2—"uniform" approximation of the ideal TS with minimal absolute error in each section; 3—"interpolation" of the ideal response

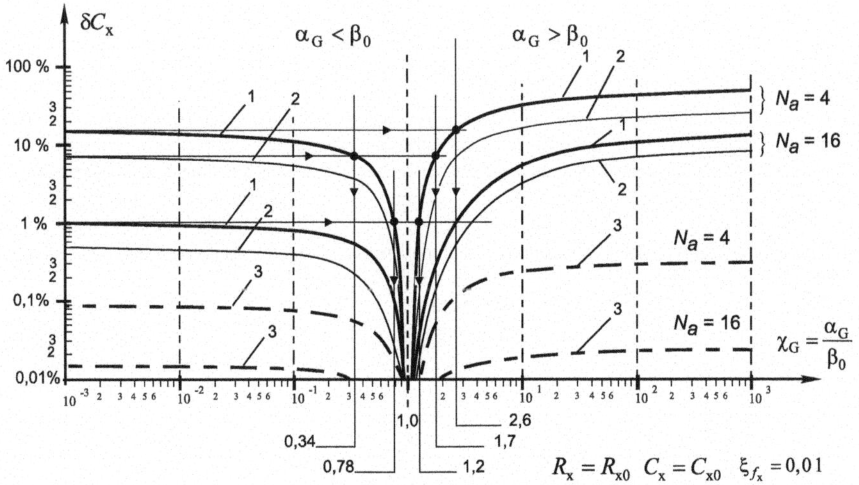

Fig. 15 Methodological error component δC_x due to the approximation method: 1—interpolation of the ideal TS; 2—"uniform" approximation of the ideal TS with minimal absolute error in each section; 3—"interpolation" of the ideal response

compared to interpolation based on the input TS shape and be of the same order as with the "uniform" approximation.

In the case of "uniform" approximation by the analytical expression of the response (in Figs. 14 and 15—thin lines 2) the estimation of the R_x,C_x parameters determination error via splines model parameters and N_a can only be obtained numerically.

4 Conclusions

1. The problem of simultaneously ensuring the generation possibility, the specified accuracy, and the requirements of a qualitative nature, that is obtaining smooth functions that satisfy on local segments the reproduced dependences, can be solved by constructing splines based on models that are adequate to real signals in electric circuits.
2. The advantage of exponential splines is the ease of generation in linear electric circuits.
3. Exponential splines are optimal for forming a class of exponential test signals that are also described by the same as spline functions, for example, in the tasks of measuring and controlling the parameters of electrical circuits using the method of zeros and poles. This allows for a limited number of approximation sections to increase the accuracy of determining the parameters of electric circuits.

References

1. Royik, O.M.: Kontrol′ i diahnostyka radioelektronnoyi aparatury na etapakh yiyi vyrobnytstva. UNIVERSUM-Vinnytsya, Vinnytsya, Ukrayina (2000). ISBN 966-641-022-2(in Ukrainian)
2. Anoshkin, V.Yu., Ginzburg, L.I., Kalyapin, V.S.: Avtomatizirovannaya sistema vnutriskhemnogo kontrolya. Elektronnaya promyshlennost' **9**, 57–59 (1985) (in Russian)
3. Likhttsinder, B.Ya.: Vnutriskhemnoye diagnostirovaniye uzlov radioelektronnoy apparatury. Tekhníka, Kiïv, SSSR (1989). ISBN 5-335-00166-6 (in Russian)
4. Royik, O.M., Arsenyuk, I.R.: Diahnostuvannya analohovykh vuzliv radioelektronnoyi aparatury. UNIVERSUM–Vinnytsya, Vinnytsya, Ukrayina (2005). ISBN 966-641-111-3 (in Ukrainian)
5. Bayda, N.P., Kuz'min, I.V., Shpilevoy, V.T.: Mikroprotsessornyye sistemy poelementnogo diagnostirovaniya REA. Radio i Svyaz', Moskva, SSSR (1987) (in Russian)
6. Martyashin, A.I., Orlova, A.V., Shlyandin, V.N.: Preobrazovateli parametrov mnogopolyusnykh elekyatricheskikh tsepey. Energoizdat, Moskva, SSSR (1981) (in Russian)
7. Martyashin, A.I., Orlova, A.V., Tsypin, B.V.: Metody poelementnogo kontrolya elektronnykh skhem. Obzornaya informatsiya. TSNIITEI priborostroyeniya. TS-5, Moskva, vol. 1 (1983) (in Russian)
8. Martyashin, A.I., Kulikovskiy, K.L., Kuroyedov, S.K., Orlova, L.V.: Osnovy invariantnogo preobrazovaniya parametrov elektricheskikh tsepey, pod red. A.I. Martyashina. Energoatomizdat, Moskva, SSSR (1990) (in Russian)

9. Tuz, Yu.M., Osadchenko, V.P.: Nekotoryye osobennosti kontrolya parametrov elektricheskikh tsepey po metodu nuley i polyusov. Resp. nauch.-tekhn. konf. Strukturnyye metody povysheniya tochnosti, chuvstvitel'nosti i bystrodeystviya izmeritel'nykh priborov i sistem, Kiı̈v, pp 127–128 (1985) (in Russian)
10. Hubar, V.I., Shumkov, YU.S., Khimichenko, B.P.: Vymiryuvannya parametriv elektrychnykh kil iz zastosuvannyam vyprobuval′nykh syhnaliv spetsial′noyi formy. Naukovi visti NTUU "KPI" **5**, 133–140 (2010). http://nbuv.gov.ua/UJRN/NVKPI_2010_5_21 (in Ukrainian)
11. Royik, O.M., Mesyura, V.I.: Syntez struktur vymiryuval′nykh peretvoryuvachiv parametriv komponentiv zamknenykh elektrychnykh kil. Visnyk VPI **4**, 5–10 (1996) (in Ukrainian)
12. Royik, O.M.: Invariantni peretvorennya parametriv elementiv skladnykh ob'yektiv. UNIVERSUM-Vinnytsya, Vinnytsya, Ukrayina (2001). ISBN 966-641-032-X (in Ukrainian)
13. Pokhodylo, Ye.V., Khoma, V.V.: Vymiryuvachi CLR z peretvorennyam "imitans-napruha", L′viv, Ukrayina: L′vivs′ka politekhnika (2011). ISBN 978-617-607-079-5 (in Ukrainian)
14. Volodarskiy, Ye.T., Gubar', V.M., Nikiforov, L.L., Tuz, Yu.M.: Sistemy avtomatizirovannogo kontrolya radioelektronnoy apparatury. Tekhníka, Kiı̈v, SSSR (1983) (in Russian)
15. Jerri, A.J.: The Shannonsampling theorem, its various extensions and applications: a tutorial review. Proc. Inst. Electr. Electron. Eng. (IEEE) **65**(11), 1565–1596 (1977). https://doi.org/10.1109/PROC.1977.10771
16. Shumkov, Yu.: Exponential splines in electric circuits' parameters measuring. In: Proceedings of the International Conference on Actual problems of Measuring Technique "Measurement-98", Kyiv, pp. 250–253 (1998)
17. Ahlberg, J.H., Nilson, E.N., Walsh, J.L.: The Theory of Splines and Their Appiications. Academic, New York, USA (1967)
18. Schumaker, L.L.: Spline Functions: Basic Theory. Wiley, New York, USA (1981)
19. Stechkin, S.B., Subbotin, Yu.N.: Splayny v vychislitel'noy matematike. Nauka, Moskva, SSSR (1976) (in Russian)
20. Zav'yalov, Yu.S., Kvasov, B.I., Miroshnichenko, V.L.: Metody splayn-funktsiy. Nauka, Moskva, SSSR (1980) (In Russian)
21. Shelevyts′kyy, I.V., Shutko, M.O., Shutko, V.M., Kolhanova, O.O.: Splayny v tsyfroviy obrobtsi danykh i syhnaliv. Vydavnychyy dim, Kryvyy Rih, Ukrayina (2008). ISBN 078-966-2915-86-0. https://doi.org/10.13140/RG.2.1.4898.9847 (in Ukrainian)
22. Späth, H.: Exponential spline interpolation. Computing **4**, 86–96 (1969)
23. Pruess, S.: Alternatives to the exponential spline in tension. Math. Comput. **33**(148), 1273–1281 (1979). https://doi.org/10.1090/S0025-5718-1979-0537971-6
24. McCartin, B.J.: Theory of exponential splines. J. Approx. Theory **66**, 1–23 (1991). https://doi.org/10.1016/0021-9045(91)90050-K
25. McCartin, B.J.: Theory, Computation, and Application of Exponential Splines. Forgotten Books, New York, USA (2018). ISBN978-0331801309
26. Unser, M., Blu, T.: Cardinal exponential splines: part I-theory and filtering algorithms. IEEE Trans. Signal Process. **53**(4), 1425–1438 (2005). https://doi.org/10.1109/TSP.2005.843700
27. Unser, M.: Cardinal exponential splines: part II-think analog, act digital. IEEE Trans. Signal Process. **53**(4), 1439–1449 (2005). https://doi.org/10.1109/TSP.2005.843699
28. Tuz, Yu.M., Shumkov, Yu.S.: Metod eksponentsial′nykh splayniv v zadachakh vymiryuvannya ta kontrolyu parametriv elektrychnykh kil. In: VII Mizhnarodna naukovo-tekhnichna konferentsiya Metrolohiya, informatsiyno-vymiryuval′ni tekhnolohiyi ta systemy (MIVT·S-2020), Kharkiv, p. 140 (2020). https://doi.org/10.24027/2306-7039.1A.2020.193279 (in Ukrainian)
29. Shumkov, Yu.S., Hrashenko, M.V., Darahan, V.C.: Formuvannya vyprobuval′nykh syhnaliv spetsial′noyi formy na osnovi eksponentsial′nykh splayniv. Mekhanika hiroskopichnykh system **35**, 30–40 (2018). https://doi.org/10.20535/0203-3771352018128611 (in Ukrainian)
30. Tsypkin, Y.Z.: Teoriya lineynykh impul'snykh sistem. Fizmatgiz, Moskva, SSSR (1963) (in Russian)
31. Makarov, I.M., Menskiy, B.M.: Tablitsa obratnykh preobrazovaniy Laplasa i obratnykh z-preobrazovaniy: drobno-ratsional'nyye izobrazheniya. Vysshaya shkola, Moskva, SSSR (1978) (in Russian)

Improving of Methods of Impedance Parameters Units Reproduction and Measurement Accuracy Increasing for Ensuring Metrological Traceability

Sergii Shevkun, Maryna Dobroliubova, and Oleksii Statsenko

Abstract A method for finding points of optimal comparing of heterogeneous impedance parameters (active resistance, electrical capacity, inductance) is developed, combinations of such points have been found to achieve the highest metrological characteristics of national standards of impedance, methods for choosing the optimal transfer scheme to minimize the measurement error are proposed. The structural-algorithmic method of increasing the accuracy of the impedance units' measures calibration by excluding cross-nominal comparing is considered. Constructive and technological methods of increasing measurement accuracy to ensure metrological traceability of impedance parameter units in the full range of values are considered. A calibration method is developed and a procedure for estimating measurements uncertainty during the calibration of measures of electrical capacitance and inductance, precision LCR-meters on state standards with traceability to international standards of electric capacity is proposed. The main factors affecting the measurement result are considered, and their contribution to measurement uncertainty is quantified. General approaches to calibration are given, measurement equations (models), as well as examples of uncertainty budget calculations, are presented. The use of the proposed methods allows reaching a new qualitative level of measurements when using existing standard equipment. The presented materials can be used by national metrological institutes, calibration and testing laboratories, conformity assessment agencies that perform measurements of electromagnetic quantities.

Keywords Standard · Comparator · Impedance · Measure of active resistance · Measure of inductance · Measure of electrical capacitance · Divider · Equinominal comparing · Uncertainty budget · Uncertainty estimation · Combined standard uncertainty · Extended uncertainty · LCR-meter · Metrological traceability

S. Shevkun
State Enterprise "Ukrmetrteststandard", Kyiv, Ukraine

M. Dobroliubova (✉) · O. Statsenko
National Technical University of Ukraine "Igor Sikorsky Kyiv Polytechnic Institute", Kyiv, Ukraine
e-mail: marina.dobroliubova@lll.kpi.ua

V. Eremenko and A. Zaporozhets (eds.), *Advanced Information-Measuring Technologies and Systems I*, Studies in Systems, Decision and Control 439,
https://doi.org/10.1007/978-3-031-40718-5_3

1 The Need for Improvement the Accuracy of Electrical Impedance Parameters Units Measurements

Electric energy is used in all areas of human life. This leads to the need for periodic diagnostics and maintenance of equipment designed for its generation, transformation and transportation. Control of the technical condition of such objects as generator units, transformer substations, power lines, switching nodes, etc. requires a series of complex, high-precision measurements to be carried out in a timely manner. First of all, it is the control of insulation parameters, which allows determining the aging degree of insulating materials, preventing power losses and avoiding equipment breakdowns. The need for trouble-free operation of power equipment indicates the acuteness of the problem of estimating the insulation condition and quality and makes it even more important.

A large number of instruments for measuring impedance parameters, such as measures of active resistance, electrical capacitance, inductance and dissipation factor, LCR-meters, bridges and impedance comparators, multimeters with the function of measuring the specified values are widely used in development, setting-up, diagnostics and during repair of apparatus and equipment in such fields as telecommunications, IT industry, energy, transport, metallurgy and manufacturing industry, telematics, security, scientific research and defense.

Modern requirements for the quality of electrical equipment, electronic circuits and their components require a significant increase in level of measurements accuracy of electrical quantities, including impedance parameters.

Measurements reliability and accuracy are achieved by calibrating measuring equipment with ensuring metrological traceability.

In order to achieve high measurements accuracy during calibration, it is necessary to use more accurate modern highly stable standards, evaluate and take into account all factors that influence the measurements results, as well as apply statistical methods when processing the measurements results.

This work is devoted to the topical issue of increasing the accuracy of measurements when calibrating the measuring equipment of electric impedance parameters units in the full range of frequencies and values of the measured quantities.

The problem in achieving high accuracy of the specified quantities measurements is the significant influence on the measurements results of the temperature and frequency dependences of the electrical measures and connecting conductors' characteristics, the heterogeneity and inconsistency of contact connections, the presence of electromagnetic interference, as well as the drift of the standards characteristics from the moment of last calibration.

The work investigates and describes methods of reproducing inductance and active resistance units with traceability to standards of electrical capacitance, methods of calibrating capacitance, inductance, and active resistance with the highest accuracy in the country, as well as methods of calibrating LCR-meters in full range of frequencies and values. A method of determining optimal comparing points when transferring a unit of physical quantity between heterogeneous impedance parameters is developed.

A collection of such points with their numerical values for achieving the highest metrological characteristics of national standards is found, methods for minimizing error and uncertainty during the calibration of impedance parameters measuring equipment with reference on heterogeneous standards are proposed by choosing the optimal scheme for transferring the size of physical quantity unit.

2 Analysis and Optimization of Transfer Schemes and Reproducing Methods of Impedance Parameters Units

The currently existing standard base of national metrological institutes, which was created over many decades, assumes the presence of a state standard for each individual unit of impedance: electrical capacitance, inductance, active resistance. Recently, in the leading countries of the world, a new generation of initial standards of impedance parameters, which are based on fundamental constants, is being created and used [1].

However, the creation and maintenance of these standards is so expensive that most developing countries cannot implement it. This problem is particularly acute in conditions of limited budgetary funding for the maintenance of the national standard base, or in the case of state self-supporting maintenance of national metrological institutes, when the costs of standards are not covered by the income from performing of metrological works using these standards. On the other hand, the competitiveness of metrological services directly depends on the realized accuracy of measurements, that is, the smaller the uncertainty of measurements achieved during calibration, the more consumers, including foreign ones, will be interested in obtaining a unit of physical quantity in this particular laboratory.

At the same time, the appearance of modern universal comparators makes it possible to use any of the specified standards to ensure measurements of impedance parameters [2]. But in this case, the process of reproducing and transferring the size of any impedance parameter unit is significantly complicated and ceases to be unambiguous. Herewith, the accuracy of reproduction and transfer of the impedance parameter unit significantly depends on the configuration and the set of possible ways of the specified process implementation.

It is proposed to solve this contradiction by finding and determining optimal ways of transferring impedance parameters units according to the criterion “highest accuracy—reasonable cost”.

Optimization of reproduction methods and transfer schemes of impedance parameters units will provide an opportunity to significantly increase these values measurements accuracy, to increase the operational and constructive indicators of the impedance parameters standard base, the competitiveness of metrological works performed in the interests of domestic and foreign customers, to ensure saving scarce financial resources for the calibration of component standards abroad, as well as more fully using the technical capabilities of the national standard base.

The purpose of this section is to highlight the methodological foundations for transferring of impedance parameter units from international standards based on fundamental physical constants to state primary standards when using as transferring standard any standard measure of capacitance, inductance or active resistance, under the condition that measurement accuracy is achieved at the level of the world's leading countries.

In order to achieve the highest accuracy of transferring the dimensions of the impedance parameters units from heterogeneous standards, the following tasks must be solved:

- developing a method of setting the points of physical quantity unit transfer between heterogeneous impedance parameters and finding their set, which will ensure the connection between impedance parameters when transferring a unit of measurement with the highest accuracy;
- developing approaches to minimize the error when implementing a scheme for transferring the impedance parameter unit size with reference to heterogeneous standards.

When solving the optimization task, it is necessary to take into account that modern technical means allow direct (with reference to the frequency standard) comparing of homogeneous or heterogeneous parameters of the standards.

Let's consider the longitudinal (with uniform parameters) transfer of measurement unit, where the functions of the comparator f and λ have the following form

$$Z_X = \alpha Z_A + \beta Z_B; \tag{1}$$

$$Z_X = (\alpha + j\beta)Z_0. \tag{2}$$

The complex resistance Z_X is described by the equation:

$$Z_X = A_X + jB_X, \tag{3}$$

where A_X and B_X—active and reactive absolute components of impedance.

Then, if the function (1) is used

$$A_X = \alpha A_0(1 + \frac{\beta B_0 tg\delta_0}{\alpha A_0}); \; B_X = \beta B_0(1 + \frac{\alpha A_0 tg\phi_0}{\beta B_0});$$
$$tg\delta_X = \frac{\alpha}{\beta} \cdot \frac{A_0}{B_0}\left(1 + \frac{\beta B_0 tg\delta_0}{\alpha A_0}\right) \times \left(1 + \frac{\alpha A_0 tg\phi_0}{\beta B_0}\right)^{-1}. \tag{4}$$

And if the function (2) is used:

- when using the standard of resistance:

$$A_X = \alpha R_0(1 + \frac{\beta tg\phi_0}{\alpha}); \ B_X = \beta R_0(1 + \frac{\alpha tg\phi_0}{\beta}); \quad (5)$$

- when using the standard of capacitance:

$$A_X = \frac{\beta}{\omega C_0}(1 + \frac{\alpha tg\delta_0}{\beta}); \ B_X = \frac{\alpha}{\omega C_0}(1 + \frac{\beta tg\delta_0}{\alpha}). \quad (6)$$

Formulas (5), (6) show that to determine any impedance parameters, it is necessary and sufficient to have standards of active or reactive parameters, which are also standards of the zero value of the loss angle tangent or the phase angle tangent.

A similar analysis was conducted for comparing of measures of heterogeneous quantities.

The conducted analysis allows to make the following conclusions about the structure of the standard base in the field of impedance parameter measurements:

1. The minimum number of standards for reproduction and measurement of impedance parameters is equal to two: standard of one of the impedance parameters and frequency standard.
2. The impedance parameter standard that is used must be both the main parameter standard and the auxiliary parameter standard (zero or a sufficiently small value of the loss angle tangent or the phase angle tangent).
3. The minimal system of standards allows certification of exemplary measures of any nature impedance according to two parameters.
4. To transfer (or reproduce) the impedance parameter unit size, it is necessary to have a set of tools that allow to compare the impedance with homogeneous or quadrature parameters.

The given description of the standard base structure is insufficient for creating an optimal metrological scheme and making a decision about the requirements for the equipment that ensures its functioning. In Fig. 1 it is shown the dynamic ranges in which the standard impedances of complex resistance individual parameters and the impedances of the corresponding standard measures are placed.

As can be seen from this figure, the ranges in which the impedances of standard measures of individual parameters are placed, as well as the impedances of the standards of these physical quantities, are significantly different. This circumstance led to the fact that the equipment for transferring the size of various units of individual impedance parameters is also significantly different and depends on the type of measured parameters and their ranges. Attempts to use existing equipment or to create new equipment to transfer unit sizes of impedance parameters according to today's requirements inevitably lead to the accumulation of expensive precision equipment.

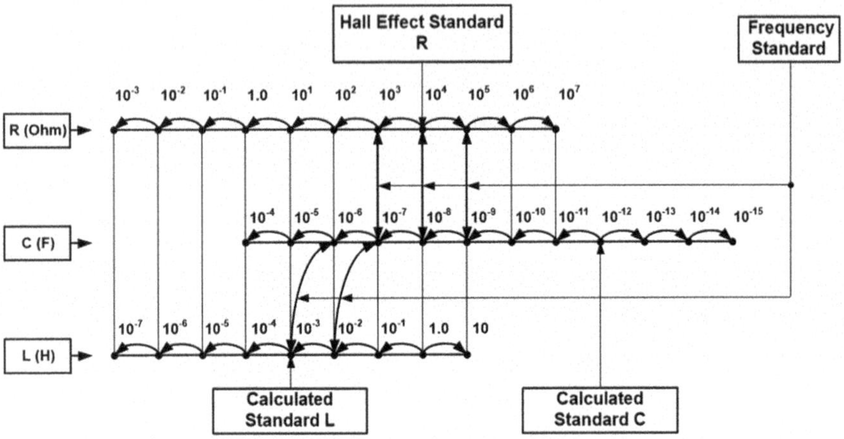

Fig. 1 The diagram of transferring the sizes of the complex resistance parameters units

Transferring the size of the impedance parameters units in the range of values is carried out using either the comparing of equal nominal standard measures, or the decimal comparing.

The technical basis for the realization of optimized transfer schemes and reproducing methods of impedance parameters units is a complex of quadrature and in-phase-opposite-phase impedance comparators placed in one casing [3].

For each frequency, it is necessary to optimize the process of quadrature transfer and to find the values of measures at which equal impedance transfer is possible. At the same time, it is desirable to use the most common measures of nominal values, multiple of 1, 2, ..., 9, 10. For some cases, it is rational to create measures of special nominal values.

Figure 2 illustrates the results of calculations for a frequency of 1 kHz when comparing capacitance and resistance measures. At a frequency of 1 kHz, three points were found at which it is possible to implement equal-impedance comparing. These points correspond to the transferring $R \leftrightarrow C$ from resistance measures with a nominal multiple of 0.2 to capacitance measures with a nominal multiple of 0.8; from resistance measure with a nominal multiple of 0.4 to capacitance measure with a nominal also multiple of 0.4; from resistance measures with a nominal multiple of 0.8 to capacitance measures with a nominal multiple of 0.2. In all these points, the relative deviation from the nominal is 0.00528. This value is large enough. With such a deviation, the comparator will not provide the maximum measurement accuracy (the maximum accuracy of the comparator is achieved with deviations $\delta \leq \pm 0.02$ %).

To eliminate this problem, after transferring the size to any of the main measures of resistance or capacitance having a nominal value, it can be corrected by adding an additional measure, the relative value of which is equal to 0.00528 of the value of the main measure. At the same time, if the additional measure is measured with an error of at least 0.01%, the total error due to this factor does not exceed 5×10^{-7}. At the

Objective Function (criterion of optimality):

$$\delta_k = \frac{r_j - z_i}{r_j} \to \min$$

System of Constraints:

$$\begin{cases} r_j \in \{0; 0.1; \ldots 1.0\}; \\ z_i = \dfrac{1}{\omega c_i}; \\ c_i \in \{0; 0.1; \ldots 1.0\}. \end{cases}$$

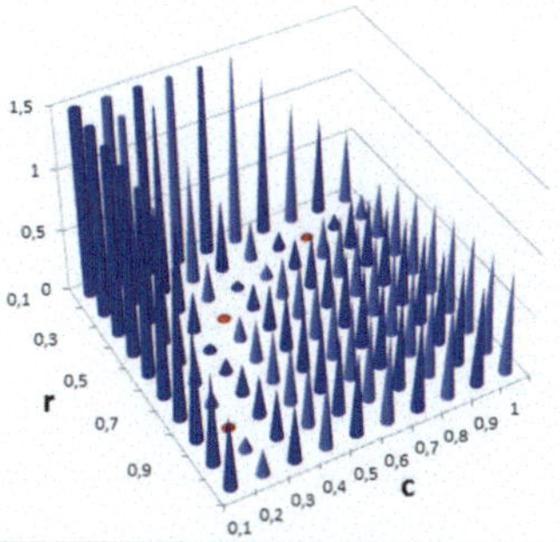

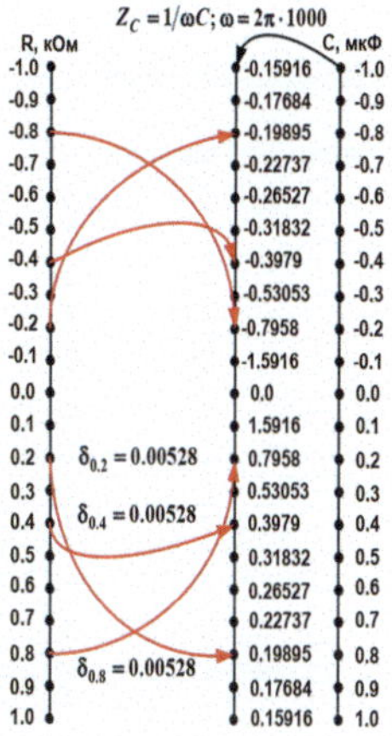

r\c	0,1	0,2	0,3	0,4	0,5	0,6	0,7	0,8	0,9	1
0,1	14,91549431	6,957747	4,305165	2,978874	2,183099	1,652582	1,273642	0,989437	0,768388	0,591549
0,2	6,957747155	2,978874	1,652582	0,989437	0,591549	0,326291	0,136821	0,005282	0,115806	0,204225
0,3	4,30516477	1,652582	0,768388	0,326291	0,061033	0,115806	0,242119	0,336854	0,410537	0,469484
0,4	2,978873577	0,989437	0,326291	0,005282	0,204225	0,336854	0,431589	0,502641	0,557903	0,602113
0,5	2,183098862	0,591549	0,061033	0,204225	0,36338	0,469484	0,545272	0,602113	0,646322	0,68169
0,6	1,652582385	0,326291	0,115806	0,336854	0,469484	0,557903	0,62106	0,668427	0,705269	0,734742
0,7	1,273642044	0,136821	0,242119	0,431589	0,545272	0,62106	0,675194	0,715795	0,747373	0,772636
0,8	0,989436789	0,005282	0,336854	0,502641	0,602113	0,668427	0,715795	0,75132	0,778951	0,801056
0,9	0,768388257	0,115806	0,410537	0,557903	0,646322	0,705269	0,747373	0,778951	0,803512	0,823161
1	0,591549431	0,204225	0,469484	0,602113	0,68169	0,734742	0,772636	0,801056	0,823161	0,840845

Fig. 2 Determination of the set of points of optimal comparing of measures R and C at 1 kHz

interval of deviations of 0.5% from the nominal, the error from the frequency dependence of modern capacitance and resistance measures can be neglected. Therefore, to eliminate the problem, the operating frequency of the comparator can be changed to the indicated deviations. Then the transfer error will be minimal.

Figure 3 illustrates the results of performed calculations for a frequency of 1 kHz when comparing resistance and inductance measures. As this figure shows, there are two points where equal-impedance comparing of inductance and resistance measurements can be made. These points correspond to the transfer $R \leftrightarrow L$ from resistance measures with a nominal multiple of 0.2 to inductance measure with a nominal multiple of 0.3, and from resistance measure with a nominal multiple of 0.5 to inductance measure with a nominal multiple of 0.8. But in the first of these points, the relative deviation from the nominal is 0.05752, and in the second—0.00531. Both of these values are large enough. To eliminate this problem, after transferring the size to the main resistance measure, which has a nominal value, it can be changed by adding an additional measure, the relative value of which is 0.05752 and 0.00531 of the main measure value, respectively. At the same time, if the additional measure is measured with an error of at least 0.02%, the total measurement error will not exceed 10^{-6}.

The previously considered transverse transfers were carried out using a quadrature bridge. Let's consider the transverse transfer of the type $C \leftrightarrow L$ with the use of an in-phase-opposite-phase comparator. The results of calculations using the given method for a frequency of 1 kHz are shown in Fig. 4. There are three points where it is possible to carry out equal-impedance comparing of capacitance and inductance measurements. These points correspond to the transfer $C \leftrightarrow L$ from capacitance measures with a nominal multiple of 0.8 to inductance measure with a nominal

Objective Function (criterion of optimality):

$$\delta_k = \frac{r_j - z_i}{r_j} \to \min$$

System of Constraints:

$$\begin{cases} r_j \in \{0; 0.1; \ldots 1.0\}; \\ z_i = \omega l_i; \\ l_i \in \{0; 0{,}1; \ldots 1{,}0\}. \end{cases}$$

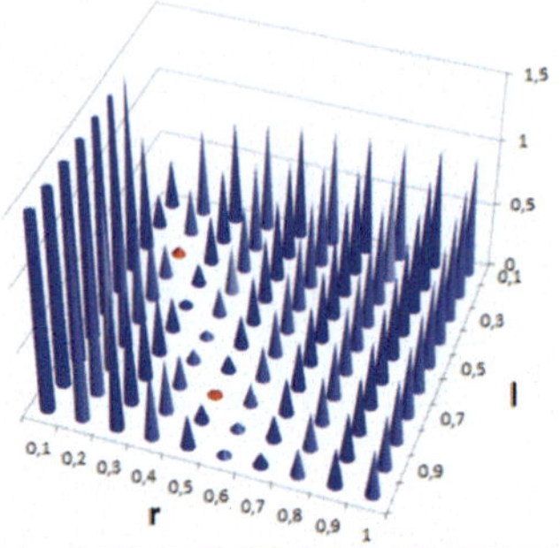

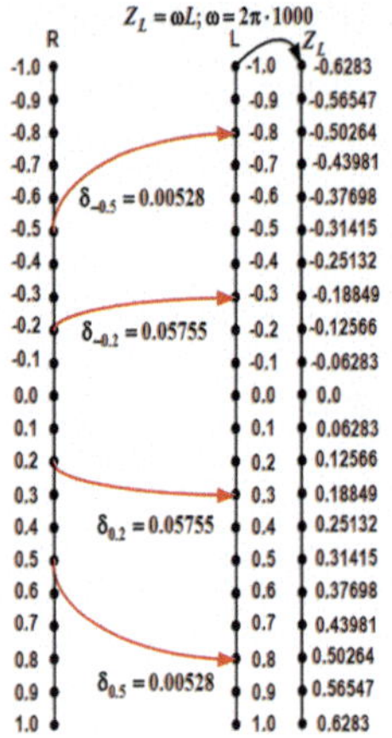

r\l	0,1	0,2	0,3	0,4	0,5	0,6	0,7	0,8	0,9	1
0,1	0,371681	0,256637	0,884956	1,513274	2,141593	2,769911	3,39823	4,026548	4,654867	5,283185
0,2	0,685841	0,371681	0,057522	0,256637	0,570796	0,884956	1,199115	1,513274	1,827433	2,141593
0,3	0,79056	0,581121	0,371681	0,162242	0,047198	0,256637	0,466077	0,675516	0,884956	1,094395
0,4	0,84292	0,685841	0,528761	0,371681	0,214602	0,057522	0,099557	0,256637	0,413717	0,570796
0,5	0,874336	0,748673	0,623009	0,497345	0,371681	0,246018	0,120354	0,00531	0,130973	0,256637
0,6	0,89528	0,79056	0,685841	0,581121	0,476401	0,371681	0,266962	0,162242	0,057522	0,047198
0,7	0,91024	0,82048	0,730721	0,640961	0,551201	0,461441	0,371681	0,281922	0,192162	0,102402
0,8	0,92146	0,84292	0,764381	0,685841	0,607301	0,528761	0,450221	0,371681	0,293142	0,214602
0,9	0,930187	0,860374	0,79056	0,720747	0,650934	0,581121	0,511308	0,441495	0,371681	0,301868
1	0,937168	0,874336	0,811504	0,748673	0,685841	0,623009	0,560177	0,497345	0,434513	0,371681

Fig. 3 Determination of the set of points of optimal comparing of measures *R* and *L* at 1 kHz

Objective Function (criterion of optimality):

$$\delta_k = \frac{z_j - z_i}{z_j} \to \min$$

System of Constraints:

$$\begin{cases} c_j \in \{0; 0.1; \ldots 1.0\}; \\ z_j = \dfrac{1}{\omega c_j}; \\ l_i \in \{0; 0.1; \ldots 1.0\}; \\ z_i = \omega l_i. \end{cases}$$

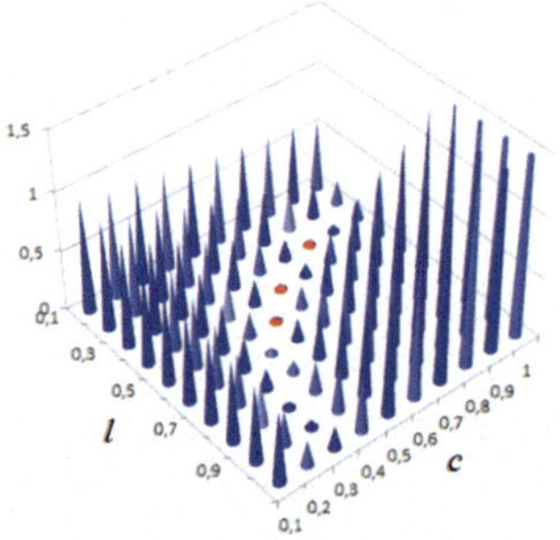

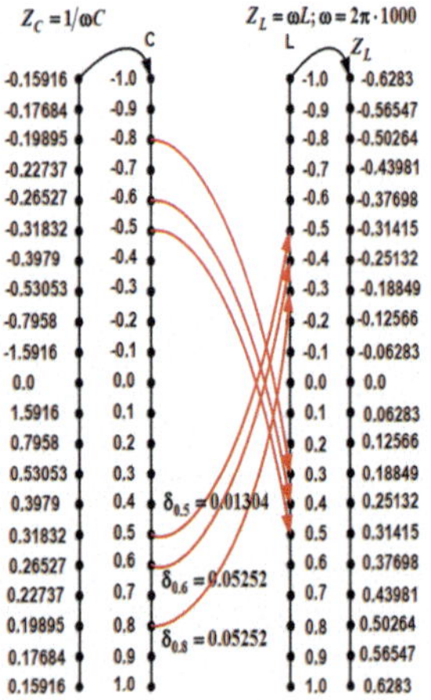

c\l	0,1	0,2	0,3	0,4	0,5	0,6	0,7	0,8	0,9	1
0,1	0,960522	0,921043	0,881565	0,842086	0,802608	0,763129	0,723651	0,684173	0,644694	0,605216
0,2	0,921043	0,842086	0,763129	0,684173	0,605216	0,526259	0,447302	0,368345	0,289388	0,210432
0,3	0,881565	0,763129	0,644694	0,526259	0,407824	0,289388	0,170953	0,052518	0,065917	0,184353
0,4	0,842086	0,684173	0,526259	0,368345	0,210432	0,052518	0,105396	0,263309	0,421223	0,579137
0,5	0,802608	0,605216	0,407824	0,210432	0,01304	0,184353	0,381745	0,579137	0,776529	0,973921
0,6	0,763129	0,526259	0,289388	0,052518	0,184353	0,421223	0,658094	0,894964	1,131835	1,368705
0,7	0,723651	0,447302	0,170953	0,105396	0,381745	0,658094	0,934442	1,210791	1,48714	1,763489
0,8	0,684173	0,368345	0,052518	0,263309	0,579137	0,894964	1,210791	1,526619	1,842446	2,158273
0,9	0,644694	0,289388	0,065917	0,421223	0,776529	1,131835	1,48714	1,842446	2,197752	2,553058
1	0,605216	0,210432	0,184353	0,579137	0,973921	1,368705	1,763489	2,158273	2,553058	2,947842

Fig. 4 Determination of the set of points of optimal comparing of measures *C* and *L* at 1 kHz

multiple of 0.3; from capacitance measure with a nominal multiple of 0.6 to inductance measure with a nominal multiple of 0.4; from capacitance measure with a nominal multiple of 0.5 to inductance measure with a nominal multiple of 0.5.

In the first and second of these points, the relative deviation from the nominal ratio of the shoulders of the comparator is 0.052518, in the third—0.01304. All these values are large. The first and second points with a relative deviation of 0.052518 can be completely excluded from consideration. In addition, it is necessary to take into account that in circulation there are measures of inductance with nominal values that are multiples of 1, 2, 3, 5, but there are no inductance measures with a nominal value that is a multiple of 8. And unlike capacitance or resistance measures, attempts to make a measure with a nominal value multiple of 8 from measures of smaller nominal values lead to significant errors.

The frequency dependence of modern inductance measures is very large and rarely known with the required accuracy. Therefore, to eliminate the problem, it is not possible to change the operating frequency of the comparator to the indicated deviations, as was proposed for the transfer of the type $R \leftrightarrow C$. It is proposed to create a capacitance measure multiple of 2.5 by means of parallel connection of capacitance multiple of 2 and 0.5. With the use of such a measure, it is possible to transfer the size of a unit from capacitance measure to decimal inductance measures. At the same time, the difference in the impedances ratio differs from the comparator shoulders ratio by about a percent and is corrected by adding the appropriate additional capacitance measure to the main measure.

On the basis of the above studies and calculations, a near-optimal diagram of the impedance units' transfer is constructed, which is shown in Fig. 5. The diagram makes it possible to transfer the size of the unit and carry out the calibration of all impedance measures in the full range of values with reference to only one of any measures.

On the basis of the conducted analysis and experimental studies, transfer schemes and reproducing methods of impedance parameter measurements units are optimized, which contribute to increasing accuracy and significantly reducing the cost of metrological works at the level of national standards.

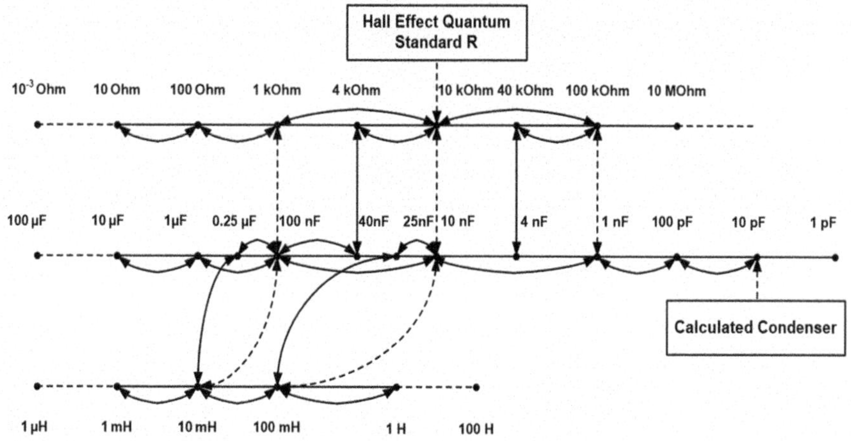

Fig. 5 Diagram of transfer of impedance units at 1 kHz

A method of finding points of physical quantity unit transfer between heterogeneous impedance parameters is developed, a set of such points is found to achieve the highest metrological characteristics. Methods of minimizing the error of transfer of the impedance parameter unit size with reference to heterogeneous standards by choosing the optimal transfer scheme are proposed and developed.

The results of the work made it possible to organize and successfully complete the international reconciliation of national standards of electrical capacitance and inductance at a frequency of 1 kHz. The final report on the topic COMET.EM-S13 and COOMET.EM-S14 was approved by all national metrological institutes and international metrological organizations, published on the KCDB website of the International Bureau of Weights and Measures (BIPM), Paris [4–6].

According to the results of international reconciliations, due to measurements accuracy increase it was adjusted the CMC lines on the BIPM website, which contributed to the official recognition of the metrological capabilities of the SE "Ukrmetrteststandard" at the international level and the involvement of a larger number of domestic and foreign customers to performing of metrological works [4].

Thus, based on the results of the research, a technical and methodological basis is theoretically justified and created for the transfer of impedance parameter units from international standards based on fundamental physical constants to national standards when using as a transferring standard any standard measure of capacitance, inductance or active resistance, providing that the accuracy of measurements is maintained at the level of the leading countries of the world.

3 Increasing the Accuracy of the Impedance Unit Size Transfer in Low-Resistance Range of Values by the Equal-Nominal Comparing Method

In the metrological attendance of modern electronic equipment used in energy, telecommunications, mechanical engineering, transport, defense and scientific research, one of the most common measurements are the measurements of such quantities as active resistance, electrical capacity, inductance and other parameters of complex resistance (impedance).

Increasing the requirements for measurement accuracy requires the use of expensive standard equipment.

Measurement of AC impedance units with the highest accuracy in Ukraine is carried out on the National Standard of Units of Electrical Capacity and Loss Factor DETU 08-06-01 and the National Standard of Units of Inductance and Loss Tangent of DETU 08-09-09, which are stored at the State Enterprise "Ukrmetrteststandart", Kyiv [7–9].

However, when transferring the size of the unit by the range from smaller to larger values and in the opposite direction, the most significant error is introduced by

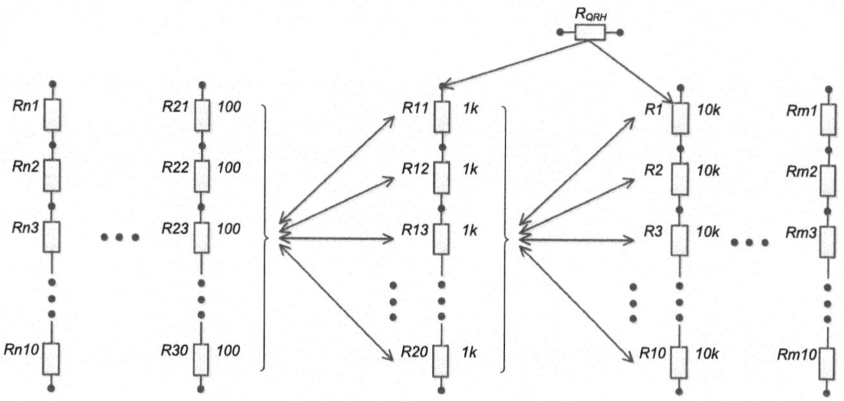

Fig. 6 Calibration of active resistance measures in the range of values using equal-nominal comparing

the autotransformer bridge (comparator included to the mentioned standards) when comparing measures of different nominal values [3, 10, 11].

The purpose of this subsection is to reveal the method of the accuracy increasing of the calibration of complex resistance units measures in the low-resistance range of values by excluding the comparing of different nominal values measures, which helps to reproduce any resistance value using a calibrated measure of only one nominal.

When comparing equal-nominal impedance measures, there is no need to use a high-precision comparator. Priority is given to such a characteristic of the comparator as its sensitivity [9]. It is also necessary to have a set of impedance measures of ten units of each nominal.

The essence of the proposed method will be considered on the example of calibration of active resistance measures in the range from 0.1 Ω to 1 MOhm. The sequence of calibration operations and their order is shown in Fig. 6.

Using equal-nominal comparing, it is possible to calibrate ten measures with a 1 kOhm nominal value with reference to a standard measure of 10 kOhm. The specified method is called the ballast (base, reference) resistor method.

For example, using the standard of active resistance on the quantum Hall effect, a 10 kOhm measure, which is obtained by connecting ten 1 kOhm measures in series, is calibrated. When choosing one of the specified measures with a nominal value of 1 kOhm as a reference resistor, each measure with a nominal value of 1 kOhm is compared with it (Fig. 7).

As a result of the comparison, the coefficients (ratio) of each measure R_n value of to the value of the reference measure R_{1ref} are determined

$$K_n = \frac{R_n}{R_{1ref}}. \quad (7)$$

A system of equations is created

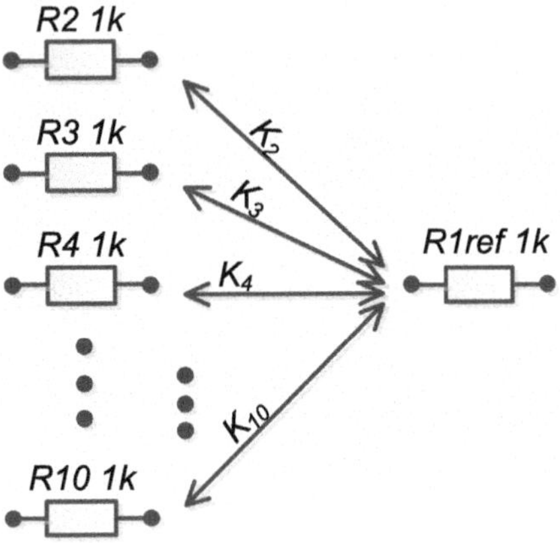

Fig. 7 An example of the ballast (base, reference) resistor method application

$$\begin{cases} R_2 = K_2 R_{1ref} \\ R_3 = K_3 R_{1ref} \\ \vdots \\ R_{10} = K_{10} R_{1ref} \\ R_{1ref} + R_2 + \ldots + R_{10} = R_{\Sigma(10\ \text{kOhm})} \end{cases}, \tag{8}$$

as a result of its solution, high-precision calculated values of each measure R_{1ref}, $R_2, \ldots, R_{10}$ with a nominal value of 1 kOhm are obtained. The specified set of measures is a group measure of active resistance with an average value

$$R_{\Sigma(1\ \text{kOhm})} = \frac{R_{1ref} + R_2 + \ldots + R_{10}}{10} = \frac{R_{\Sigma(10\ \text{kOhm})}}{10}. \tag{9}$$

The next step is to calibrate the measures with a nominal value of 100 Ω. For this purpose, a set of ten measures with a nominal value of 100 Ω is created, which is compared by comparing with a group measure of 1 kOhm nominal value. Further on, the ballast resistor method is used again and the calculated value of each measure with a nominal value of 100 Ω is found using formulas similar to (7)–(9).

Using this method, all measures are successively calibrated in the direction of decreasing nominal values from 100 to 0.1 Ω.

The calibration of resistance measures from 10 kOhm in the direction of increasing ratings is carried out by equal-nominal comparing of the 10 kOhm standard measure with each of the ten measures with a nominal value of 10 kOhm, which form the corresponding assembly of measures. This specified assembly will be a high-precision measure with a calculated value of 100 kOhm

$$R_{\Sigma(100\ \text{kOhm})} = R_{1(10\ \text{kOhm})} + R_{2(10\ \text{kOhm})} + \cdots + R_{10(10\ \text{kOhm})}. \tag{10}$$

In the same way

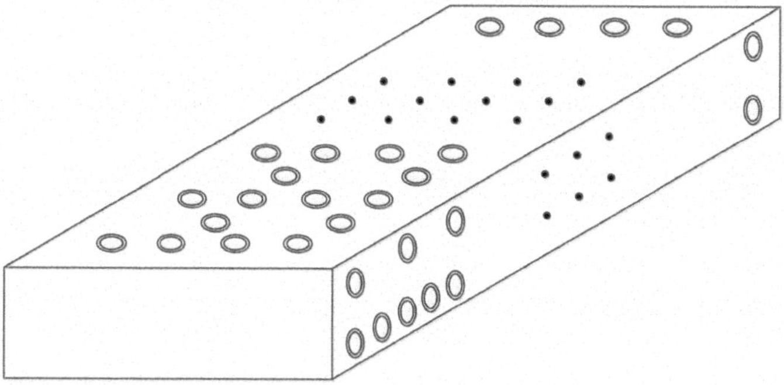

Fig. 8 Appearance of the divider

$$R_{\Sigma(1\ \text{MOhm})} = R_{1(100\ \text{kOhm})} + R_{2(100\ \text{kOhm})} + \cdots + R_{10(100\ \text{kOhm})}. \quad (11)$$

The described method of equal-nominal comparing has limitations for use in the area of high impedances. This is explained by the fact that during testing of high-resistance measures combined into an assembly, parasitic current leakage to the case through the middle point of the divider acquires significant values. Therefore, calibration of meters with a nominal value of more than 1 MOhm is impractical.

Based on the performed research results, it is proposed to create a divider for the active resistance unit's values. A typical divider is four assemblies, each of which is formed by series connection of high-precision resistance measures with 4-terminal connection of each measure. Each of the four assemblies is formed by the serial connection of ten measures with nominal values of 0.1 Ω, 10 Ω, 1 kOhm and 100 kOhm, respectively. The specified assemblies are combined in one metal case. The appearance of the divider is shown in Fig. 8.

The connection scheme of measures in the divider is shown in Fig. 9.

The measurements equivalent scheme of each assembly of divider is shown in Fig. 10.

To measure the resistance values of the resistor Z_{01}, taking into account the resistance of the connecting conductors Z_1', it is necessary to connect the current generator at the points I_1 and I_3, and the voltage indicator—at the points U_1 and U_2; to measure the resistance values of the resistor Z_{02}, taking into account the resistance of the connecting conductors $Z_2' + Z_2''$, it is necessary to connect the current generator at the points I_2 and I_5, and the voltage indicator—at the points U_2 and U_3; to measure the resistance of the resistor Z_{03}, taking into account the resistance of the connecting conductors $Z_3' + Z_3''$, it is necessary to connect the current generator at points I_4 and I_7, and the voltage indicator—at points U_3 and U_4.

The use of the described divider allows calibration of impedance measures in the full range of values from 0.1 Ω to 1 MOhm with high productivity. At the same time, only equal-nominal comparing is used.

The reviewed equal-nominal comparing method makes it possible to significantly increase the accuracy of the complex resistance units measures calibration

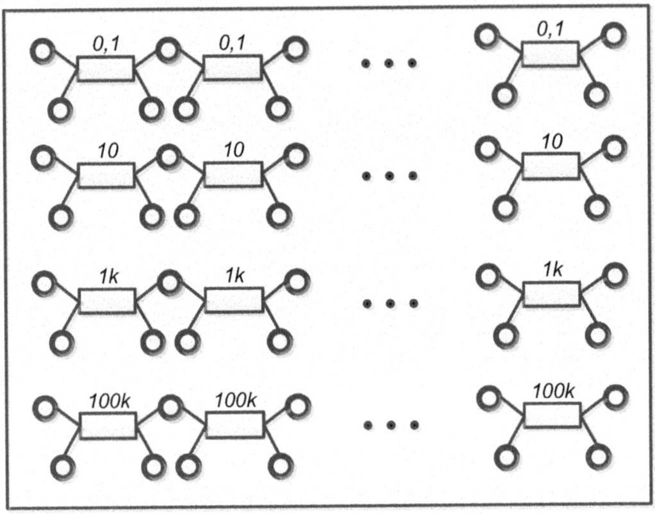

Fig. 9 The connection scheme of measures in the divider

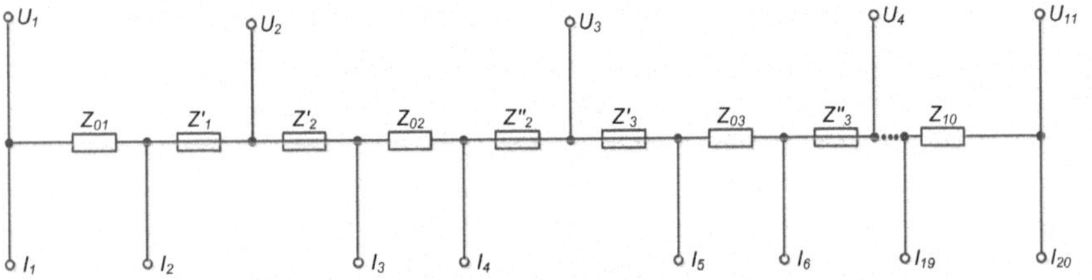

Fig. 10 Equivalent scheme of the divider assembly

in the low-resistance range of values by eliminating the comparing of measures of different nominal values and reach a new qualitative level of measurements when using existing standard equipment. Using a sufficient set of measures, it is possible to combine them constructively into one block and, with certain switchings and connections, ensure high productivity of metrological works.

The equal-nominal comparing method was used in the evaluation of the uncertainty components introduced by the autotransformer bridge during international reconciliations of origin standards of electrical capacitance and inductance in the themes COOMET EM-S13 (544/UA/12) and COOMET EM-S14 (584/UA/12). This method will be used in further international reconciliations.

4 Constructive and Technological Methods of Increasing the Accuracy of Measurements to Ensure Metrological Traceability of Impedance Parameters Units in the Range of Values

The modern development of radio-electronic and electrical measuring equipment imposes increased requirements on the accuracy of electrical impedance units' measurements. At the same time, the task of achieving high accuracy while ensuring the traceability of impedance units' measurements in the range of values becomes significant. The highest accuracy is achieved with equal-nominal comparing. Decimal dividers are used for equal-nominal comparing during successive calibration of measures in a range of values [12].

However, when using decimal dividers on AC, the following problems arise:

- violation of connection four- parity in the area of low-resistance impedances due to changes in the geometry of the current lines at the potential electrode when varying current sources;
- difference in the values of the temperature coefficient of resistance in the materials of the connecting conductor and the measure;
- violation of the connection four-parity when measuring the connection parameters of several measures (assembly) due to mutual induction;
- violation of the connection four-parity due to the presence of a distributed capacitance between the connecting elements of the divider measures and the common shield of the measure.

The purpose of this subsection is to analyze and describe the developed constructive and technological methods for improving the accuracy of measurements when transferring the sizes of impedance parameters units over a range of values.

The National Standard of Inductance DETU 08-09-09 [3, 7–9, 11] includes a comparator, which, with equal-nominal impedance comparing, introduces to the measurements an uncertainty component of no more than 1×10^{-8}. Therefore, an urgent task is to bring the accuracy of the measurement result closer to this value.

To transfer the size of the impedance unit of in the range of values for a long time, dividers of various constructions are used. The most widely used for this purpose are Haymon dividers [13], which are now commercially available and in the range of average DC resistance values allows reduction of the transfer error to units by 10^{-9}. Unfortunately, in a wider range of values and on AC, their metrological characteristics are significantly lower. This is due to the fact that the Haymon dividers are fundamentally rebuilt using switches.

In [12], a method is considered for ensuring the traceability of measurements in the range of values by means of equal-nominal comparing using non-switched voltage dividers, which are 10 (or 11) series-connected equal-nominal impedance measures. The use of such a technique on alternating current greatly simplifies the process of calibrating impedance measures in a range of values, however, it also has

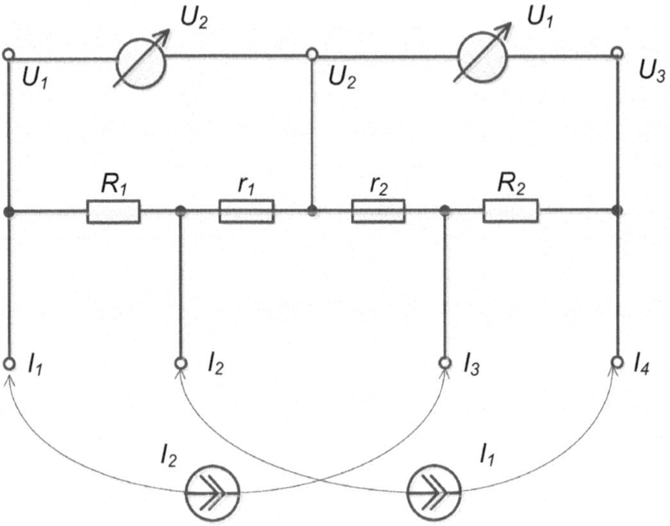

Fig. 11 The equivalent circuit of the voltage divider, taking into account the contribution of the connecting conductors: R_1, R_2—comparable measures; r_1, r_2—connecting conductors; I_1, I_2—current sources; U_1, U_2—vector voltmeters

limitations. This subsection is devoted to an analysis of these limitations, as well as an estimation of the resulting errors.

To analyze the resulting errors, let's consider a simplified divider structure that can be used for calibration with binary measurement ranges or can be part of a decimal divider. The equivalent circuit of such a divider, taking into account the contribution of the resistances of the connecting conductors, is shown in Fig. 11.

On Fig. 11 the connection of current generators and comparator voltmeters during the calibration process is shown. In the process of measurement, not the ratios of the resistances of the measures used R_1/R_2 are determined, but the ratios of the resistances of these measures together with the resistances of the connecting wires $(R_1 + r_1)/(R_2 + r_2)$.

Some characteristic error sources are considered below, which in this case limit the accuracy of the measurements.

4.1 Violation of the Connection Four-Parity in the Region of Low-Resistance Impedances due to a Change in the Geometry of the Current Lines at the Potential Electrode When Varying Current Sources

When measuring a single impedance or the sum of impedances, the configuration of the current lines in the region of the potential electrode will be different (Figs. 12,

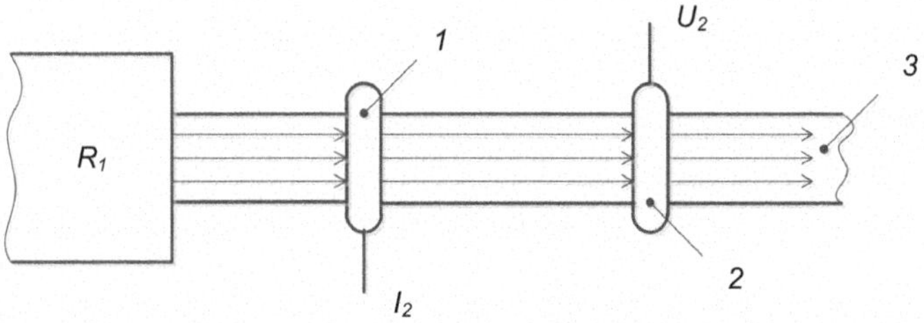

Fig. 12 Scheme of connecting current and potential contacts to the conductor: 1—current contact; 2—potential contact; 3—supply conductor; R_1—examined measure

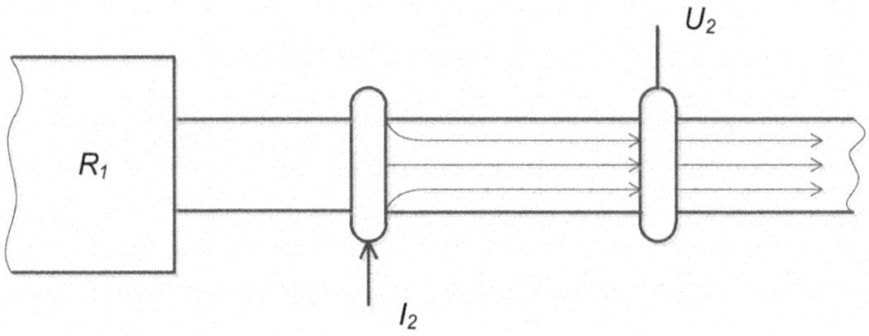

Fig. 13 Configuration of current lines when applying current to a contact I_2

13). This leads to a difference in the measured voltage of the examined signal, which introduces an additional error into the measurement result.

The task of minimizing the error is to achieve, by means of constructive and technological methods, the constancy of the current lines configuration, regardless of the connection point and the choice of current source [14].

The collinearity of the current lines at the point of the potential electrode connection U_2 should not depend on the point of the measuring signal current supply (I_1 or I_2).

If the current lines in the conductor are collinear, the resistance of the conductor is determined by the formula

$$r_{10} = 4\rho\frac{l}{\pi d^2}, \tag{12}$$

where l and d—conductor length and diameter, ρ—its resistivity.

Studies show that if this requirement is not met, the equivalent resistance of the conductor

$$r_{eq} = r_{10}\left(1 - e^{-\pi\frac{l}{d}}\right). \tag{13}$$

The value $e^{-\pi(l/d)}$ is the measurement error of the connecting wire resistance.

In the examined divider, the quantities $R_1 + r_1$ and $R_2 + r_2$ are used as measures. The temperature instability δR_t of such a composite measure is determined by the obvious formula

$$\delta R_t = \left(TKR_m + TKR_c \frac{r_1}{R_1} \right) \Delta t, \tag{14}$$

where TKR_m и TKR_c—temperature coefficients of measure and connecting wire, Δt—temperature instability in the thermostat where the measurements are usually performed.

Comparison of formulas (13) and (14) shows that for reduction of the first component of the error, it is necessary to increase the ratio l/d, at the same time for reduction of the second component of the error, the opposite is necessary—to reduce the specified ratio.

Thus, it is necessary to determine the optimal ratio l/d experimentally and by calculation, while putting forward requirements for the temperature instability of the thermostat.

An analysis of the total error shows that its minimum value is achieved at $l/d = -\ln(TKR_c \cdot \Delta t) \cdot \pi$. This minimum error decreases with decreasing of Δt.

4.2 *Difference of Temperature Coefficient Between Conductor and Measure*

Let's assume the examined measure has a nominal value of 1 Ω, and the connection wire made of copper has a resistance of 1 mOhm. The measure is in a thermostat with temperature instability of 0.01 °C. With the temperature instability of the examined measure $\delta R_{0T} = 10^{-8}$, the temperature instability of the connecting copper conductor, taking into account the error of the thermostat, will be $\delta R_{wT} = 4 \times 10^{-8}$, that is, the temperature instability of the conductor-measure complex will be 4–5 times higher than the temperature instability of the measure itself due to the connecting conductors. The contribution to the temperature instability of the conductor-measure complex is 4 times greater than the contribution of the measure itself. It follows from this that the temperature stability of the measure-conductor pair will be achieved when the ratio of nominal value R and value r is the same as the ratio of their TKR, that is, approximately 1:4000.

4.3 Violation of the Connection Four-Parity When Connecting Several Measures (Assemblies) Due to Mutual Induction

The voltage divider circuit, taking into account the influence of mutual magnetic induction, is shown in Fig. 14.

On low-resistance measuring ranges, two loops of the measuring circuit appear. Each loop geometrically has a different area. The mutual influence of the magnetic fields of these loops of the circuit introduces a parasitic inductance, which appears due to the inter-circuit mutual induction, and also introduces an error in the measured voltage (15), (16).

$$\Delta U = j\omega M_{\Sigma} I_1, \quad (15)$$

$$j\omega M_{\Sigma} = \frac{\Delta U}{I_1} = j\omega L_{eq}, \quad (16)$$

where ΔU –parasitic voltage induced in a potential circuit due to the flow of current through the current circuit; M_{Σ}—interloop mutual magnetic induction; M_1, M_2, M_3—magnetic induction of loops R_1, R_2 и $R_2 + R_2$; L_{eq}—equivalent value of inductance introduced by mutual magnetic induction.

Interloop mutual magnetic induction M_{Σ} creates additional inductance of the resistance measure.

Accordingly, the error introduced by parasitic inductance, which appears as a result of interloop mutual induction, will be different:

$$M_1 + M_2 \neq M_{\Sigma}. \quad (17)$$

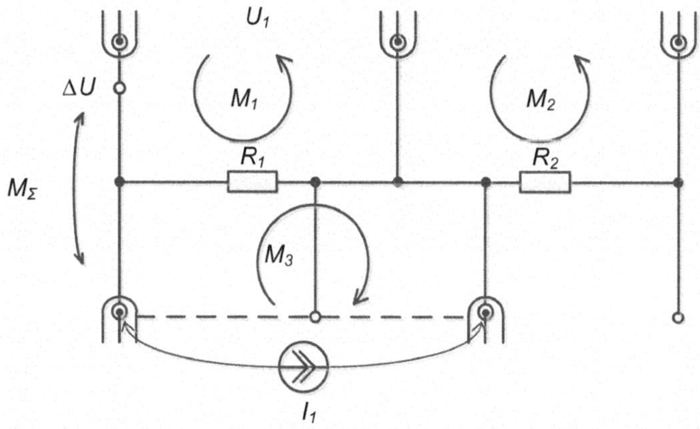

Fig. 14 Voltage divider circuit taking into account the influence of mutual magnetic induction

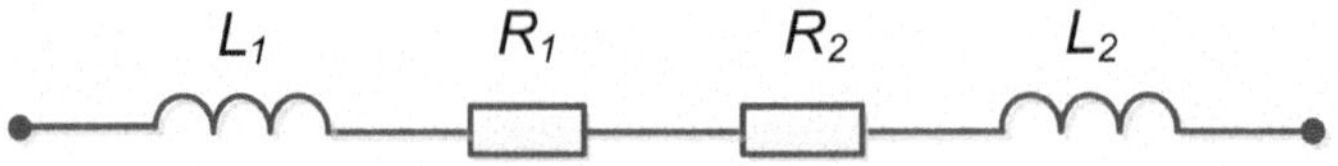

Fig. 15 The equivalent circuit of the voltage divider, taking into account the influence of mutual magnetic induction

The equivalent circuit of the voltage divider, taking into account the influence of mutual magnetic induction, is shown in Fig. 15.

When a measurement of $R_1 + R_2$ is made, the current loop and its configuration take on a different form.

Wherein

$$R_\Sigma = R_1 + R_2 + j\omega(L_1 + L_2) = (R_1 + R_2)\left[1 + j\frac{\omega(L_1 + L_2)}{R_1 + R_2}\right]. \tag{18}$$

4.4 Violation of the Four-Parity Due to Distributed Capacitance Between the Conductors and the Common Shield

During measuring parameters of two measures R_1 or R_2 using comparator with a four-pair object connection, the conductivity between the casing and the center connection does not affect the measurement result. But when the total resistance of the chain of measures $R_1 + R_2 + \ldots$ is measured, then the influence of parasitic capacitances, their conductivities, is detected and significantly affects the measurement result (Fig. 16).

Equivalent resistance of a circuit consisting of only two elements:

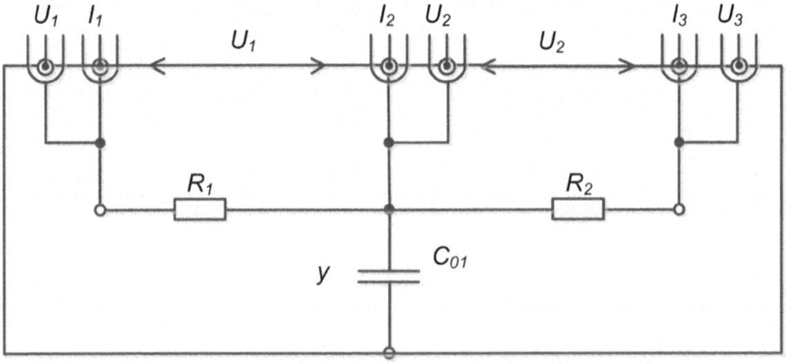

Fig. 16 The equivalent circuit of the voltage divider, taking into accounts the influence of parasitic capacitances

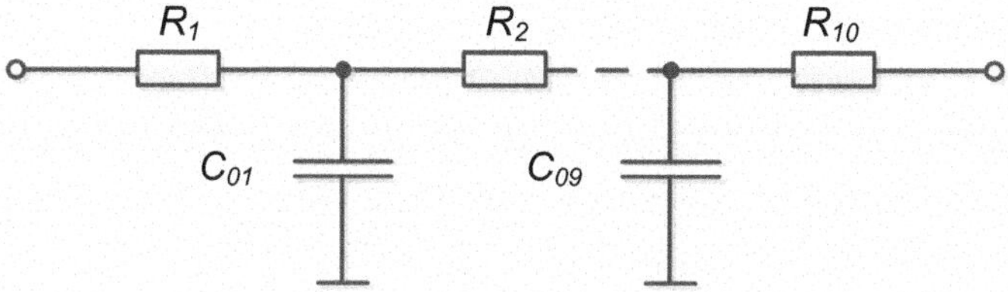

Fig. 17 Equivalent circuit of a voltage divider of ten resistors, taking into account the parasitic capacitances of the conductor-shield

$$R_{eq} = R_1 + R_2 + R_1 R_2 y, \tag{19}$$

where y—parasitic complex conduction.

At low frequencies y is just conduction, at high frequencies $y = j\omega C$.

$$R_{eq} = R_1 + R_2 + jR_1 R_2 y\omega C = (R_1 + R_2)(1 + j(R_1 R_2/(R_1 + R_2))\omega C), \tag{20}$$

where $\frac{R_1 R_2}{R_1 + R_2}\omega C$—loss tangent.

If a divider of ten resistors is examined (Fig. 17), then the formula takes the form:

$$R_1 + R_2 + R_1 R_2/(1/j\omega C_{01}) = R_1 + R_2 + j\omega C_{01} R_1 R_2 = 2R\left(1 + \frac{1}{2} j\omega C_{01} R\right). \tag{21}$$

As a result of the research, constructive and technological methods were proposed to improve the accuracy of measurements to ensure the metrological traceability of the units of impedance parameters in the range of values.

These methods allow solving following metrological problems: take into account the difference in the resistance temperature coefficient of the connecting conductors and the measure; take into account the violation of the connection four-clamp and four-parity due to changes in the geometry of the contacts, the mutual induction of the measuring circuits, as well as the presence of a distributed parasitic capacitance.

The use of the proposed approaches makes it possible to experimentally evaluate all parasitic capacitances and inductances of measuring circuits during the comparing of standard measures of impedance parameters, evaluate the errors and uncertainties that they introduce separately and in total, and also take into account and compensate them by computational methods.

5 Estimation of Measurement Uncertainty in the Calibration of Precision Measures of Electrical Capacitance with Reference to the Standard Measure of Capacitance of the Same Nominal Value on the National Standard of Units of Electrical Capacitance and Dissipation Factor

The technical basis of the system for ensuring the unity of measurements in the state is the national standard base, the development level of which determines the pace of scientific and technical progress in all sectors of the economy, industry, science and defense. The accuracy of impedance parameters measurements plays an important role in maintaining proper characteristics and improving technical means of telecommunications, energy, transport, etc.

In order to maintain the standard base at a high technical level, it is necessary to carry out systematic research and reconciliation with the standards of other countries for the international recognition of the results of measurements and tests, and to improve it constantly in accordance with state development programs.

The process of international integration requires the harmonization of the domestic regulatory normative document base with international documents in the field of metrology and metrological activity, which concerns, first of all, the presentation of measurement results in terms of uncertainty theory.

Thus, according to the requirements of international and interstate normative documents and the requirements of DSTU 3231:2007, it is envisaged to evaluate measurement uncertainty when reproducing and transferring units of physical quantities by state and secondary standards [15].

In 2001, at the State Enterprise "All-Ukrainian State Scientific-Production Center for Standardization, Metrology, Certification and Protection of Consumer Rights" (SE "Ukrmetrteststandard"), the National Standard of Units of Electrical Capacitance and Dissipation Factor DETU 08-06-01 was created and successfully operated [15].

Similar standards were created in the SE "Ukrmetrteststandart" for such national metrological institutes as GUM (Poland) and NIST (USA).

The purpose of this subsection is with the use of a modern approach to processing measurement results to develop a methodology for evaluating the measurements uncertainty when calibrating electrical capacitance measures with the reference on the standard capacitance measure of the same nominal value on the National Standard of Units of Electrical Capacitance and Dissipation Factor DETU 08-06-01.

The appearance of the standard is shown in Fig. 18.

The standard includes a complex of measuring equipment:

- set of standard thermostatic electrical capacitance measures Andeen Hagerling AH11A of 10 pF nominal value (4 units);
- set of standard thermostatic electrical capacitance measures Andeen Hagerling AH11A of 100 pF nominal value (4 units);

Fig. 18 Appearance of the national standard of units of electrical capacitance and dissipation factor DETU 08-06-01

- set of transient standard measures of electrical capacitance with nominal value from 1 pF to 1 μF (6 units);
- set of standard measures of active resistance with nominal value from 0.1 Ω to 10 MOhm (8 units);
- the standard comparator with a personal computer;
- the standard of frequency and time SCHV-74;
- electronic counter frequency meter CHZ-54.

The state standard provides reproduction of measurement units at a frequency of 1000 Hz.

The standard comparator from DETU 08-06-01 makes it possible to transfer the sizes of impedance parameters units in the range of values determined by a set of measures of active resistance and electrical capacitance [3].

The standard thermostatic electrical capacitance measures Andeen Hagerling AH11A form a group two-nominal measure.

The appearance of the group two-nominal measure is shown in Fig. 19.

Standard measures from this group measure have been continuously studied for more than 20 years. Thanks to the studies conducted in the national metrological institutes of the leading countries of the world (BIPM, France; PTB, Germany; NIST, USA; NPL, Great Britain; SMI, Czech Republic; GUM, Poland), as well as studies conducted in the SE "Ukrmetrteststandart", the value of the electrical capacitance of the AH11A standard measurements is known with an expanded uncertainty $U(C_{AH}) = 8.0 \times 10^{-8}$ pF at the coefficient of coverage $k = 2$. Constant studies of the specified measures allow evaluating and taking into account the drift of their main characteristics.

Fig. 19 Appearance of the group two-nominal measure of four measures Andeen-Hagerling type AH11A

When calibrating an electrical capacitance measure with reference to standard measure of the same nominal, the value of the electrical capacitance measure C_X being calibrated is obtained from the expression

$$C_X = C_S + \Delta C_C + \Delta C_{QC} + \Delta C_{Tx} + \Delta C_{SC} + \Delta C_{CE} + \Delta C_\gamma, \qquad (22)$$

where C_S—the value of the electrical capacitance of the standard measure specified in the calibration certificate; ΔC_C—the difference between the value of the electrical capacitance of the standard measure specified in the calibration certificate and the displayed value of the electrical capacitance of the measure being calibrated; ΔC_{QC}—the deviation caused by the discreteness of the comparator readings; ΔC_{Tx}—correction caused by the temperature dependence of the measure being calibrated; ΔC_{SC}—correction for the sensitivity of the comparator; ΔC_{CE}—error of comparing; ΔC_γ—correction caused by the drift of the value of the electrical capacitance of the standard measure since the last calibration.

Tables 1 and 2 show the calculations of the uncertainty budget (C_X), which are used when calibrating electrical capacitance measures with nominal values of 10 pF and 100 pF, respectively.

The expanded uncertainty in Table 1 is obtained from the expression

Table 1 Uncertainty budget (C_X) for electrical capacitance measures with a nominal value of 10 pF

Quantity X_i	Estimate x_i	Standard uncertainty $u(x_i)$	Probability distribution	Method of evaluative (A, B)	Sensitivity coefficient c_i	Uncertainty contribution $u_i(y)u_i(y)$
C_S	10.00001166 pF	4.0E-07 pF	Normal	B	1	4.0E-07 pF
ΔC_C	0.0000051 pF	6.1E-07 pF	Normal	A	1	6.1E-07 pF
ΔC_{QC}	0	5.77E-07 pF	Rectang.	B	1	5.77E-07 pF
ΔC_{Tx}	0	4.1E-07 pF	Normal	B	1	4.1E-07 pF
ΔC_{SC}	0	1.07E-08 pF	Rectang.	B	1	1.07E-08 pF
ΔC_{CE}	0	5.00E-07 pF	Normal	B	1	5.00E-07 pF
ΔC_Y	0.0000001 pF	5.08E-07 pF	Normal	B	1	5.08E-07 pF
C_X	10.0000169 pF					
		Combined standard uncertainty			u_c	1.24E-06 pF
		Effective degrees of freedom			ν_{eff}	>200, $k = 2$
		Expanded uncertainty ($p \approx 95\%$)			U	2.48E-06 pF

$$U = k \cdot u(C_X) = 2 \cdot\ 1.24\text{E} - 06\text{pF}\ = 2.48\text{E} - 06\text{pF}.$$

The expanded uncertainty in Table 2 is obtained from the expression

$$U = k \cdot u(C_X) = 2 \cdot\ 1.28\text{E} - 05\text{pF}\ = 2.56\text{E} - 05\text{pF}.$$

Ensuring the stability of the 1000 Hz frequency at which the measurements are performed, as well as the slight dependence of the electrical capacitance of the measures on the frequency, allows not taking into account the frequency dependence of the standard measure and the measure being calibrated.

6 Ensuring the Traceability of Measurements of Electrical Capacitance and Active Resistance

The metrological traceability of the National Standard of Units of Electrical Capacitance and Dissipation Factor DETU 08-06-01 is ensured by periodic calibration of Andeen-Hagerling AH11A thermostatic electrical capacitance standard measures with a nominal value of 10 pF (3 units) and 100 pF (3 units), which are part of

Table 2 Uncertainty budget (C_X) for electrical capacitance measures with a nominal value of 100 pF

Quantity X_i	Estimate x_i	Standard uncertainty $u(x_i)$	Probability distribution	Method of evaluative (A, B)	Sensitivity coefficient c_i	Uncertainty contribution $u_i(y)u_i(y)$
C_S	100.0001038 pF	4.0E-06 pF	Normal	B	1	4.0E-06 pF
ΔC_C	0.0000251 pF	8.2E-06 pF	Normal	A	1	8.2E-06 pF
ΔC_{QC}	0	5.77E-07 pF	Rectang.	B	1	5.77E-07 pF
ΔC_{Tx}	0	4.1E-06 pF	Normal	B	1	4.1E-06 pF
ΔC_{SC}	0	1.07E-07 pF	Rectang.	B	1	1.07E-07 pF
ΔC_{CE}	0	7.90E-06 pF	Normal	B	1	7.90E-06 pF
ΔC_{γ}	0.0000001 pF	8.50E-07 pF	Normal	B	1	8.50E-07 pF
C_X	100.0001290 pF					
		Combined standard uncertainty			u_c	1.28E-05 pF
		Effective degrees of freedom			ν_{eff}	>200, $k = 2$
		Expanded uncertainty ($p \approx 95\%$)			U	2.56E-05 pF

the standard, in the national metrological institutes of the leading countries of the world. The last regular calibration was carried out in the period from November 20 to December 7, 2020 at the International Bureau of Weights and Measures (BIPM), Paris.

The traceability of the values of the BIPM standard capacitors capacitance is provided by a sequence of measurements using a chain of impedance bridges to the calculated capacitor with reference on the values of such fundamental physical constants as the von Klitzing constant $R_K = h/e^2 = 25{,}812.80746\ \Omega$ using fixed numerical values of the Planck constant h and elementary charge e.

Calibration results for 10 pF capacitance measures at the International Bureau of Weights and Measures (BIPM), Paris, are given in Table 3. Measurement method is comparison with a standard capacitor of 100 pF when using a coaxial bridge for 2-pin connection of two impedances.

Calibration results for 100 pF capacitance measures are given in Table 4. Measurement method is comparison with a standard capacitor of 10 pF when using a coaxial bridge for 2-pin connection of two impedances.

The uncertainty u for C given in Tables 3 and 4 is the result of the sub-summation of the contribution from the measure instability observed during the measurement period and the contribution from BIPM measurement uncertainty and traceability.

All uncertainties given in the certificates are standard estimated uncertainties without taking into account the application of the coverage factor k.

Table 3 Calibration results of Andeen-Hagerling AH11A 10 pF measures

Factory no. of measure AH11A	01261	01263	01327
Factory no. of measure casing AH1100	00084	00086	00108
Nominal temperature in the laboratory, °C	23		
The average temperature of the measure, °C	30.7	31.1	30.5
The average value of the measure temperature drift «DRIFT (PPM)»	0.048	−0.050	0.014
Nominal voltage on the measure, V	100		
Nominal measurement frequency, Hz	1592		
BIPM estimate of the coefficient of capacitance dependence on voltage, pF/V	0.6×10^{-9}	0.6×10^{-9}	0.2×10^{-9}
Standard uncertainty of the coefficient of capacitance dependence on voltage, pF/V	2.0×10^{-9}		
BIPM estimate of the relative change in capacitance with frequency going from 1000 to 1592 Hz: $(C_{1592} - C_{1000})/C_{1000}$	-9.2×10^{-8}	-5.4×10^{-8}	-6.0×10^{-8}
Standard uncertainty of the relative change in capacitance with frequency going from 1000 to 1592 Hz	6.0×10^{-8}		
The obtained value of the measure capacitance C at 1592 Hz and 100 V, pF	10.00001074	10.00001104	9.99999022
Contribution to uncertainty u due to measure instability	0.2×10^{-8}	0.2×10^{-8}	0.1×10^{-8}
Contribution to uncertainty u due to BIPM measurement and traceability uncertainties	4.0×10^{-8}		
Relative standard uncertainty u for C	4.0×10^{-8}		

In order to ensure the traceability of active resistance measurements, the calibration of active resistance measures (9 units) from DETU 08-06-01 is carried out annually at the National Scientific Center "Institute of Metrology", Kharkiv. The last regular calibration was carried out in September 2022.

Calibrations were carried out on the National State Primary Standard of Electrical Resistance NDETU EM-06-2021 by the direct method according to MKU 08-674:2021. Metrology. Resistance measures. Calibration methodology on the NDETU standard EM-06-2021. Measurements made during calibration are traceable to SI units in accordance with the capabilities of NSC "Institute of Metrology" published in the KCDB BIPM database. The results of calibration of active resistance measures are shown in Table 5.

The expanded uncertainty is obtained as the product of the standard uncertainty and the coverage coefficient $k = 2$ corresponding to the 95% confidence level for the assumption of a normal distribution of the measurement results.

Based on the results of the research, a methodology was developed that allows to estimate the combined standard and extended uncertainty of measurements when

Table 4 Calibration results of Andeen-Hagerling AH11A 100 pF measures

Factory no. of measure AH11A	01260	01262	01328
Factory no. of measure casing AH1100	00084	00086	00108
Nominal temperature in the laboratory, °C	23		
The average temperature of the measure, °C	30.7	31.1	30.5
The average value of the measure temperature drift «DRIFT (PPM)»	0.006	−0.060	−0.084
Nominal voltage on the measure, V	10		
Nominal measurement frequency, Hz	1592		
BIPM estimate of the coefficient of capacitance dependence on voltage, pF/V	0.4×10^{-7}	0.0×10^{-7}	0.2×10^{-7}
Standard uncertainty of the coefficient of capacitance dependence on voltage, pF/V	2.0×10^{-7}		
BIPM estimate of the relative change in capacitance with frequency going from 1000 to 1592 Hz: $(C_{1592} - C_{1000})/C_{1000}$	-2.0×10^{-8}	-35.8×10^{-8}	-3.5×10^{-8}
Standard uncertainty of the relative change in capacitance with frequency going from 1000 to 1592 Hz	6.0×10^{-8}		
The obtained value of the measure capacitance C at 1592 Hz and 100 V, pF	100.0001018	100.0000942	100.0001174
Contribution to uncertainty u due to measure instability	0.2×10^{-8}	0.4×10^{-8}	0.3×10^{-8}
Contribution to uncertainty u due to BIPM measurement and traceability uncertainties	4.0×10^{-8}		
Relative standard uncertainty u for C	4.0×10^{-8}		

calibrating measures of electrical capacitance on the National Standard of Units of Electrical Capacitance and Dissipation Factor DETU 08-06-01 with reference to the standard capacitance measure of the same nominal value. The use of this methodology allows DETU 08-06-01 to be involved in the international reconciliation of origin standards of electrical capacitance of national metrological institutes of the leading countries of the world.

The practical value lies in ensuring the unity of measurements of electrical capacitance and dissipation factor in the country.

The National Standard of Units of Electrical Capacitance and Dissipation Factor is intended for reproducing, storing and transferring the size of the unit to both working standards and high-precision measuring equipment—standard measures of electrical capacitance, impedance and amittance meters, LCR- and RLC-meters, which, in their in turn, transfer the measurement unit to the working measuring equipment of electrical capacitance, loss tangent and dissipation factor, multimeters, combined

Table 5 Results of calibration of active resistance measures

No.	Type of measure	Serial no.	Certificate, no.	Temperature, °C	Calibration point, Ohm	Measured value, Ohm	Absolute extended uncertainty (k = 2), Ohm
1	P4022M1	04	UA01 №2195 from 21.09.2022	20.3	1×10^7	0.999981×10^7	1.9×10^2
2	P4016	821	UA01 №2196 from 21.09.2022	20.6	1,000,000	999,999.7	5.6
3	P4015	098	UA01 №2197 from 21.09.2022	20.6	100,000	100,000.44	0.18
4	MAC-2	010	UA01 №2198 from 05.09.2022	20.2	100	99.997543	0.000023
5	MAC-2 10 Ом	006	UA01 №2199 from 05.09.2022	20.2	10	9.9997817	0.0000035
6	MAC-2 1 Ом	006	UA01 №2200 from 05.09.2022	20.2	1	1.00000126	0.00000034
7	MAC-2 0,1 Ом	006	UA01 №2201 from 05.09.2022	20.2	0.1	0.100000891	0.000000067
8	MAC-2	003	UA01 №2202 from 05.09.2022	20.2	1000	999.99966	0.00028
9	MAC-2	002	UA01 №2203 from 05.09.2022	20.2	10,000	9999.8631	0.0039

devices, etc., which are widely used in such fields as: energy, radio electronics, utilities, construction, transport, telecommunications, metallurgy, machine and aircraft construction industry, security industry, as well as scientific research and defense.

More than 10 million measuring devices are used in Ukraine, which receive a unit of measurement from the National Standard of Units of Electrical Capacitance and Dissipation Factor.

Meters of impedance, total resistance of the phase-zero loop, can belong to the scope of legally regulated metrology, provided that the measurement results are used during the control of the working conditions safety.

The completed set of works with the standard ensures stability, metrological traceability of the measurement results in the above-mentioned fields and in relation to the measuring equipment used, including, compliance estimation in accordance with the requirements of the Technical Regulation of legally regulated measuring equipment.

The results of the research of Andeen-Hagerling AH11A 10 pF and 100 pF standard capacitance measures confirm that the metrological characteristics of the National Standard of Units of Electrical Capacitance and Dissipation Factor DETU 08-06-01 are within the permissible values and correspond to the data specified in the passport for the standard.

The developed standard examination methods and results of performed research correspond to the world level and contribute to improving the quality and competitiveness of domestic products, ensuring the conditions for innovative development of the national economy, as well as ensuring the functioning of a modern standard base adequate to the requirements of a market economy, harmonized with international rules and requirements; protection of the interests of consumers or third parties in obtaining reliable measurement results.

7 Transfer of the Electrical Capacitance Unit Size by the Range of Values Using the National Standard of Units of Electrical Capacitance and the Dissipation Factor

The transfer of the unit size of capacitance over the value range is carried out with the use of the universal automated precision comparator included in the DETU 08-06-01 [11]. When comparing of electrical capacitance measures the comparator has two transfer ratio values: 1:1 or 1:10. Using only these two transfer ratio values, it is possible to realize the transferring of the unit size of capacitance by consecutive calibrations of capacitance measures in the wide range of values toward both high and low impedance.

An example of transferring of the unit size of capacitance over the value range during the calibration of capacitance measures with the nominal value of 10 nF with reference on the standard capacitance measure of 100 pF and with the use of intermediate capacitance measure of 1 nF is represented in Fig. 20.

The estimation of uncertainty in the calibration of capacitance measures is carried out according to the measurement model (equation) [16]:

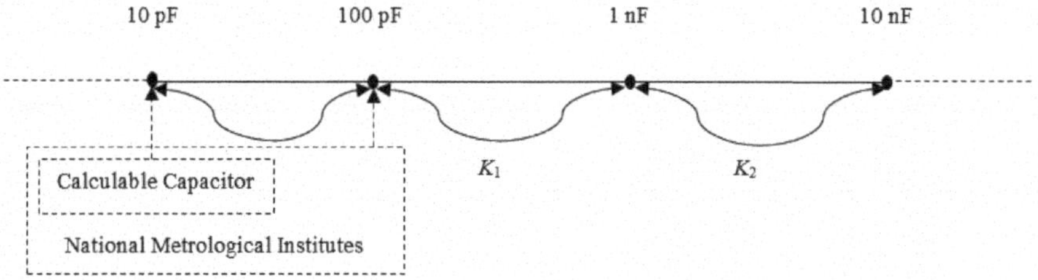

Fig. 20 The transfer of the unit size of capacitance over the range of values

$$C_X = \frac{C_{100\ \text{pF}}}{K_1 K_2}, \tag{23}$$

where $C_{100\ \text{pF}} = C_S + \Delta C_{TS} + \Delta C_{fS} + \Delta C_{\gamma S}$, C_S—value of capacitance of a standard measure with the nominal value of 100 pF, indicated in a calibration certificate; $C_{100\ \text{pF}}$—the actual value of the capacitance of the standard measure with nominal value of 100 pF; ΔC_{TS}—correction for temperature dependence of standard measures; ΔC_{fS}—correction for frequency dependence of standard measures; $\Delta C_{\gamma S}$—correction for a drift of standard measures from the moment of the last calibration; K_1—the transfer coefficient of the comparator when calibrating an intermediate capacitance measure of 1 nF from a set of thermostatic measures SA 5200RC with reference to a standard measure of electrical capacitance with a nominal value of 100 pF:

$$K_1 = \frac{C_{1\ \text{nF}}}{C_{100\ \text{pF}}}, \tag{24}$$

where K_2—the transfer factor of the comparator in the calibration of the intermediate capacitance measure C_X with the nominal value of 10 nF with reference to the intermediate capacitance measure with the nominal value of 1 nF:

$$K_2 = \frac{C_X}{C_{10\ \text{nF}}}. \tag{25}$$

The example of the uncertainty budget of measurements of the capacitance value in the calibration of the measure C_X is presented in Tables 6 and 7 [17].

Calculation of the relative total standard uncertainty $w(C_X)$ and relative expanded uncertainty $W(C_X)$ during the transferring of the size of the physical quantity from the capacitance measure with the nominal value of 100 pF to the calibrated capacitance measure with the nominal value of 10 nF is carried out in a relative form by the formulas (26), (27):

Table 6 Budget of uncertainty when determining the actual value of the standard measure $C_{100\ \text{pF}}$

Quantity X_i	Estimate x_i	Standard uncertainty $u(x_i)$	Probability distribution	Method of evaluative (A, B)	Sensitivity coefficient c_i	Uncertainty contribution $u_i(y)u_i(y)$
C_S	100.0001038 pF	4.0E-06 pF	Normal	B	1	4.0E-06 pF
ΔC_{TS}	0	4.1E-06 pF	Normal	B	1	4.1E-06 pF
ΔC_{fS}	0 pF	1.20E-07 pF	Normal	B	1	1.20E-07 pF
$\Delta C_{\gamma S}$	0.0000001 pF	8.50E-07 pF	Normal	B	1	8.50E-07 pF
$C_{100\ \text{pF}}$	100.0001039 pF					
		Combined standard uncertainty			u_c	5.79E-06 pF
		Effective degrees of freedom			ν_{eff}	>200, $k = 2$
		Expanded uncertainty ($p \approx 95\%$)			U	1.16E-05 pF

Table 7 Uncertainty budget when calibrating a measure C_X

Quantity X_i	Estimate x_i	Standard uncertainty $w(x_i)$	Probability distribution	Method of evaluative (A, B)	Sensitivity coefficient p_i	Uncertainty contribution $p_i \cdot w(x_i)p_i \cdot w(x_i)$
$C_{100\ \text{pF}}$	100.0001039 pF	5.79E-06	Normal	B	1	5.79E-06
k_1	0.099996	1.20E-07	Normal	A	−1	−1.20E-07
k_2	0.100002	8.50E-07	Normal	A	−1	−8.50E-07
C_X	10,000.2104 pF					
		Relative combined standard uncertainty			$w(C_X)$	4.09E-06
		Effective degrees of freedom			ν_{eff}	>200, $k = 2$
		Relative expanded uncertainty ($p \approx 95\%$)			$W(C_X)$	8.18E-06

$$w(C_X) = \sqrt{w^2(C_S) + \sum_{i=1}^{N} p_i^2 w_i^2(x_i)}. \tag{26}$$

$$W(C_X) = kw(C_X) \tag{27}$$

The value of standard uncertainty of the factors K_1, K_2 takes into account:

- the deviation caused by the comparator quantization error;
- correction for the sensitivity of the comparator and the error of comparing.

The values of the factors K_1, K_2 and their uncertainties are indicated in the comparator certificate, but these factors can be refined for every point of the measurement range by the comparison of pre-calibrated measures.

It should be noted that in the total standard measurement uncertainty of the calibration result, it is also necessary to take into account the frequency dependence of the transfer factor of the comparator, which has a substantial influence on the measurement result during the calibration of capacitance measures.

However, during the measurements, the frequency drift is negligible and the measurement uncertainty is about the value of 1×10^{-10}. Thus, the uncertainty component introduced by the frequency dependence can be neglected.

The measured value of capacitance of the calibrated measure with the nominal value of 10 nF at the measurement temperature of (22 ÷ 24) °C and relative humidity of (30 ÷ 45) % at the frequency of the examined signal of 1 kHz made up 100,002,104 nF ± 8.18 μF/F.

CMC of NMIs of countries are published as pdf-files in the Annex C of the BIPM Key Comparison Database (KCDB) in the form of tables [4]. The above-mentioned values of measurement uncertainties correspond to the data published in the KCDB for Ukraine in the range of capacitance values from 10 pF to 10 nF.

The developed calibration procedure allows to estimate the relative total standard and relative extended measurement uncertainty when calibrating the standard capacitance measures using the National standard of the units of electrical capacitance and dissipation factor (DETU 08-06-01).

The estimation of the uncertainty values of electrical capacitance measurements in a wide range of capacitance values according to the developed method is confirmed by published CMC for Ukraine in the range of capacitance values from 10 pF to 10 nF at frequencies of 1 and 1.592 kHz.

The use of this technique makes it possible to estimate the measurement uncertainty of the capacity values of standard measures during international comparisons.

8 Uncertainty Estimation in the Calibration of Inductance Measures with Traceability to Capacitance Standards

In 2001–2009, the National standards of the units of electrical capacitance, inductance and dissipation factor were created and successfully are used till now in State Enterprise "All-Ukrainian State Scientific and Production Centre for Standardization, Metrology, Certification and Protection of Consumer" (SE "Ukrmetrteststandard"). The appearance of the National Standard of the Units of Inductance and Dissipation Factor is shown in Fig. 21.

Fig. 21 The national standard of the units of inductance and dissipation factor DETU 08-09-09

The standards are created on the base of the universal automated precision comparator [7], which allows realizing the calibration in the total range of values: high impedance with reference on the low impedance, capacitance with reference on the resistance, inductance with reference on the capacitance.

Similar standards were created in SE "Ukrmetrteststandard" for such national metrology institutes as BelGIM (Belarus), GUM (Poland) and NIST (USA).

The purpose of the section is to present the main points of the estimation of uncertainty in the calibration of inductance measures based on a modern approach to the processing of measurement results using the National standards of the units of electrical capacitance, inductance and dissipation factor.

To calibrate working standards of inductance, it is necessary to provide traceability from a standard capacitance measure of 100 pF or 10 pF with the appropriate calibration certificate, where calibration uncertainty should be specified [18, 19].

Such measures are included to the National Standard of the Units of Electrical Capacitance and Dissipation Factor DETU 08-06-01 as group of four precision capacitance measures Andeen-Hagerling AH11A (4×100 pF and 4×10 pF). Due to the calibration of the specified measures conducted at NIST (USA), PTB (Germany), BIPM (France) and NPL (Great Britain), as well as own constant research, the value of the electrical capacitance of all capacitance measures AH11A is known with the expanded uncertainty $U(C_{AH}) = 8.0 \cdot 10^{-8}$ pF with the probability $P = 0.95$ at the coverage factor $k = 2$.

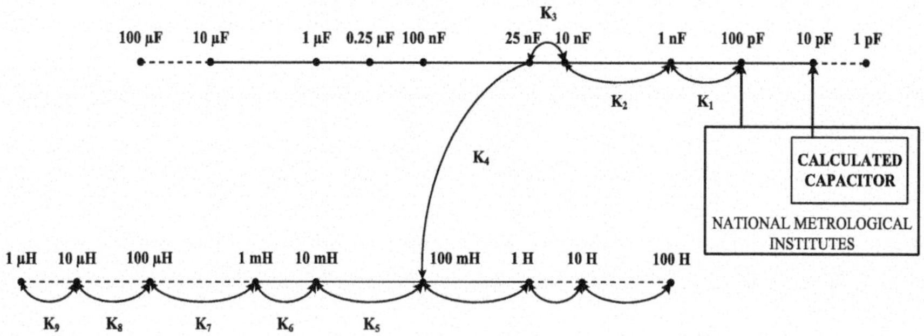

Fig. 22 The transferring of the unit size of inductance over the value range with the basis on the calculated standard capacitor

A working frequency of the standard is 1 kHz and 1.592 kHz. The calibration of the capacitance measure AH11A on the same frequencies was conducted on national standards of other countries, which are based on the use of fundamental physical constants.

The transferring of the unit size of inductance over the value range from capacitance measures is carried out with the use of the universal automated precision comparators included in the standards. The comparators have the following transfer ratio values: 1:1, 1:10 or 10:1. Using these transfer (comparison) ratio values only, it is possible to realize the transferring of the unit size of inductance from capacitance measures by consecutive calibrations in the total range of values toward both high and low impedance. An example of transferring of the unit size of inductance over the value range from capacitance measures in the calibration of inductance measure with the nominal value of 1 µH based on the standard capacitance measure of 100 pF and with the use of a series intermediate standard thermostated capacitance and inductance measures is represented in Fig. 22.

The measurements model (equation) that follows from the measurement scheme:

$$L_X = \frac{K_5 K_6 K_7 K_8 K_9}{\omega^2 K_1 K_2 K_3 K_4 C_{100\text{ pF}}}, \tag{28}$$

$$C_{100\text{ pF}} = C_S + \Delta C_{TS} + \Delta C_{fS} + \Delta C_{\gamma S}, \tag{29}$$

K_1—transfer coefficient of the comparator when calibrating the electrical capacitance measure 1 nF with the reference to the standard capacitance measure AH11A with the nominal value of 100 pF

$$K_1 = \frac{C_{1\text{ nF}}}{C_{100\text{ pF}}}; \tag{30}$$

K_2—transfer coefficient of the comparator when calibrating the electrical capacitance measure 10 nF with the reference to the intermediate capacitance measure with the nominal value of 1 nF, for the set of thermostated capacitance measures CA 5200RC

$$K_2 = \frac{C_{10\ \mathrm{nF}}}{C_{1\ \mathrm{nF}}}; \tag{31}$$

K_3—transfer coefficient of the comparator when calibrating the electrical capacitance measure 25 nF with the reference to the intermediate capacitance measure with the nominal value of 10 nF, for the set of thermostated capacitance measures CA 5200RC

$$K_3 = \frac{C_{25.33\ \mathrm{nF}}}{C_{10\ \mathrm{nF}}}; \tag{32}$$

K_4—transfer coefficient of the comparator when calibrating the inductance measure 100 mH with the reference to the intermediate capacitance measure 25.33 pF and frequency ω; ω—operating frequency of the test signal 6279.897 rad/s (1 kHz) on which measurements are made;

K_5—transfer coefficient of the comparator when calibrating the inductance measure 10 mH with the reference to the intermediate inductance measure 100 mH

$$K_5 = \frac{L_{10\ \mathrm{mH}}}{L_{100\ \mathrm{mH}}}; \tag{33}$$

K_6—transfer coefficient of the comparator when calibrating the inductance measure 1 mH with the reference to the intermediate inductance measure 10 mH

$$K_6 = \frac{L_{1\ \mathrm{mH}}}{L_{10\ \mathrm{mH}}}; \tag{34}$$

K_7—transfer coefficient of the comparator when calibrating the inductance measure 100 μH with the reference to the intermediate inductance measure 1 mH

$$K_7 = \frac{L_{100\ \mu H}}{L_{1\ \mathrm{mH}}}; \tag{35}$$

K_8—transfer coefficient of the comparator when calibrating the inductance measure 10 μH with the reference to the intermediate inductance measure 100 μH

$$K_8 = \frac{L_{10\ \mu H}}{L_{100\ \mu H}}; \tag{36}$$

K_9—transfer coefficient of the comparator when calibrating the inductance measure 1 μH with the reference to the intermediate inductance measure 10 μH

$$K_9 = \frac{L_{1\ \mu H}}{L_{10\ \mu H}}; \tag{37}$$

$C_{100\ \text{pF}}$—actual value of the Andeen-Hagerling AN11A standard capacitance measure with the nominal value of 100 pF, taking into account the time drift of the main parameter, as well as other influential factors;

C_S—value of the Andeen-Hagerling AN11A standard capacitance measure indicated in the calibration certificate;

ΔC_{TS}—correction for the temperature dependence of the standard capacitance measure AH11A;

ΔC_{fS}—correction for the frequency dependence of the standard capacitance measure AH11A;

$\Delta C_{\gamma S}$—correction for the drift of the standard measure AH11A since the last calibration.

At the same time, it should be taken into account that, in accordance with the requirements of the customer, the calibration of inductance measures should be carried out either with two-pair or three-pair connection of measures.

An example of calculating the uncertainty budget of the actual capacitance value C_X of the AH11A standard measure is given in Table 8 [17].

The example of the uncertainty budget calculation when calibrating the inductance measure L_X with the reference to the standard capacitance measure $C_{100\ \text{pF}}$ AH11A is presented in Table 9.

Calculation of the relative total standard uncertainty $\omega(L_X)$ and relative expanded uncertainty $W(L_X)$ when transferring the unit size of the physical quantity from the standard capacitance measure AH11A with the nominal value of 100 pF to the calibrated inductance measure L_X with the nominal value of 1 μH is carried out in a relative form by the formulas (38), (39)

$$\omega(L_X) = \sqrt{\omega^2(C_{100\ \text{pF}}) + \sum_{i=1}^{N} p_i^2 \omega_i^2(x_i)}, \tag{38}$$

Table 8 Uncertainty budget of the actual capacitance value C_X of the AH11A reference measure

Quantity X_i	Estimate x_i	Standard uncertainty $u(x_i)$, pF	Probability distribution	Method of evaluation (A, B)	Sensitivity coefficient c_i		Uncertainty contribution $c_i u(x_i)$, pF
C_S	100.00002	3.7E-05	Normal	A	1		3.7E-05
ΔC_{TS}	0	7.1E-07	Normal	B	1		7.1E-07
ΔC_{fS}	0	4.1E-07	Normal	B	1		4.1E-07
$\Delta C_{\gamma S}$	−1.10E-06	5.8E-07	Normal	B	1		5.8E-07
$C_{100\ \text{pF}}$	100.0000189						
		Combined standard uncertainty, pF				$u(C_X)$	3.70E-05
		Effective degrees of freedom				ν_{eff}	>200, $k = 2$
		Expanded uncertainty ($p \approx 95\%$)				$U(C_X)$	7,403E-05

Table 9 Uncertainty budget when calibrating an inductance measure L_X with reference to standard capacitance measure $C_{100\ pF}$ AH11A

Quantity X_i	Estimate x_i	Relative standard uncertainty $\omega(x_i) = u(x_i)/\lvert x_i\rvert$	Probability distribution	Method of evaluative (A, B)	Sensitivity coefficient p_i	Relative uncertainty contribution $p_i\omega(x_i)$
$C_{100\ pF}$	100.000019 pF	3.83E-07	Normal	B	−1	−3.83E-07
ω	6283.185370 rad/c	1.00E-07	Normal	B	−2	−2.00E-07
K_1	10.000245	1.20E-06	Normal	A	−1	−1.20E-06
K_2	10.000239	1.20E-06	Normal	A	−1	−1.20E-06
K_3	2.532112	1.50E-06	Normal	A	−1	−1.50E-06
K_4	10.100035	7.00E-06	Normal	A	−1	−7.00E-06
K_5	0.099970	1.50E-06	Normal	A	1	1.50E-06
K_6	0.100041	1.50E-06	Normal	A	1	1.50E-06
K_7	0.099982	2.00E-06	Normal	A	1	2.00E-06
K_8	0.099999	3.00E-06	Normal	A	1	3.00E-06
K_9	0.100001	5.00E-06	Normal	A	1	5.00E-06
L_X	0.990337 μH					
		Relative combined standard uncertainty			$\omega(L_X)$	9.84E-06
		Effective degrees of freedom			ν_{eff}	>200, $k = 2$
		Relative expanded uncertainty ($p \approx 95\%$)			$W(L_X)$	3.46E-05

$$W(L_X) = k\omega(L_X). \tag{39}$$

In the value of the standard uncertainty of the coefficients $K_1, K_2, \ldots, K_9$, the following are taken into account: deviation due to the comparator quantization error; comparator sensitivity correction, readout resolution correction and comparing error. The values of the coefficients $K_1, K_2, \ldots, K_9$ are given in the passport of the comparator, but can be specified for each point of the measurement range by comparing pre-calibrated standard measures. It should be noted that in the total standard uncertainty of the calibration result, it is also necessary to take into account the frequency dependence of the comparator gain. Frequency instability has a significant effect on the result of measurements when calibrating both the capacity measurements and the measures of inductance. However, during the measurements, the frequency drift is negligible and the measurement uncertainty is about the value of 1×10^{-10}. Thus, the components of uncertainty introduced by the frequency dependence can be neglected.

The measured value of inductance of the calibrated standard measure with the nominal value of 1 μH at the measurement temperature of (22 ÷ 24) °C and relative humidity of (30 ÷ 45) % at the frequency of the examined signal of 1 kHz is equal to 0.990337 μH ± 3.46 · 10^{-5}.

This measurement result corresponds to the published CMC-lines in the table of calibration and measurement capabilities of SE "Ukrmetrteststandard" on the website of KCDB BIPM.

The developed calibration procedure allows estimating the relative total standard and relative extended measurement uncertainties when calibrating the standard inductance measures using the National standards of the units of electrical capacitance, inductance and dissipation factor. The use of this calibration procedure allows estimating the measurement uncertainty of the inductance value of standard measures during international comparisons.

9 Calibration Method of Precision LCR-Meters

One of the most important tasks for all fields of science and technology related to the use of electrical measurements is the measurement of inductance, capacitance and active resistance across a wide range of frequencies and values.

The high accuracy and traceability of impedance parameter measurements is essential in such fields as electrical engineering, electronics, machine tools, instrumentation, construction, transportation, telecommunications, metallurgy, mechanical and aeronautical industries, security, and also research and defense [17, 20–23].

Currently, a large number of high-precision standard and working impedance measuring instruments are widely used, namely: precision Keysight (Agilent) E4980 AL LCR meters of various modifications, IET/QuadTech 7600 Plus, GW INSTEK LCR-6300, Applent Instruments AT810A, MHC-1100 and MOE-1200, etc. Combined devices and multimeters have become the most widely used in the creation, commissioning and operation of various electronic equipment. They require calibration over a wide frequency range and range of values of inductance, electrical capacitance, active resistance.

The purpose of the research is to develop methods of calibration of LCR-meters of precision and other impedance parameter meters, as well as to create a unified approach to determining the main components of the measurement model, uncertainty budget and design of calibration results [24].

The vast majority of calibrated precision LCR meters have the following metrological characteristic:

- measurement ranges at the main frequency of 1 kHz:
  - active resistance *R*—from 1 · 10E-5 to 1 · 10E11 Ohm;
  - electrical capacitance *C*—from 1 · 10E-17 to 10 F;
  - inductance *L*—from 1 · 10E-10 to 1 · 10E8 H;
- loss tangent $tg\delta$—from ± 1 · 10E-6 to ± 1;
- operating frequency range—from 0.5 to 1000000 Hz;

- extended relative uncertainty of measurement of specified parameters at the fundamental frequency and at direct current depending on the range is from 1 · 10E-3 to 1 · 10E-5.

Calibration of precision LCR meters in Ukraine is performed with the help of highly stable measures of inductance, electrical capacitance and active resistance, which are part of the National standard of units of inductance and loss tangent, as well as of National standard of units of electrical capacitance and dissipation factor.

These measures have metrological traceability to international standards on fundamental physical constants by calibration or participation in international intelligence [5, 25–29].

Calibration of LCR-meters begins with an external inspection and performance check (testing), which are carried out in accordance with the operational documentation for the LCR-meter. During the initial calibration, electrical strength and insulation resistance tests are performed.

Before using the LCR-meter, the operating frequency and the measuring voltage are set and its automatic self-calibration is performed. After that, the initial parameters are determined on the selected measurement ranges—the active and reactive components of the resistance of connecting cables and terminal devices. It should be taken into account that the initial parameters at different measurement ranges are of a different nature and may differ significantly.

Also, it is necessary set the substitution scheme (parallel/serial) for the measurement object with control elements of the LCR-meter. In the case of incorrect selection of the measurement object substitution scheme, the value of the main parameter being measured may differ greatly from the actual value.

Before connecting to a LCR meter to reactive measurements object it must be discharged—shorten the contacts for a few seconds.

As a rule, the calibration of LCR-meters is carried out according to the values of: active resistance R, electrical capacitance C, inductance L. Sometimes customers who need to use the LCR-meter in the most accurate measurements will require calibration according to the operating frequency.

9.1 Calibration of a Precision LCR Meter for Measuring Active Resistance

For calibration of the LCR-meter for active resistance measurements, the standard measures from the set of MAC-2 with nominal values of 0.1 Ω, 1 Ω, 10 Ω, 100 Ω, 1 kOhm, 10 kOhm, P4015 with nominal value 100 kOhm, P4016 with nominal value 1 MOhm, P4022M1 with nominal value 10 MOhm are used. The measurement scheme is shown in Fig. 23.

During calibration, n independent measurements of the resistance R_S are carried out, which is reproduced by the standard measure. The average value $\overline{R}_{LCR}$ is obtained from the readings of the LCR meter.

Fig. 23 Scheme of measurements of active resistance during calibration of a precision LCR-meter

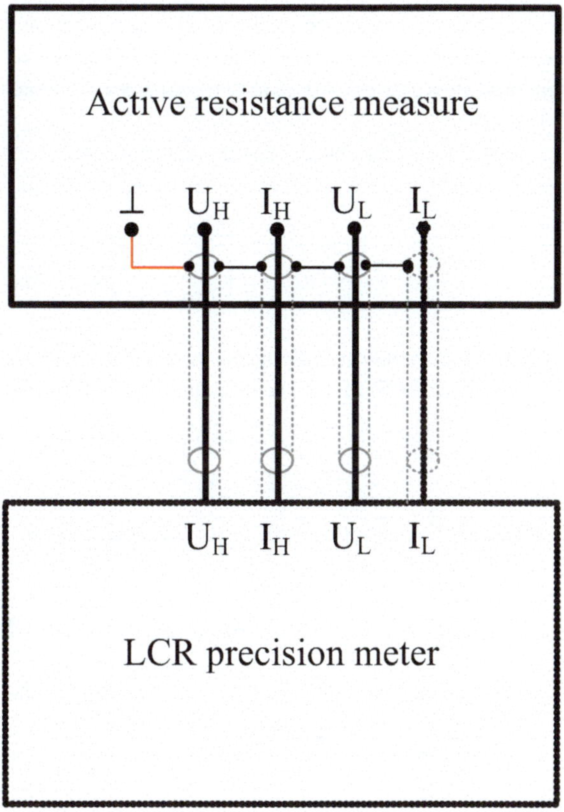

The equation (model) of measurements has the following form:

$$\Delta R_{LCR} = R_{LCR} - R_S + \Delta R_{Sf} + \Delta R_{S\gamma} + \Delta R_{QLCR} + \Delta R_{ST} + \Delta R_{TLCR} + \Delta R_l,$$

where ΔR_{LCR}—deviation of readings of LCR meter being calibrated from the true value of the active resistance reproduced by the standard measure; $\overline{R}_{LCR}$—average value of the active resistance displayed on the screen of the LCR-meter during n measurements of the active resistance of the standard measure; R_S—value of the active resistance of the standard measure from the calibration certificate; ΔR_{Sf}—correction for frequency dependence of the standard measure of active resistance; $\Delta R_{S\gamma}$—correction to drift of the standard measure of active resistance since the last calibration; ΔR_{QLCR}—correction for the quantization error (sampling) of LCR meter being calibrated; ΔR_{ST}—correction for the temperature dependence of the standard measure; ΔR_{TLCR}—correction for the temperature dependence of the calibrated LCR meter when measuring active resistance; ΔR_l—correction for parasitic active resistance of connecting cables and contacts.

Based on this measurements model, an example of calculating the uncertainty budget when calibration of LCR-meter for active resistance measurements using standard measure of 100 Ω at a frequency of 1 kHz may have the form shown in Table 10.

At the same time, the standard uncertainty of type A, associated with the random component of R_{LCR} estimate, is determined by the formula:

$$u_A(R_{LCR}) = \sqrt{\left(\frac{1}{n(n-1)}\right)\Sigma_{j=1}^{n}(R_{LCRi} - \overline{R}_{LCR})^2}.$$

The total standard uncertainty of the of active resistance measurements when calibrating of LCR meter is calculated by the formula:

$$u_\Sigma(\Delta R_{LCR}) = \sqrt{u_A^2(R_{LCR}) + \Sigma_{i=1}^{N} u_{Bi}^2(\Delta R_{LCR})}.$$

Extended uncertainty with coverage factor $k = 2$ and confidence probability $p \approx 95\%$:

$$U = k \cdot u_\Sigma(\Delta R_{LCR}) = 2 \cdot u_\Sigma(\Delta R_{LCR}).$$

It should be noted that the given calculations of the calibration result correspond to only one measurement point—100 Ω at a frequency of 1 kHz. At the same time, the calibration of the meter, which is a precision LCR-meter, must be carried out in

Table 10 Uncertainty budget when calibrating a precision LCR-meter for measuring active resistance of 100 Ω at 1 kHz

Quantity X_i, Ohm	Estimate x_i, Ohm	Standard uncertainty $u(x_i)$, Ohm	Probability distribution	Method of evaluative (A, B)	Sensitivity coefficient c_i	Uncertainty contribution $u_i(y)$, Ohm
$\overline{R}_{LCR}$	99.9989	1.45E-04	Normal	A	1	1.45E-04
R_S	99.9977	1.75E-05	Normal	B	−1	−1.75E-05
ΔR_{Sf}	−0.0461	1.10E-05	Normal	B	1	1.10E-05
$\Delta R_{S\gamma}$	0.00001	1.00E-05	Normal	B	1	1.00E-05
ΔR_{QLCR}	0	5.77E-05	Rectang.	B	1	5.77E-05
ΔR_{ST}	0	5.00E-05	Rectang.	B	1	5.00E-05
ΔR_{TLCR}	0	1.00E-04	Rectang.	B	1	1.00E-04
ΔR_l	0	1.00E-04	Normal	B	1	1.00E-04
ΔR_{LCR}	0.0012					
		Combined standard uncertainty				2.17E-04
		Effective degrees of freedom			ν_{eff}	>200, $k = 2$
		Expanded uncertainty ($p \approx 95\%$)				4.34E-04

the full range of frequencies and values. Exceptions are cases when the calibration range is narrowed at the request of the customer.

Therefore, calibration must be carried out at several points (at least three) of each sub-range. It is advisable to present the calibration result in the form of a graph, which shows the dependence of the deviation of the LCR-meter readings from the actual value of the measured value with the associated measurement uncertainty.

This graph is obtained by approximating the calibration results at the corresponding calibration points [30]. In areas of the range where there may be extremes of the graph, or significant non-linearity, the number of calibration points should be increased.

9.2 Calibration of a Precision LCR-Meter for Measuring Inductance

Let's consider the calibration of a precision LCR-meter for measuring inductance. The standard termostated inductance measures from the National standard of the units of inductance and loss tangent with a nominal value of 1 mH, 10 mH, 100 mH, as well as standard inductance measures P593, P596, P5101–P51015 with a nominal value of 0.01 μH–100 mH and others are used as a standard during calibration. The measurement schemes for 2-pin and 3-pin connection are shown in Figs. 24 and 25.

As an example, we'll consider the calibration of a precision LCR-meter for measuring inductance using a thermostated standard measure P5113 with a nominal value of 100 mH as a standard [31].

The equation (model) of measurements has the following form:

$$\Delta L_{LCR} = \overline{L}_{LCR} - L_S + \Delta L_{Sf} + \Delta L_{S\gamma} + \Delta L_{QLCR} + \Delta L_{ST} + \Delta L_{TLCR} + \Delta L_l,$$

where ΔL_{LCR}—deviation of the readings of LCR-meter being calibrated from the true value of the inductance reproduced by standard measure; $\overline{L}_{LCR}$—average value of the readings of the LCR-meter when measuring the inductance of the standard measure; L_S—value of inductance of the standard measure from the calibration certificate; ΔL_{Sf}—correction for frequency dependence of the standard measure of inductance; $\Delta L_{S\gamma}$—correction to drift of the standard measure of inductance since the last calibration; ΔL_{QLCR}—correction for the discreteness of the readings of the calibrated LCR-meter; ΔL_{ST}—correction for the temperature dependence of standard measure; ΔL_{TLCR}—correction for the temperature dependence of the calibrated LCR-meter; ΔL_l—correction for the inductance of connecting cables and contacts.

Based on this measurement model, an example of calculating the uncertainty budget when calibrating a precision LCR meter for measuring inductance using a standard measure with a nominal value of 100 mH at a frequency of 1 kHz can have the form given in Table 11.

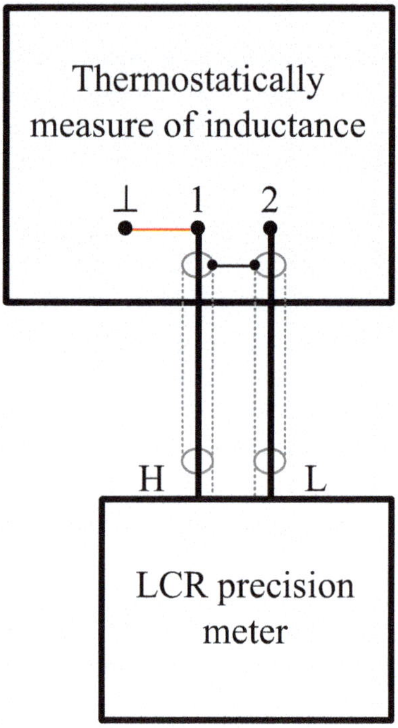

Fig. 24 2-pin connection of the standard measure

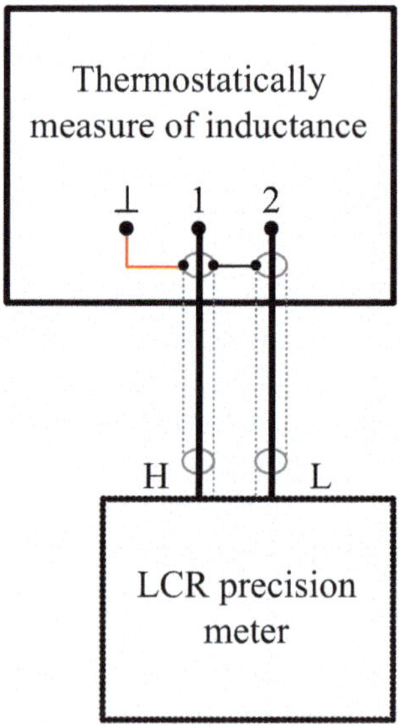

Fig. 25 3-pin connection of the standard measure

Table 11 Uncertainty budget when calibrating a precision LCR meter for measuring inductance of 100 mH at 1 kHz

Quantity X_i	Estimate x_i, mH	Standard uncertainty $u(x_i)$, mH	Probability distribution	Method of evaluation (A, B)	Sensitivity coefficient c_i	Uncertainty contribution $u_i(y)$, mH
$\overline{L}_{LCR}$	100.1100	3.52E-04	Normal	A	1	3.52E-04
L_S	100.0082	4.00E-04	Normal	B	−1	−4.00E-04
ΔL_{Sf}	0.0001	1.00E-05	Normal	B	1	1.00E-05
$\Delta L_{S\gamma}$	0	1.10E-05	Normal	B	1	1.10E-05
ΔL_{QLCR}	0	5.77E-05	Rectang.	B	1	5.77E-05
ΔL_{ST}	0	3.60E-05	Rectang.	B	1	3.60E-05
ΔL_{TLCR}	0	1.00E-04	Rectang.	B	1	1.00E-04
ΔL_l	0	1.00E-04	Normal	B	1	1.00E-04
ΔL_{LCR}	0.1018					
		Combined standard uncertainty				5.56E-04
		Effective degrees of freedom			ν_{eff}	>200, $k = 2$
		Expanded uncertainty ($p \approx 95\%$)				1.11E-03

Type A standard uncertainty associated with a random component of the estimate L_{LCR} is determined by the formula:

$$u_A(L_{LCR}) = \sqrt{\left(\frac{1}{n(n-1)}\right)\Sigma_{j=1}^{n}(L_{LCRi} - L_{LCR})^2}.$$

The total standard uncertainty of inductance measurements when calibrating LCR-meter is calculated by the formula:

$$u_\Sigma(\Delta L_{LCR}) = \sqrt{u_A^2(L_{LCR}) + \Sigma_{i=1}^{N} u_{Bi}^2(\Delta L_{LCR})}.$$

Extended uncertainty with coverage factor $k = 2$ and confidence probability $p \approx 95\%$:

$$U = k \cdot u_\Sigma(\Delta L_{LCR}) = 2 \cdot u_\Sigma(\Delta L_{LCR}).$$

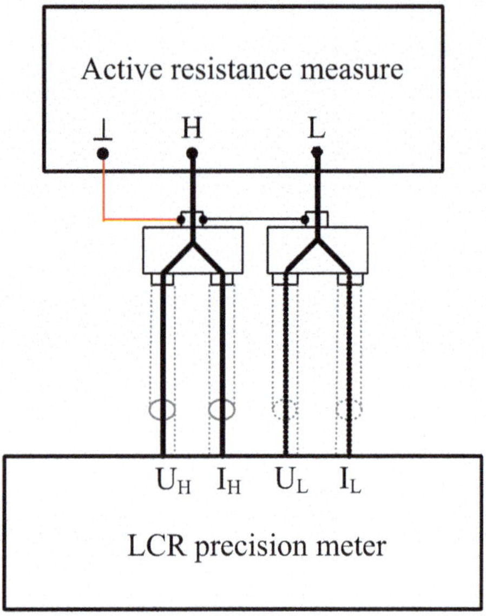

Fig. 26 Scheme of measurements of electrical capacitance when calibrating precision LCR-meter

9.3 *Calibration of a Precision LCR-Meter for Measuring Electrical Capacitance*

Let's consider the calibration of a precision LCR-meter for measuring electrical capacitance. As a standard, Andeen-Hagerling type AH11A thermostated standard measures of electrical capacitance with a nominal value of 10 pF and 100 pF from the National standard of units of electrical capacitance and dissipation factor, a thermostated set of capacitances and resistances, as well as standard capacitance measures P597 with a nominal value of 1 pF, 10 pF, 100 pF and others. The measurement scheme is shown in Fig. 26.

For example, we'll consider the calibration of a precision LCR-meter for measuring electrical capacity using the Andeen-Hagerling thermostated electrical capacitance standard measure of electrical capacitance of the AH11A type with a nominal value of 100 pF as a standard.

The equation (model) of measurements has the following form:

$$\Delta C_{LCR} = \overline{C}_{LCR} - C_S + \Delta C_{Sf} + \Delta C_{S\gamma} + \Delta C_{QLCR} + \Delta C_{ST} + \Delta C_{TLCR} + \Delta C_l,$$

where ΔC_{LCR}—deviation of the readings of the calibrated LCR-meter from the true value of the capacitance reproduced by the standard measure; $\overline{C}_{LCR}$—average value of the readings of the LCR-meter for measuring the capacitance of the standard measure; C_S—value of the standard measure capacitance from the calibration certificate; ΔC_{Sf}—correction of frequency dependence of the standard measure of capacitance; $\Delta C_{S\gamma}$—correction to drift of the standard capacitance measure since the last calibration; ΔC_{QLCR}—correction for the quantization error of the calibrated LCR-meter; ΔC_{ST}—correction for the temperature dependence of the standard measure; ΔC_{TLCR}—correction for the temperature dependence of the calibrated LCR-meter; ΔC_l—correction for parasitic capacitance of connecting cables and contacts.

Based on this measurement model, an example of calculating the uncertainty budget when calibrating a precision LCR-meter for measuring capacitance using a standard measure with a nominal value of 100 pF at a frequency of 1 kHz can have the form given in Table 12.

Type A standard uncertainty associated with a random component of the estimate C_{LCR} is determined by the formula:

$$u_A(C_{LCR}) = \sqrt{\left(\frac{1}{n(n-1)}\right)\Sigma_{j=1}^{n}(C_{LCRi} - C_{LCR})^2}.$$

The total standard uncertainty of the capacitance measurements when calibrating LCR-meter is calculated by the formula:

$$u_\Sigma(\Delta C_{LCR}) = \sqrt{u_A^2(C_{LCR}) + \Sigma_{i=1}^{N} u_{Bi}^2(\Delta C_{LCR})}.$$

Extended uncertainty with coverage factor $k = 2$ and confidence probability $p \approx 95\%$:

$$U = k \cdot u_\Sigma(\Delta C_{LCR}) = 2 \cdot u_\Sigma(\Delta C_{LCR}).$$

Table 12 Uncertainty budget when calibrating a precision LCR meter for measuring capacitance of 100 pF at 1 kHz

Quantity X_i	Estimate x_i, pF	Standard uncertainty $u(x_i)$, pF	Probability distribution	Method of evaluative (A, B)	Sensitivity coefficient c_i	Uncertainty contribution $u_i(y)$, pF
$\overline{C}_{LCR}$	100.00008	1.47E-05	Normal	A	1	1.47E-05
C_S	100.00014	4.00E-05	Normal	B	−1	−4.00E-05
ΔC_{Sf}	0.00008	1.00E-05	Normal	B	1	1.00E-05
$\Delta C_{S\gamma}$	−0.00001	4.00E-05	Normal	B	1	4.00E-05
ΔC_{QLCR}	0	5.77E-05	Rectang.	B	1	5.77E-05
ΔC_{ST}	0	1.90E-04	Rectang.	B	1	1.90E-04
ΔC_{TLCR}	0	1.00E-04	Rectang.	B	1	1.00E-04
ΔC_l	0	1.00E-04	Normal	B	1	1.00E-04
ΔC_{LCR}	−0.00007					
		Combined standard uncertainty				2.51E-04
		Effective degrees of freedom			ν_{eff}	>200, $k = 2$
		Expanded uncertainty ($p \approx 95\%$)				5.01E-04

When calibrating the LCR-meter, calculations of the measurement result and the uncertainty budget are carried out in a similar manner based on values such as electrical capacitance, active resistance, and other impedance parameters.

As a result of the conducted research, a method of calibrating precision LCR-meters based on national standards of units of electrical capacitance, inductance and loss tangent is developed. The method fully allows obtaining measurement results and carrying out uncertainty calculations when calibrating LCR-meters, multimeters and combined devices for such values as inductance, electrical capacitance and active resistance in the full range of frequencies and values. The method can be used when conducting international reconciliations of the original standards of national metrological institutes. The results of the research will contribute to the organization of metrological support for measurements of impedance parameters in all branches of science and technology in the country.

10 Conclusions

On the basis of the conducted analysis and experimental studies, transfer schemes and reproducing methods of impedance parameter measurements units are optimized, which contribute to increasing accuracy and significantly reducing the cost of metrological works at the level of national standards.

A method of finding points of physical quantity unit transfer between heterogeneous impedance parameters is developed, a set of such points is found to achieve the highest metrological characteristics. Methods of minimizing the error of transfer of the impedance parameter unit size with reference to heterogeneous standards by choosing the optimal transfer scheme are proposed and developed.

The results of the work made it possible to organize and successfully complete the international reconciliation of national standards of electrical capacitance and inductance.

According to the results of international reconciliations, due to measurements accuracy increase it was adjusted the CMC lines on the BIPM website, which contributed to expansion of official metrological capabilities and strengthening the recognition of the SE "Ukrmetrteststandard" at the international level. This contributed to the involvement of a larger number of domestic and foreign customers to performing of metrological works.

Thus, based on the results of the research, a technical and methodological basis is theoretically justified and created for the transfer of impedance parameter units from international standards based on fundamental physical constants to national standards when using as a transferring standard any standard measure of capacitance, inductance or active resistance, providing that the accuracy of measurements is maintained at the level of the leading countries of the world.

Constructive and technological methods of increasing measurement accuracy to ensure metrological traceability of impedance parameter units in the full range of values are considered.

These methods allow solving following metrological problems: take into account the difference in the resistance temperature coefficient of the connecting conductors and the measure; take into account the violation of the connection four-clamp and four-parity due to changes in the geometry of the contacts, the mutual induction of the measuring circuits, as well as the presence of a distributed parasitic capacitance.

The use of the proposed approaches makes it possible to experimentally evaluate all parasitic capacitances and inductances of measuring circuits during the comparing of standard measures of impedance parameters, evaluate the errors that they introduce separately and in total, and also take into account and compensate them by computational methods.

The equal-nominal comparing method described in the paper makes it possible to significantly increase the accuracy of the calibration of complex resistance units' measures in the low-ohm range of values by eliminating the comparing of measures of different nominal values. This makes it possible to reach a new qualitative level of measurements when using existing standard equipment. Using a sufficient set of measures, it is possible to combine them structurally into one block and, with certain switchings and solutions, ensure high accuracy and productivity of metrological work.

Based on the results of the research, a methodology was developed that allows to estimate the combined standard and extended uncertainty of measurements when calibrating measures of electrical capacitance on the National Standard of Units of Electrical Capacitance and Dissipation Factor DETU 08-06-01. The use of this methodology allows DETU 08-06-01 to be involved in the international reconciliation of origin standards of electrical capacitance of national metrological institutes of the leading countries of the world.

The equal-nominal comparing method was used in the evaluation of the uncertainty components introduced by the autotransformer bridge during international reconciliations of origin standards of electrical capacitance and inductance in the themes COOMET EM-S13 (544/UA/12) and COOMET EM-S14 (584/UA/12). This method will be used in further international reconciliations.

Materials are provided on the estimation of measurement uncertainty during the calibration of electrical capacitance measures on the National Standard of Units of Electrical Capacitance and Dissipation Factor. The main factors affecting the measurement result are considered.

The method of calibrating measures of inductance on the National Standard of Ukraine of Units of Inductance and Loss Tangent with uncertainty estimation and traceability to international standards of electric capacitance is disclosed.

Methods for evaluating metrological characteristics during calibration of precision LCR-meters using National standards of units of electrical capacitance, inductance and loss tangent are presented. General approaches to the calibration of LCR-meters are given, equations (models) of measurements during calibration are presented, as well as examples of uncertainty budget calculation. The components of the calibration uncertainty budget are disclosed in detail. The given methods and approaches to uncertainty estimation allow calibration of a variety of LCR-meters by such values as inductance, capacitance and active resistance in a wide frequency range and range of values. The materials can be used by laboratories that carry out calibration, testing and conformity assessment of measuring equipment of electromagnetic quantities.

The practical value of the developed methods and approaches lies in ensuring the unity of measurements of impedance parameters units in the country in such areas as energy, radio electronics, communal sphere, construction, transport, telecommunications, metallurgy, machine and aircraft construction industry, security industry, and also scientific research and defense. Meters of impedance, the total resistance of the phase-zero loops belong to the sphere of legally regulated metrology, when the measurement results are used during the safety control of working conditions.

The results of the research of Andeen-Hagerling AH11A 10 pF and 100 pF stamdard capacitance measures confirm that the metrological characteristics of the National Standard of Units of Electrical Capacitance and Dissipation Factor DETU 08-06-01 are within the permissible values and correspond to the data specified in the passport for the standard.

The developed standard research methods and their results correspond to the world level and contribute to improving the quality and competitiveness of domestic products, ensuring the conditions for innovative development of the national economy, as well as ensuring the functioning of a modern standard base, adequate to the requirements of the market economy, harmonized with international rules and requirements of the international standards.

References

1. Klitzing, V.K., Dorda, G., Pepper, M.: New method for high-accuracy determination of the fine-structure constant based on quantized hall resistance. Phys. Rev. Lett. **45**, 494–497 (1980). https://doi.org/10.1103/PhysRevLett.45.494
2. Delahaye, F.: AC-bridges at BIMP. BNM-LCIE, 1–6 (1998)
3. Surdu, M.N., Ahmadov, A.A., Ahmadov, S.A., Kursin, S.N., Lameko, A.L., Muharovskij, M.Y.: Precizionnyj komparator dlya vosproizvedeniya edinicy induktivnosti i peredachi ee razmera v diapazone znachenij. Ukr. Metrol. J. **4**, 14–22 (2008). (in Russian)
4. The BIPM key comparison database (KCDB 816) – Access mode. http://kcdb.bipm.org/
5. Velychko, O., Shevkun, S.: Final report on COOMET supplementary comparison of capacitance at 10 pF and 100 pF (COOMET.EM-S13) (2013)
6. Velychko, O., Shevkun, S.: Mizhnarodni zvirennia v ramkakh COOMET natsionalnykh etaloniv odynyts elektrychnoi yemnosti nominalom 10 i 100 pF. Metrolohiia ta prylady **2**, 3–8 (2015). (in Ukrainian). http://nbuv.gov.ua/UJRN/mettpr_2015_2_3
7. Shevkun, S., Surdu, M., Velichko, O.: Evaluation of uncertainty of inductance measures on state primary standard of the unit of inductance. In: XVIII TC04 IMEKO Symposium and IX International Congress on Electrical Metrology «Metrologia-2011» (2011). ISBN: 978-85-86920-08-0. ID 83589
8. Shevkun, S.M., Surdu, M.M., Velychko, O.M., Dobroliubova, M.V., Shumkov, Yu.S.: Analiz pokhybky vymiriuvannia nerivnovahy universalnoho komparatora na osnovi avtotransformatornoho mosta v natsionalnykh etalonakh odynyts parametriv impedansu. Metrolohiia ta prylady **2 II**(40), 257–261 (2013). (in Ukrainian)
9. Velychko, O.M., Surdu, M.M., Shevkun, S.M., Dombrovskyi, M.H., Dobroliubova, M.V.: Otsinka nevyznachenosti rezultativ kalibruvannia rivnonominalnykh mir elektrychnoi

yemnosti na Derzhavnomu etaloni odynyts elektrychnoi yemnosti ta faktora vtrat. Inf. Process. Syst. **3**(110), 161–163 (2013). (in Ukrainian)
10. Kibble, B.P., Rayner, G.H.: Coaxial AC bridges (1984)
11. Surdu, M., Lameko, A., Surdu, D., Kursin, S., Mukharovsky, M., Akhmadov, A., Shevkun, S.: Accurate universal set of automatic comparators for impedance parameters units reproduction and transfer. In: XVIII TC04 IMEKO Symposium and IX International Congress on Electrical Metrology «Metrologia-2011» (2011). ISBN: 978-85-86920-08-0. ID 83536
12. Shevkun, S.M., Surdu, M.M., Velychko, O.M., Dobroliubova, M.V.: Pidvyshchennia tochnosti peredachi rozmiru odynytsi impedansu v nyzkoomnomu diapazoni znachen metodom rivnonominalnoho komparuvannia. Metrolohiia ta prylady **1 II**(45), 243–246 (2014). (in Ukrainian)
13. Lisowski, M., Krawchuk, K.: Insulation resistance influence on high resistance Hamon transfer accuracy. Metrol. Meas. Syst. **1 XVI**, 33–45 (2009)
14. Shevkun, S.M., Surdu, M.M., Velichko, O.M., Dobrolyubova, M.V.: Konstruktivno-tehnologicheskie metody povysheniya tochnosti izmerenij dlya obespecheniya metrologicheskoj proslezhivaemosti edinic parametrov impedansa v diapazone znachenij. In: XXIV International Scientific Symposium «Metrology and Metrology Assurance 2014», pp. 64–70 (2014). (in Russian)
15. DSTU 3231:2007 Metrolohiia. Etalony odynyts vymiriuvan derzhavni, pervynni ta vtorynni. Osnovni polozhennia, poriadok rozroblennia, zatverdzhennia, reiestratsii, zberihannia ta zastosuvannia (2008). (in Ukrainian)
16. Velychko, O., Surdu, M., Shevkun, S., Dobroliubova, M.: Transfer of the units size by the range of values with using the state primary standard of the units of electrical capacitance and dissipation factor. In: XXVII International Scientific Symposium «Metrology and Metrology Assurance 2017», pp. 40–44 (2017)
17. Velychko, O., Shevkun. S.: Support of metrological traceability of capacitance measurements in Ukraine. East.-Eur. J. Enterp. Technol. **3/9**(87), 4–10 (2017). https://doi.org/10.15587/1729-4061.2017.101897
18. Shevkun, S., Velychko, O., Surdu, M., Dobroliubova, M.: Estimation of uncertainty in calibration of inductance measures by using the state primary standards of the units of electrical capacitance, inductance and dissipation factor. In: XXVIII International Scientific Symposium «Metrology and Metrology Assurance 2018», pp. 35–38 (2018)
19. Shevkun, S.M., Velichko, O.M., Surdu, M.M., Dobrolyubova, M.V., Dombrovskij, M.G.: Metodika kalibrovki mer induktivnosti na Gosudarstvennom etalone Ukrainy edinic induktivnosti i tangensa ugla poter. In: XXIII International Scientific Symposium «Metrology and Metrology Assurance 2013», pp. 454–458 (2013). (in Russian)
20. DSTU ISO/IEC 17025:2017, General requirements for the competence of testing and calibration laboratories (2017). (in Ukrainian)
21. Velychko, O., Shevkun. S.: A support of metrological traceability of inductance measurements in Ukraine. East.-Eur. J. Enterp. Technol. Inf. Control. Syst. **5/9**(89), 12–18 (2017). https://doi.org/10.15587/1729-4061.2017.109750
22. Velychko, O., Shevkun. S., Gordiyenko, T., Dobrolyubova, M.: Metrological traceability of impedance parameter measurements in Ukraine. East.-Eur. J. Enterp. Technol. Inf. Control. Syst. **4/9**(94), 43–49 (2018). https://doi.org/10.15587/1729-4061.2018.139689
23. Velychko, O., Shevkun, S., Dobroliubova, M.: Features of calibration of precision LCR meters. Sens. Transducers **237**(9–10), 171–177 (2019)
24. ISO/IEC Guide 98–3:2008, Uncertainty of measurement. Part 3. Guide to the expression of uncertainty in measurement (2008).
25. Velychko, O., Shevkun. S.: Final report on COOMET supplementary comparison of inductance at 10 mH and 100 mH at 1 kHz. COOMET.EM-S14. Metrol. **53**(1A), Technical Supplement 01009 (2016)
26. Dierikx, E., et al.: Final report on the supplementary comparison EURAMET.EM-S26: inductance measurements of 100 mH at 1 kHz. EURAMET Project 816. Metrol. **49**, 01002 (2012)

27. Velychko, O., Akhmadov, O.: Final report on COOMET key comparison of capacitance at 10 pF. COOMET.EM-K4. Metrol. **54**(1A), 01005 (2017)
28. Velychko, O., Akhmadov, O.: Final report on COOMET supplementary comparison of capacitance at 100 pF. COOMET.EM-S4. Metrol. **54**(1A), 01006 (2017)
29. Velychko, O., Shevkun, S., Bartholomew, J., Alrobaish A.: Final report on GULFMET supplementary comparison of inductance at 10 mH and 100 mH at 1 kHz. GULFMET. EM-S4. Metrol. **56**(1A), 01013 (2019)
30. International vocabulary of metrology – Basic and general concepts and associated terms (VIM) (2021)
31. Kuzmenko, Iu., Velychko, O., Shevkun, S., Dobroliubova, M.: Estimation of Uncertainty In Calibration Of Precision LCR-meters on the state primary standard of units of inductance and tangent angles of losses. In: 5th International Conference on Sensors and Electronic Instrumentation Advances (SEIA'19), pp. 331–332 (2019)

Implementation of Information and Measurement Systems at the Base of Specialized Internet Protocols

Sergiy Bogomazov and Nazar Povorozniuk

Abstract The object of research is network microcontroller systems for collecting and processing experimental data. The subject of the study is the organization of distributed microcontroller systems for collecting measurement information with remote Internet access to microserver networks of virtual intelligent sensors based on specialized web protocols. The purpose of the work is the development and research of a complex of hardware and software tools for building network microcontroller systems for collecting and processing experimental information based on intelligent sensors according to IEEE-1451 standards with support for web protocols and embedded Java technologies. The research method is theoretical and experimental based on the analysis and synthesis of hardware and software of microcontroller network systems for remote collection of experimental data. An approach to the organization of experimental data collection and processing systems for remote collection and processing of experimental data as microcontroller networks of intelligent sensors of IEEE-1451 standards based on embedded Java machines with support for specialized web technologies, MQTT (Message Queue Telemetry Transport) and web-REST services. Structures of intelligent sensors for studying the composition of substances have been developed. A basic set of hardware and software tools for microcontroller modules of intelligent sensors has been developed. Existing standards of industrial networks are used for communication between logical sensors, MQTT protocol, and REST web services are used for communication between intelligent sensors. The network processor NCAP (Network Capable Application Processor) of intelligent sensors is implemented based on built-in Java machines. This made it possible to ensure the platform independence of software solutions from hardware implementation, to use the advantages of network Java technologies, to significantly reduce the cost of remote Internet access to equipment, and to reduce the mass and size indicators and energy consumption of systems compared to computer architectures.

S. Bogomazov (✉) · N. Povorozniuk
National Technical University of Ukraine "Igor Sikorsky Kyiv Polytechnic Institute", Kyiv, Ukraine
e-mail: sbogmzv@gmail.com

V. Eremenko and A. Zaporozhets (eds.), *Advanced Information-Measuring Technologies and Systems I*, Studies in Systems, Decision and Control 439,
https://doi.org/10.1007/978-3-031-40718-5_4

Keywords Microcontroller networks · MQTT protocol · Hardware and software · Intelligent sensor · Remote data collection · Web services works

1 Organization of Distributed Data Collection Systems Based on Network Intelligent Sensors

1.1 Features of the Use of Intelligent Sensors in Distributed Data Collection Systems

To collect experimental data and manage complex technical objects or processes with a dynamically changing state, distributed computer systems capable of solving the tasks of collecting and processing experimental data in real-time are widely used. When creating distributed systems, it is necessary to ensure the possibility of applying distributed intelligent methods of management, distribution of calculations, and intelligent processing of information. Today, this also applies to low-level control devices. Intelligent nodes of control systems are not only industrial computers and controllers, but also sensors and executive devices. Complex, functionally saturated systems for collecting and processing technological information require the use of converters capable of solving additional tasks, in addition to providing information about the level of signals or about turning on and off equipment elements [1]. A typical structure of a network of intelligent sensors is shown in Fig. 1.

The improvement of modern electronic devices is due, first of all, to the high rate of development of microelectronics. The constant decrease in the cost of microprocessor elements and the rapid growth of their functional capabilities allow these components to be built into smaller and smaller products. The appearance of a new generation of sensors, which have received the name of intelligent sensors, is connected with the change of the element base of electronic signal processing devices in primary converters. An analog of the term intelligent sensor is the term "smart sensor", which usually means a sensor with integrated electronics (analog–digital converter, microprocessor, digital signal processor, system on a crystal, etc.) as well as the implementation of a digital interface and network communication protocols.

Thus, the ability to connect it to a wired or wireless network is considered an integral function of an intelligent sensor. Accordingly, in addition to the presence of self-diagnostic and self-learning functions, such a converter must perform self-identification functions, which are required to access it in a network of other sensors. In addition to the ability to connect the sensor to the network, the presence of a network interface allows you to perform configuration, operation mode selection, and diagnostics from a remote workplace, which provides advantages both in operation and in the cost of their maintenance. The term "intelligent" is used in a narrow sense to refer to devices that, due to the use of information processing in them, acquire new functional capabilities.

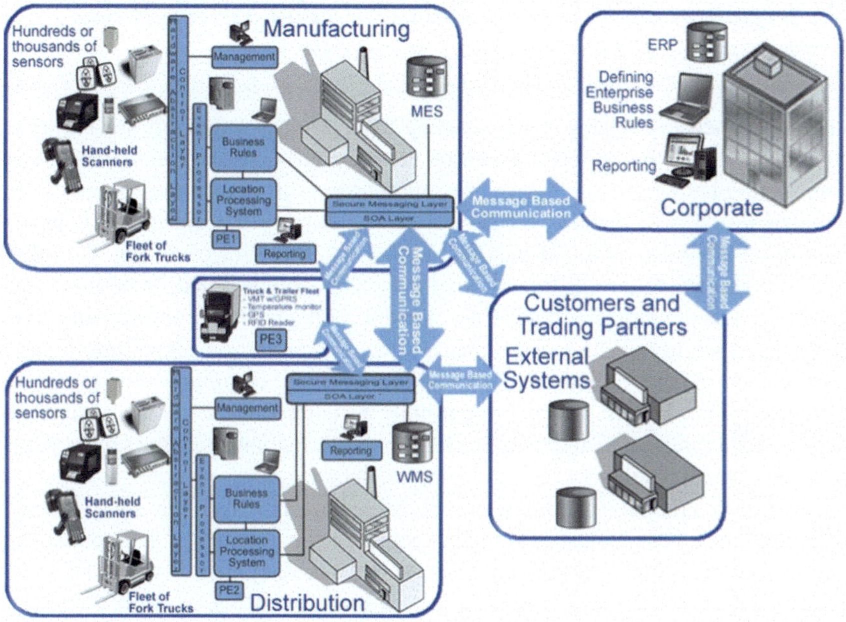

Fig. 1 The structure of the network of intelligent sensors

Intelligence in sensors, like in humans, manifests itself in many different forms. Sensors owe their intelligence to microprocessor technologies. The microprocessor is the brain of the sensor, which allows the device to "learn" the conditions in which it operates. The sensor is capable of processing large amounts of information at high speed. Thanks to microprocessors, today the user has sensors that are easy to install, configure and use.

A distinctive feature of intelligent sensors is the digital processing of signals directly from the output of the primary converter. This guarantees high accuracy and stability of its characteristics in all permissible measurement ranges, as well as low sensitivity to external disturbances. Digital signal processing and the ability to upgrade the software allow you to implement various functions of converting controlled values with further improvement of the characteristics and functions of the sensor.

Intelligent sensors can be significantly different from what we are used to understanding by the term "sensor". Today, these are rather specialized controllers that receive the signal from the primary converter and deal with its processing and can exchange information with other intelligent nodes of the automation system through digital communication channels. The only thing that continues to "dilute" them with ordinary sensors is their small size.

1.2 Organization of Network Data Collection Systems Based on the IEEE 1451 Group of Standards

In intelligent sensors, analog signals coming from primary converters are amplified and converted into digital form. Based on these signals and calibration data stored in the non-volatile memory device, the microprocessor adjusts the values of the measured (or related) quantity received from the primary transducer and converts them into the required units of measurement. Thus, the error caused by the influence of temperature, zero drift, and other reasons is compensated. In addition, the microprocessor monitors the state of the elements of the primary converter and evaluates the probability of the measurement result. The processed digital information is transmitted using a digital interface and implemented communication protocols to the user, who, in turn, has the opportunity to configure the sensor parameters (measurement range, etc.) and request additional information about the sensor status and measurement results.

To manage complex technological processes with a dynamic state change, distributed networks of sensors are being acquired for monitoring and controlling parameters in real-time. A significant obstacle in the development of wired and wireless networks of intelligent sensors is the lack of a single network standard. Currently, dozens of types of various interfaces (RS-485, HART, USB, 4-20 mA, IEEE-488) and industrial networks (ProfiBus, Fieldbus, DeviceNet, Interbus, CANbus, Modbus, LIN) are used. In the current situation, sensor manufacturers face the need for a complex choice of the type of digital interface and communication protocol, since the production of the same type of intelligent sensors for each of the currently popular networks is economically unprofitable.

With the emergence of the IEEE 1451 group of standards, the situation began to change for the better due to the unification of the interface between the intelligent sensor and the network [2–4]. These standards, whose purpose is to accelerate the process of transition from the use of individual sensors to the use of networks of sensors, are divided into several groups that define the hardware and software principles of connecting intelligent sensors to the network. IEEE 1451 standards describe two classes of devices: STIM (Smart Transducer Interface Module) and NCAP (Network Capable Application Processor) [5–10].

The IEEE 1451.1 standard defines a single interface for connecting to a network of intelligent sensors and contains the specifications of the NCAP module, which performs the functions of communication between the external network and the STIM of the intelligent sensor. The IEEE 1451.2 standard defines a digital interface for connecting the STIM intelligent converter module to the network adapter. In addition, it also defines the concept of TEDS (Transducer Electronic Data Sheet)—electronic specification of the sensor, which provides self-identification of the device in the network. The TEDS specification has been widely distributed and includes data such as model code, serial number, date of issue and calibration data, units of measurement, and date of calibration. The TEDS specification allows for automatic sensor

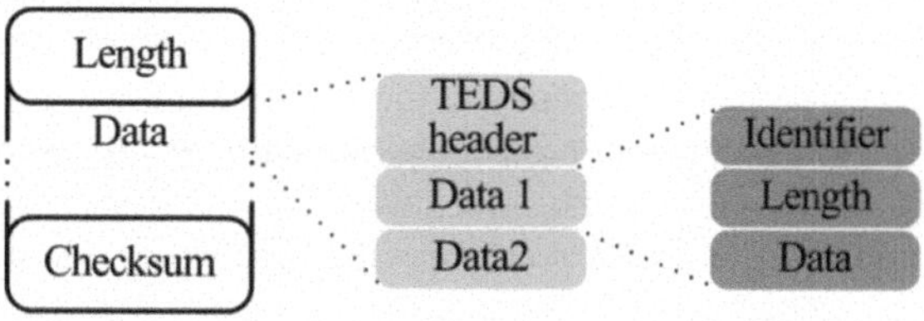

Fig. 2 Structure of TEDS

configuration, similar to the Plug-and-Play technology used for personal computer device identification and self-configuration.

At the same time, the need to create and maintain a database with information about each sensor disappears, which facilitates their operation and replacement. One of the main goals of connecting sensors in a network is to provide access to the information they provide to user-level software, which is independent of specific types of sensors and ways of organizing the network. Thus, the network of sensors can be perceived as an intermediate level between the sensors themselves and the instrument tasks solved with their help.

The standards of the IEEE 1451 group are universal and can be used in various areas, for example, for intelligent cameras or environmental protection systems. The IEEE 1451 group of standards defines a network consisting of intelligent sensors. This network contains a TIM (Transducer Interface Module) and one NCAP network processor. The TIM module may contain one or more sensitive elements. Each built-in sensitive element forms one measurement channel.

The TIM module connects to the NCAP module, which in turn controls the TIM module. The TIM module contains electronic documentation—TEDS [11]. A typical TEDS structure is shown in Fig. 2 [12].

The TEDS memory contains electronic documents describing the TIM module. Meta TEDS describes in detail the available and available channels in TIM, which are described in more detail in the Channel TEDS of each channel. Channel TEDS describes information related to the physical layer (WPAN, Wi-Fi and Bluetooth, RFID, etc.). TEDS also contains a specially designated place for storing user information (a byte array, the length of which depends on the manufacturer). The data that can be used in the calibration process (linear parameters and calibration coefficients) for all channels are recorded in the corresponding, separate calibration documentation for each channel—Calibration TEDS. Frequency characteristics (tables with points consisting of frequencies, amplitudes, and phases) are recorded in frequency documentation—Frequency TEDS. Transfer functions (numerator and denominator of coefficients or parameters of zeros and poles) are recorded in the appropriate documentation for the transfer function—Transfer Function TEDS [5].

The standard also allows TIM manufacturers to create their own TEDS (custom data structures) for each sensor channel. TEDS has a flexible structure and can represent physical units that cannot be described using the international SI system of quantities. Each record in the TEDS memory is divided into areas: the first area, which contains the length of the TEDS memory, the second area, which contains the data directly, and an area that stores the checksum used to verify the integrity of the data. The typical structure of the TIM module is shown in Fig. 3 [12].

The data transfer protocol between the NCAP module and the TIM module does not depend on the physical layer. The transfer process is initiated by the NCAP module, which addresses a specific TIM with a command. When the TIM receives and recognizes a command, the command is processed and, if requested by the command, a response is sent to the NCAP module. The data exchange sequence diagram is presented in Fig. 4 [12].

The packet with the command sending NCAP to the TIM consists of the following parts: the address of a specific channel within one TIM; the group number to which

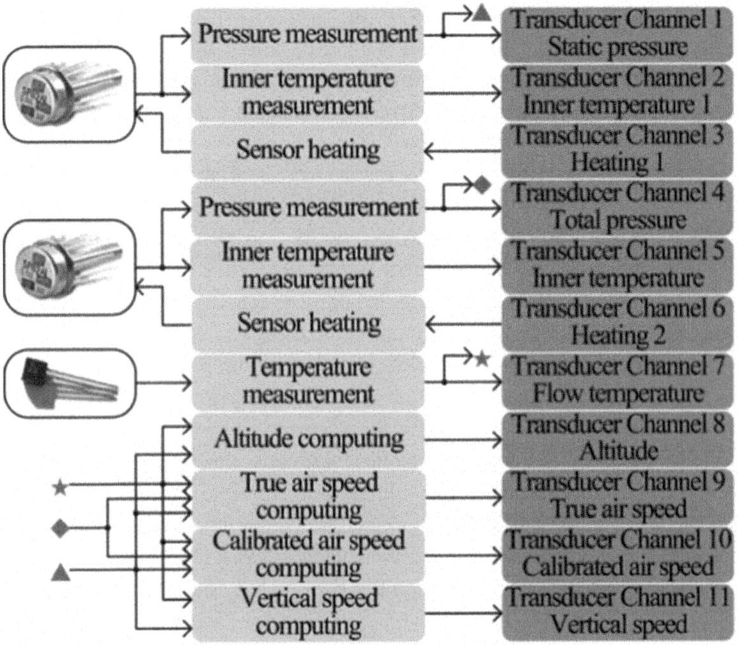

Fig. 3 A typical structure of a TIM module

Fig. 4 Data exchange sequence diagram between TIM and NCAP

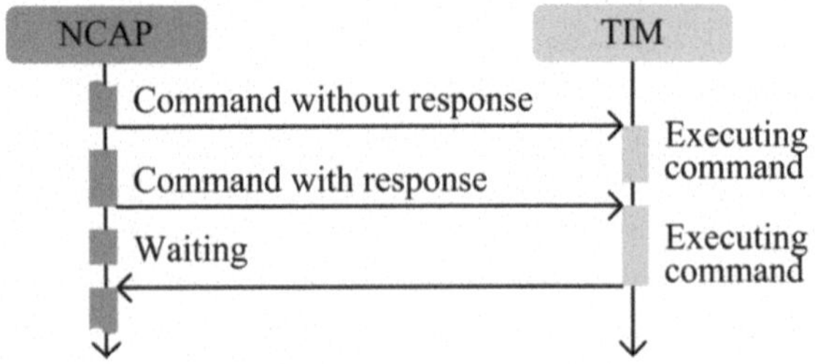

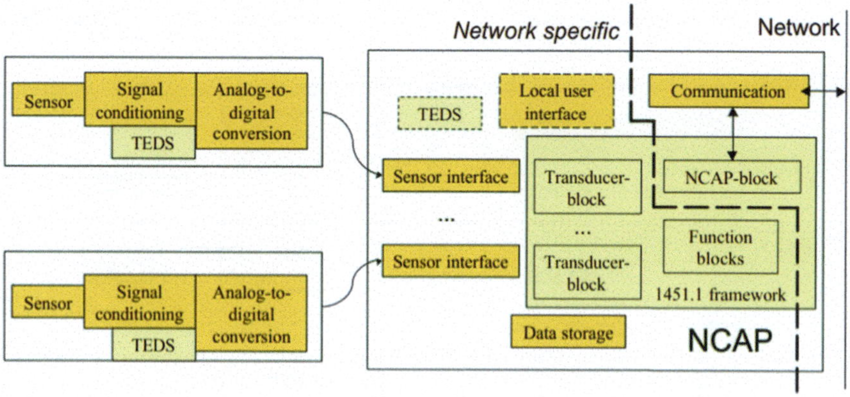

Fig. 5 Structure of TIM and NCAP modules according to standard 1451.1

the channel belongs; the number of the group to which the team belongs; metadata about the length of the sent information; data to execute the command (optional).

The response consists of the following parts: a flag indicating whether or not the command is executed; metadata about the length of the response; data resulting from command execution.

The IEEE 1451.1 "Network Capable Application Processor Information Model" standard defines a single object model for connecting intelligent converters to the network and contains interface specifications. The structure of the TIM and NCAP modules according to the IEEE 1451.1 standard is shown in Fig. 5 [13].

The IEEE 1451.2 standard "Transducer to Microprocessor Communication Protocol and TEDS Formats" defines a TII (Transducer Independent Interface) digital interface for connecting a TIM module with a digital output to a microprocessor network module. Transducer Independent Interface (TII)—an independent interface of the transducer, which defines: an initialization function that initiates the reading and recording of transducer readings; method of bitwise transmission; the byte structure of the record for transmission from NCAP to TIM; a byte record structure for transmission from TIM to NCAP; data frame transfer protocol.

The structure of the TIM and NCAP modules according to the IEEE 1451.2 standard is shown in Fig. 6 [13].

The IEEE 1451.4 standard defines methods for encoding the information recorded in TEDS for a wide range of sensors, which reduces the amount of memory required to store TEDS. The structure of the TIM and NCAP modules of the IEEE 1451.4 standard is shown in Fig. 7 [13].

An example of a message sent from NCAP to STIM is shown in Fig. 8 [12].

Meta TEDS is a TIM electronic specification. It contains information that is available to all channels and stores information that is necessary to gain access to any of the channels. This information is permanent and read-only. An example of the Meta TEDS structure is shown in Fig. 9.

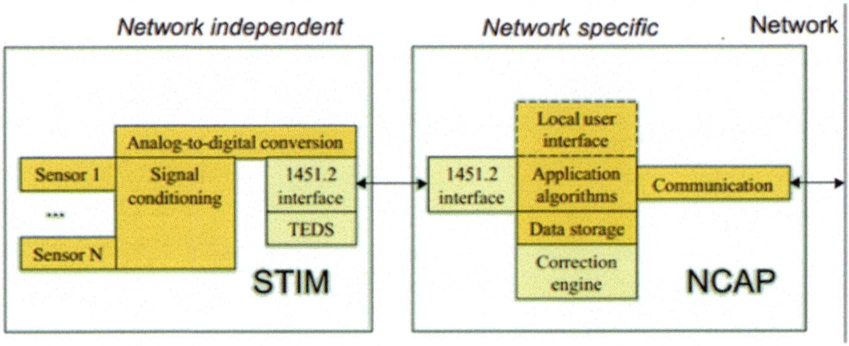

Fig. 6 Structure of TIM and NCAP modules according to standard 1451.2

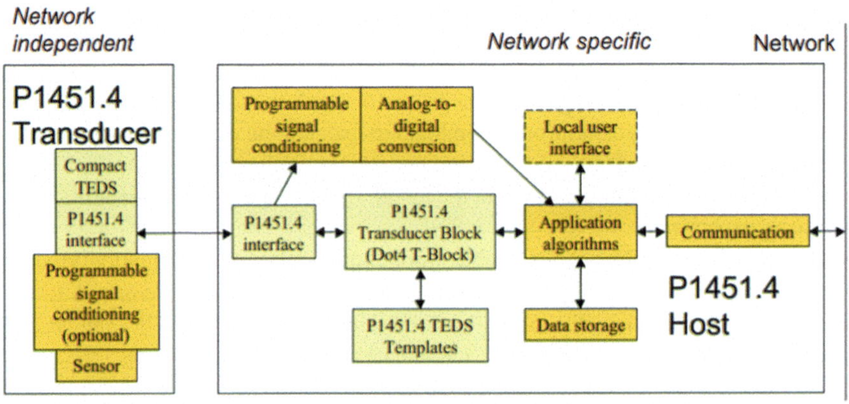

Fig. 7 Structure of TIM and NCAP modules according to standard 1451.4

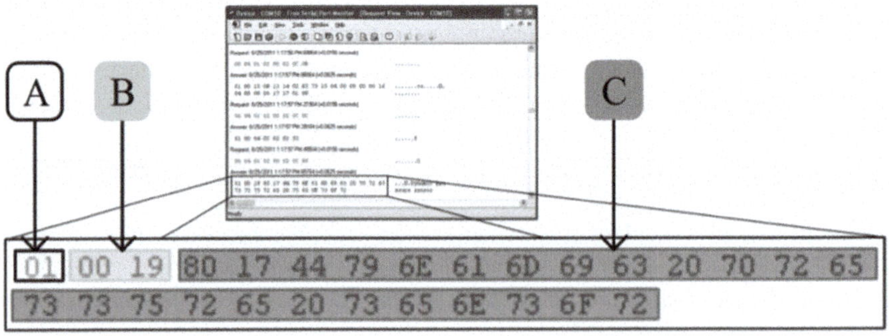

Fig. 8 An example of the structure of a message sent from NCAP to STIM

Fig. 9 A typical meta TEDS structure

MSB LSB	Field definition
00 00	Total length
00 24	
03 04	Header
00 01	
01 01	03 04 00 01 01 01
04 0A	UUID
81 C0	04 0A 81 C0 F9 74 48 81 F5 62 2E 78
F9 74	
48 81	
F5 62	
2E 78	
0A 04	Operational timeout
3F 00	0A 04 3F 00 00 00
00 00	
0C 04	Self-test timeout
C0 A0	0C 04 C0 A0 00 00
00 00	
0D 02	Number of channels
00 01	0D 02 00 01
F8 82	Checksum

Channel TEDS is an electronic measurement channel specification that stores information about a specific measurement channel. The information is permanent and read-only.

Calibration TEDS is an electronic calibration specification that stores information used by the correction module. TEDS data can be accessed both for reading and writing.

Frequency response TEDS is an optional electronic specification of frequency characteristics of a measurement channel, which stores frequency and phase data for a specific measurement channel. It can be available for both reading and writing.

Transfer function TEDS is an optional electronic specification of the transfer function of the corresponding measurement channel, which stores the coefficients of the transfer function. TEDS data can be accessed both for reading and writing.

The end user's application-specific TEDS is an optional electronic specification of an arbitrary format, which is used to store data about system users. TEDS data can be accessed both for reading and writing.

Industry extensions TEDS is an optional electronic specification of any format (the format is specified by the developer) that stores the data necessary for the TEDS developer. TEDS data can be both read and write accessible.

The implementation of data collection systems using TEDS provides the following opportunities and advantages:

- automation of setting up data collection systems, which traditionally required manual entry of sensor parameters, such as connection scheme, range, sensitivity, or pre-recording of all these data in software code;
- implementation of automatic configuration of sensors and simplification of coordination with other electronic equipment;

- reduction of time spent on starting the system, which was spent on checking the correspondence of the serial numbers of the sensors with the numbers of the connecting cables and on checking the correctness of all connections;
- the need to track the number of the measuring channel to which the necessary converter is connected, which is achieved thanks to TEDS, has disappeared;
- the ability to dynamically change the information in TEDS without stopping the system or partially disabling some modules has been implemented;
- the ability of the sensor to independently and automatically monitor the need for calibration and initiate the recalibration process has been implemented.

Thus, the standards of the IEEE 1451 group provide opportunities for quick reconfiguration of the system when replacing modules (Plug-and-Play), which is achieved with the help of electronic specifications of sensors. They also define a digital interface for accessing these specifications and reading data from sensors. A set of logical read and write functions is also defined for accessing the electronic specifications of the sensors. The implementation of this standard allows you to abandon the traditional practice of accounting for the use of sensors and significantly reduces specific maintenance costs for one measurement channel of data collection, their verification and analysis in multi-channel measurement systems.

1.3 Development of Hardware and Software of the STIM Module of Intelligent Sensors of the IEEE 1451 Standard

The functions of the STIM module of the demonstration intelligent sensor were implemented based on the MSP-430G2 microcontroller device with the MSP340G2553 microcontroller (Texas Instruments). The software and hardware of the NCAP module are implemented based on a Java microserver created based on a Cubieboard single-board computer with the Linux—Cubian operating system developed based on the Debian operating system.

One of the advantages of this operating system is the built-in developer kit for the Java programming language—Java Development Kit (JDK). Also, this operating system has built-in support for I/O port drivers, which makes it easier to set up the development environment. This operating system has an SSH server, which, by default, is set to port 36,000, which provides the possibility of remote development and deployment of NCAP module software based on Cubieboard (Fig. 10).

The software of the STIM module is implemented in the embedded C language, the software of the NCAP module is implemented in the Java programming language. A UART serial data transmission interface is used as a data transmission interface between NCAP and STIM. For data exchange, the Java Simple Serial Connector (JSSC) library was used, which implements data exchange through a serial exchange interface.

Fig. 10 Appearance of the demonstration system

Fig. 11 Appearance of the STIM module

The software has been developed for the MSP-430G2 microcontroller device, which implements the functions of the STIM module following IEEE 1451 standards (Fig. 11).

The module software consists of the following files:

- main.c—contains the main program execution script;
- main.h—header file to the file main.c;
- functions.c—contains data transfer and initialization functions;
- functions.h—header file to the functions.c file;
- teds.h—contains a description of Meta and Channel TEDS;
- tii.c—contains data exchange functions between TIM and NCAP;
- tii.h is the header file to the tii.c file.

The main() function is the entry point, from which the execution of the program begins. It consists of sequentially calling two functions: the TIM initialization function—initSTIM() and the waiting function—goToSleep().

Function USCI0RX_ISR() (Fig. 12).

The given function is an interruption handler that occurs in the case of UART to the UART to the TIM module of one data byte. Teams coming to TIM from NCAP have a length of three bytes. After receipt from NCAP to TIM 3 bytes, TIM sends a 4-by-standing meta-response to NCAP. It consists of 3 bytes of the request and the 1st byte of validity. The validation byte indicates the correctness of the request sent

```
#pragma vector=USCIAB0RX_VECTOR
__interrupt void USCI0RX_ISR(void) {

  if(readRequest() == NCAP_REQUEST_RECIVED) {

    writeFrame(Ncap_Request, 3);

    validateRequest();
    sendPreResponse(validRegistr);

    if(validRegistr == VALID) {
      doRequestedFunction(Ncap_Request[2]);
    }

    clearValidRegistr();
    clearNcapRequest();
  }
}
```

Fig. 12 Tripping handler to transmit data from NCAP to TIM

Fig. 13 Appearance of the STIM module

to the TIM request. In the case when the request is correct, the processing of this request begins.

The request consists of 3 bytes. The first two bytes are the channel ID, and the last byte is the function that should be caused for a certain channel. The request is valid when the request contains channel ID and ID function (Fig. 13), which is implemented in this TIM.

In case of a successful request, the validity register (Validregistr) remains zero. Thus, if the validity register does not contain zero, then the request did not pass the validation.

The doRequestedFunction(…) function from the main.c file (Fig. 14).

A function compares the command number sent from NCAP to the processing number that must be performed according to the NCAP request. The developed implementation of the IEEE-1451 standard has four available commands: readAvailbleChannels(…)—read information about all available channels within one TIM; readMetaTeds()—read Meta-TEDS; readChannelTeds()—read Channel-TEDS; singleMeasurement()—make measurements.

initStim() from main.c: The purpose of this feature is to initialize TIM before starting work. That is, this function encapsulates the challenges of four functions of initialization: stopWDT()—a function to stop the guard timer; configUART()—a function for UART configuration; configADC()—a function for ADC configuration; configDiode() is a function for GPIO configuration.

```
void doRequestedFunction(char functionNumber) {
  switch(functionNumber) {
    case 0x00:
      readAvailableChannels(Channel_Name, Channel_Id, CHANNELS_AMOUNT);
      break;
    case 0x01:
      readMetaTeds();
      break;
    case 0x02:
      readChannelTeds();
      break;
    case 0x03:
      singleMeasurement();
      break;
  }
}
```

Fig. 14 Calling team functions from NCAP

```
void readMetaTeds(void) {
  calculateChekSum(META_TEDS, META_TEDS_BYTE_SIZE - 2);
  writeFrame(META_TEDS, META_TEDS_BYTE_SIZE);
}

void readChannelTeds(void) {
  calculateChekSum(CHANNEL_TEDS, CHANNEL_TEDS_BYTE_SIZE - 2
  writeFrame(CHANNEL_TEDS, CHANNEL_TEDS_BYTE_SIZE);
}
```

Fig. 15 Functions for meta-teds and channel-teds transmission

The calculateChekSum(…) function: the purpose of this feature is to calculate the control amount as required by IEEE 1451.

Functions readMetaTeds() and readChannelTeds() (Fig. 15).

Functions send TEDS tables recorded in TIM to NCAP: respectively Meta-TEDS and Channel-TEDS. The answer sent to NCAP consists of 2 parts: the meta-answer and the answer itself. A response may not be sent in the event of an invalid request from NCAP. In all other cases, the answer is mandatory. The response format consists of 3 parts: response length (4 bytes); data block (0… 2^{30} bytes); checksum (2 bytes).

Function readAvailableChannels(…) (Fig. 16).

This function sends the list of available TIM channels to NCAP. Data block format: N—number of channels (1 byte); the name of the Nth channel encoded in ASCII characters; Id of the Nth channel (2 bytes).

The singleMeasurement() function (Fig. 17).

This function initiates the measurement by calling the measureTemp(…) function and generates the measurement result that is sent to the NCAP module.

The measureTemp(…) function (Fig. 18).

This function receives data. The measurement starts with the activation of the ADC, which starts the conversion. After the conversion, the result is in the

```
void readAvailableChannels(char* name, char* id, char amountofChannels) {
  char length[4] = {0x00, 0x00, 0x00, 0x0E};
  char checkSum[2] = {0xFF, 0xF1};

  writeFrame(length, 4);
  writeByte(amountofChannels);
  writeByte(0x04);
  writeFrame(name, 4);
  writeFrame(id, 2);
  writeFrame(checkSum, 2);
}
```

Fig. 16 A function to transmit a list of available TEDS channels

```
void singleMeasurement(void) {
  char measurement[2] = {0x00, 0x00};
  char length[4] = {0x00, 0x00, 0x00, 0x08};
  char checkSum[2] = {0xFF, 0xF7};

  measureTemp(measurement);

  writeFrame(length, 4);
  writeFrame(measurement, 2);
  writeFrame(checkSum, 2);
}
```

Fig. 17 Function for transferring the measurement result

```
void measureTemp(char* result) {
  int sample;

  ADC10CTL0 |= ENC + ADC10SC;
  while ( ADC10CTL1 & ADC10BUSY );
  sample = ADC10MEM;

  result[1] = sample;    //correction
  result[0] = sample >> 8;
}
```

Fig. 18 A function for reading data from an ADC

ADC10MEM register. Since the ADC has a resolution of 10 bits, and the information to the NCAP is transmitted byte by byte (that is, by 8 bits), it is necessary to convert a 10-bit integer into two 8-bit integers.

Function writeFrame() (Fig. 19).

The function accepts a pointer to a byte array and the number of elements in the array to be transferred.

```
void writeFrame(char *bytePtr, char size) {
  int i = 0;
  for(; i < size; i++) {
    writeByte(bytePtr[i]);
  }
}
```

Fig. 19 A function for transferring an array of bytes

1.4 Implementation of the NCAP Module of Intelligent Sensors According to the IEEE 1451 Standard Based on a Single-Board Java Microserver Cubieboard

A library for supporting the functions of the NCAP module according to the IEEE 1451 standard has been developed for a Java machine based on a single-board Cubieboard computer (Fig. 20).

The developed NCAP Java library consists of the following packages: *.ncap.channels—contains classes for working with TIM channels; *.ncap.teds—contains classes for working with TEDS; *.ncap.tii—contains classes for organizing information transfer through TII; *.ncap.util—contains auxiliary classes for working with the library; *.ncap—contains the main class for working with the library.

The class diagram of the library is shown in Fig. 21.

The Msp430TimManager class coordinates the interaction of library classes. This class provides the ability to obtain references to classes that implement IEEE 1451 standard functions.

This class encapsulates three fields: tii—a reference to the class that implements methods for exchanging information with TIM; functionManager—a reference to a class that provides access to methods implemented in TIM; channelManager—a reference to a class that provides information about available channels in the TIM.

The getTii() method: This method returns an instance of the Msp430Tii class by reference to the Tii interface that this class implements. In case the instance has not yet been created, the method creates a new instance.

The getFunctionManager() and getChannelManager() methods are similar to the getTii() method, with the difference that they return instances of the Msp430FunctionManager and Msp430ChannelnManager classes, respectively.

The *.ncap.tii package contains classes and interfaces that implement information exchange through the abstract layer of data exchange between TIM and NCAP—the TII interface. The class diagram for this package is shown in Fig. 22.

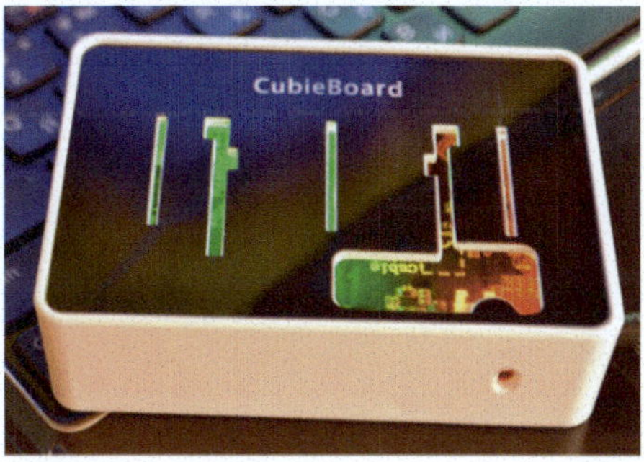

Fig. 20 Appearance of the NCAP module

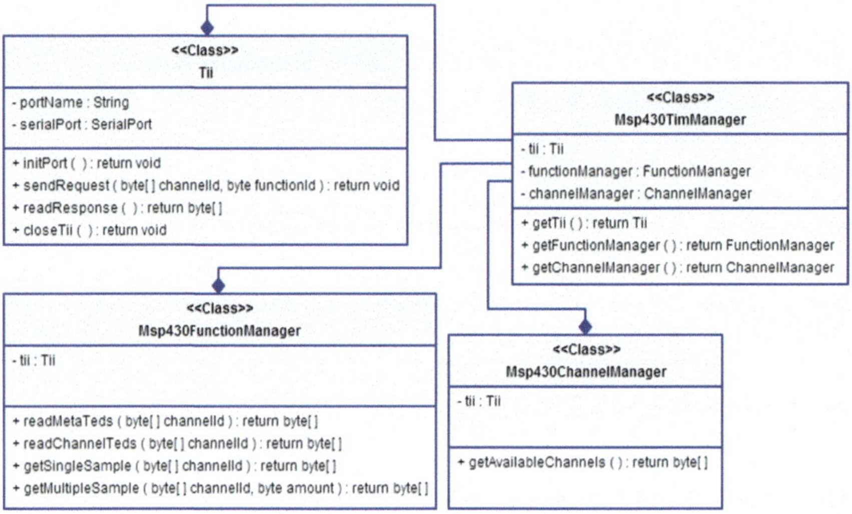

Fig. 21 NCAP Java library class diagram

Fig. 22 Class diagram of the *.ncap.tii package

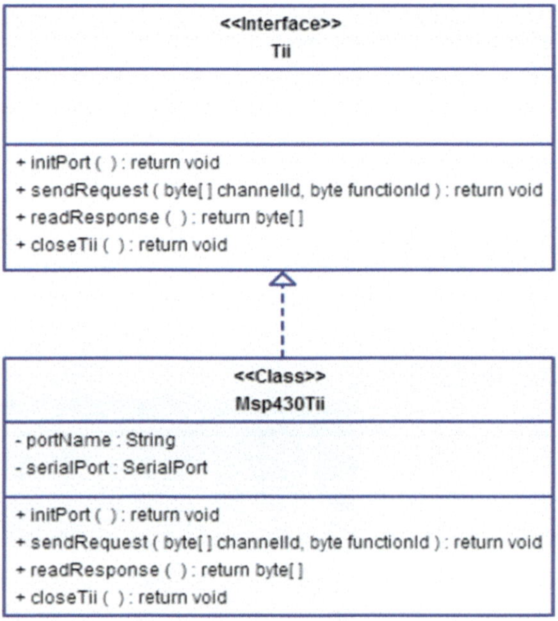

The Msp430Tii class implements the functions of the TII Java interface for exchanging information with the TIM module. Since the implementation involves data transfer through a low-level UART data transfer interface, this class uses the capabilities provided by the JSSC library.

The initPort() method opens a UART connection to the TIM.

The sendReuest() method sends a command to the TIM, which consists of the channel number id and function number id passed to the method as parameters (Fig. 23).

The readResponse() method (Fig. 24).

The method reads the response from the TIM into a byte array. If an incorrect (not valid) request is sent to the TIM, the TIM, in turn, informs the NCAP about this by sending a meta-response to the NCAP. Thanks to this, the method can notify the system about the transmission of an incorrect command to the TIM.

The *.ncap.channels package contains classes and interfaces for providing information about available channels in the TIM module. The class diagram of the package is shown in Fig. 25.

The Msp430ChannelsManager class implements a method for obtaining information about all available channels in the TIM.

Method getAvailableChannels() (Fig. 26).

```
      public void sendRequest(byte[] channelId, byte functionId) throws
SerialPortException {
            if(channelId != null) {
              if(channelId.length == CHANNEL_ID_LENGTH) { byte[] outPut = new
      byte[3]; outPut[0] = channelId[0]; outPut[1] = channelId[1];
      outPut[2] = functionId;
                  serialPort.writeBytes(outPut);
            } else {
              System.out.println("Error: channelId.length != 2");
                  }
              } else {
              System.out.println("Error: channelId == null");
         }
      }
```

Fig. 23 A method for passing a command from NCAP to TIM

```
      public byte[] readResponse() throws SerialPortException,
SerialPortTimeoutException {
      byte[] result = null;
      byte[] metaAnswer = serialPort.readBytes(META_RESPONSE_LENGTH,
TIMEOUT_IN_MILLS);

      if(metaAnswer[3] == VALID_REQUEST) {
      byte[] length = serialPort.readBytes(4, TIMEOUT_IN_MILLS);
      byte[] data = serialPort.readBytes(length[3] - 4, TIMEOUT_IN_MILLS)
      result = mergeTwoArray(length, data);
      } else {
      System.out.println("InValid request from TIM");
      }

      return result;
      }
```

Fig. 24 A method for reading a response from a TIM

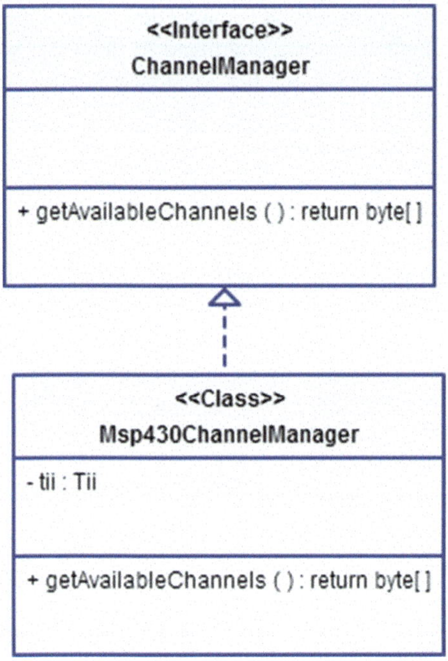

Fig. 25 Class diagram of the *.ncap.channels package

```
      public byte[] getAvailableChannels() throws SerialPortException,
SerialPortTimeoutException {
      byte[] generalChannelsId = {GENERAL_CHANNELS_ID, GENERAL_CHANNELS_ID};

      tii.sendReuest(generalChannelsId, META_TEDS_FUNCTION_ID);
      byte[] answer = tii.readResponse();

      return answer;
      }
```

Fig. 26 A method for obtaining information about available TIM channels

This method sends a command to the TIM to read all available channels. TIM sends a response to NCAP in the form of a byte array. The byte array contains the names and Ids of the available channels.

The *.ncap.teds package contains interfaces and classes that provide access to all functions supported in TIM. The class diagram of the package is shown in Fig. 27.

Msp430FunctionManager class—supports the following TIM functions: Meta TEDS reading function; Channel TEDS reading function; the function of receiving data from sensors.

Method readMetaTeds() and readChannelTeds() (Fig. 28).

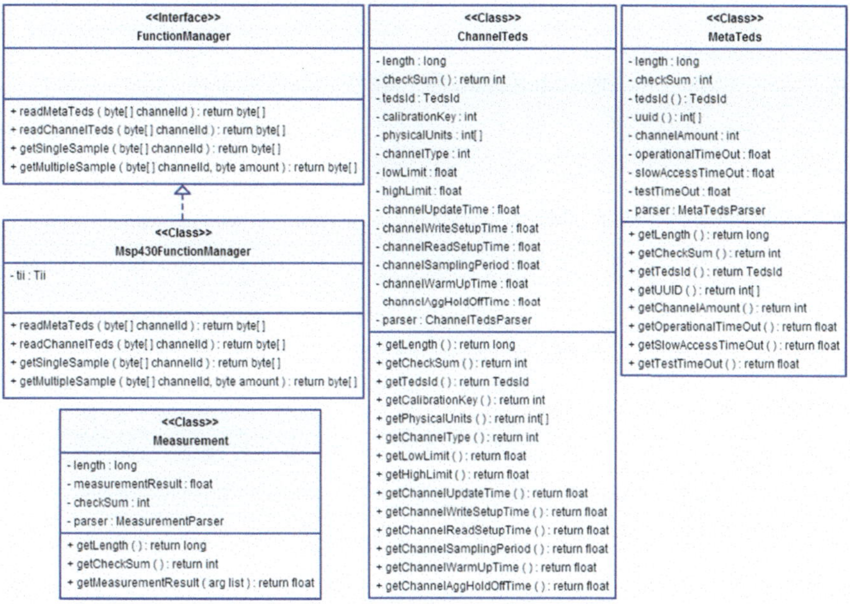

Fig. 27 Class diagram of the *.ncap.teds package

```
	public byte[] readMetaTeds(byte[] channelId) throws SerialPortException,
SerialPortTimeoutException {
	tii.sendReuest(channelId, META_TEDS_FUNCTION_ID);
	byte[] answer = tii.readResponse();

	return answer;
	}

	public byte[] readChannelTeds(byte[] channelId) throws SerialPortException,
SerialPortTimeoutException {
	tii.sendReuest(channelId, CHANNEL_TEDS_FUNCTION_ID);
	byte[] answer = tii.readResponse();

	return answer;
	}
```

Fig. 28 Methods for obtaining TEDS tables

The purpose of both methods is to receive electronic documentation from TIM—TEDS. The algorithm consists of two parts:

- send a command to the TIM with the Id of the corresponding function;
- get a byte array with the required TEDS from TIM.

The getSingleSample() method receives sensor data from TIM (Fig. 29).

The *.ncap.util package contains auxiliary classes for the operation of the NCAP library. The class diagram of this package is shown in Fig. 30.

```
    public byte[] getSingleSample(byte[] channelId) throws SerialPortException,
SerialPortTimeoutException {
    tii.sendReuest(channelId, SINGLE_MEASUREMENT_FUNCTION_ID);
    byte[] answer = tii.readResponse();
    return answer;
    }
```

Fig. 29 Method for obtaining measurement results

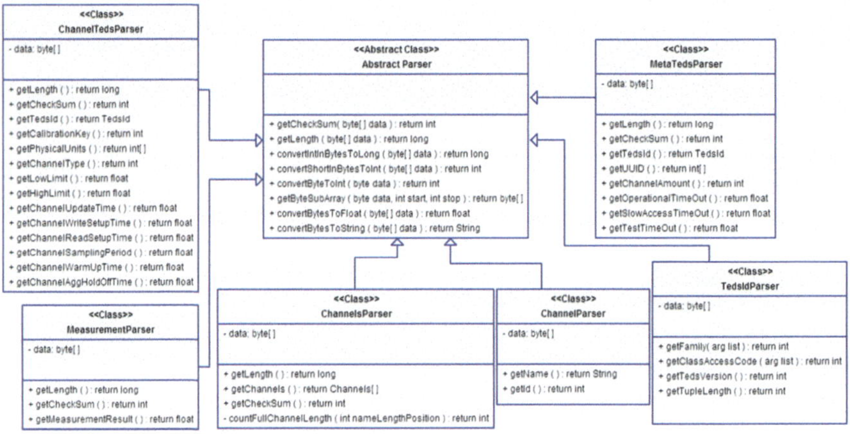

Fig. 30 Class diagram of the *.ncap.util package

The AbstractParser class is abstract. Its main purpose is to implement the functions of converting byte arrays into primitive data types of the Java programming language.

According to the IEEE 1451 standard, all primitive types must be unsigned, but the Java programming language has no unsigned primitive types. This problem is solved by converting to a type with a wider range. For example, the type unsigned int, whose values are in the range from 0 to 2^{32}—1, is converted to the type long, whose values are in the range from -2^{63} to 2^{63}—1.

The convertIntInBytesToLong() method (Fig. 31).

```
protected long convertIntInBytesToLong(byte[] data, int highBytePosition) {
    long result = convertByteToInt(data[highBytePosition]);
    for (int i = 1; i < INT_AMOUNT_OF_BYTES; i++) {
        result <<= SHIFT_BY_ONE_BYTE;
        result |= convertByteToInt(data[highBytePosition + i]);
    }
    return result;
}
```

Fig. 31 A method for converting types

```
protected int getCheckSum(byte[] data) {
      int checkSumHighByte = data.length - 2;
      return convertShortInBytesToInt(data, checkSumHighByte);
}

protected long getLength(byte[] data) {
      int lengthHighBytePosition = 0;
      return convertIntInBytesToLong(data, lengthHighBytePosition);
}
```

Fig. 32 Methods for providing the checksum and length of a data frame

This method converts an array, which consists of four bytes and is a byte representation of an unsigned integer type—int, into a Java programming language type—long.

The getCheckSum() and getLength() methods (Fig. 32).

These methods provide the checksum and length of the received data frame defined by the IEEE 1451 specification and are used to form each data frame that is transmitted from the TIM to the NCAP. Thus, their implementation in the parent class makes it possible to simplify the descendant class.

The software of the NCAP module allows the user to visually familiarize himself with information about its parameters. The output window with the list of measurement channels available in TIM is shown in Fig. 33.

The output window with the contents of the TEDS meta-electronic specification is shown in Fig. 34.

The output window with measurement data is shown in Fig. 35.

The output window with the contents of the channel TEDS electronic specification is shown in Fig. 36.

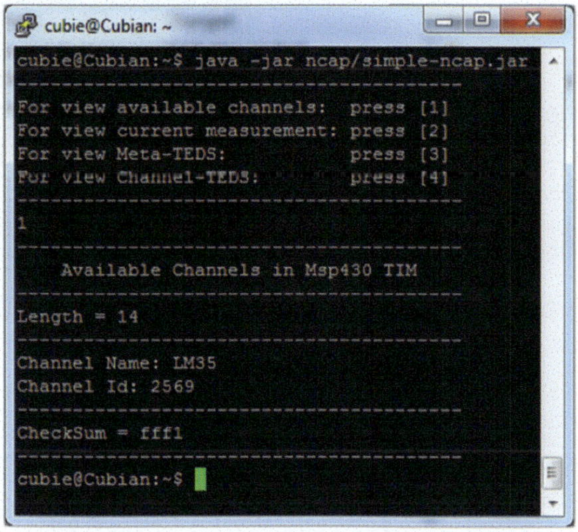

Fig. 33 A terminal window with a list of available channels in TIM

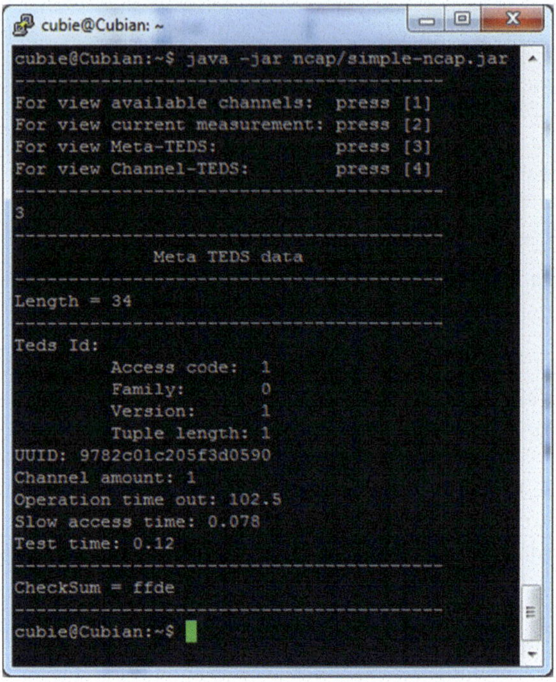

Fig. 34 Terminal window with meta TEDS

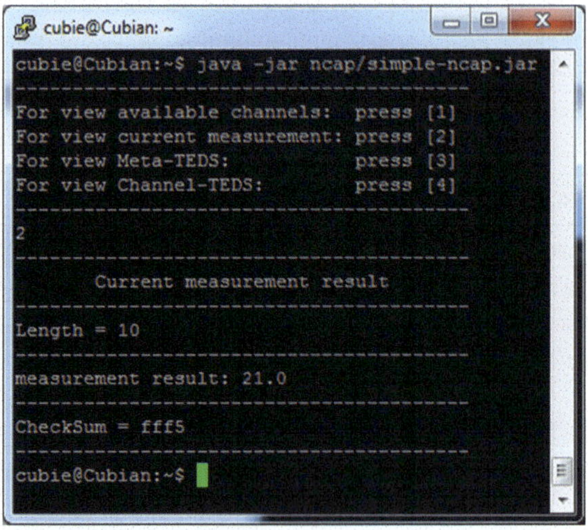

Fig. 35 Terminal window with measurement results

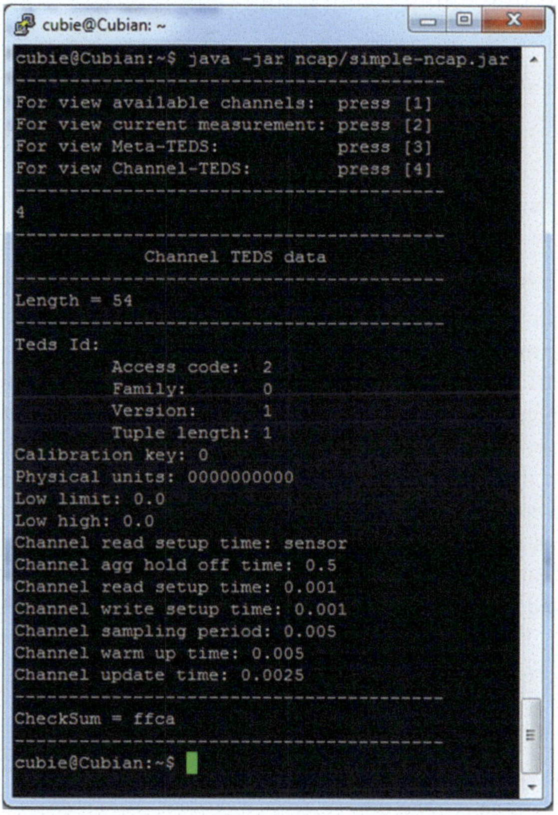

Fig. 36 Terminal window with channel TEDS

2 Development of Network Intelligent Sensors Based on the Specialized Web Protocol MQTT

2.1 Peculiarities of Using the MQTT Protocol in Network Data Collection Systems

MQTT (Message Queue Telemetry Transport) is a messaging protocol that implements the "publish/subscribe" model and is intended for communication between computerized devices. The protocol was proposed in 1999, and its authors are believed to be Andy Stanford-Clark of IBM and Arlen Nipper of Arcom. The protocol was originally created for sensors that monitor the condition of pipes, but later its scope was expanded and it was used in many built-in applications and smartphones [14–18]. For example, the Facebook social network uses this protocol for messaging (Facebook Messenger) [19].

The working principle of the MQTT protocol is shown in Fig. 37.

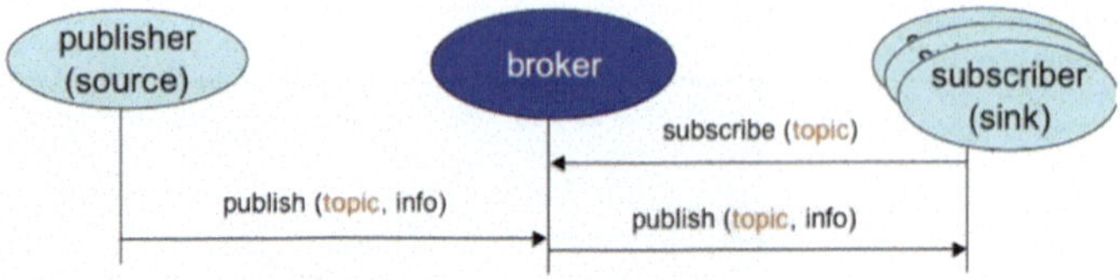

Fig. 37 The working principle of the MQTT protocol

There are currently two versions of the MQTT specification: MQTT v3.1—the core specification for TCP/IP-based networks, and MQTT-S v1.2 for sensors and embedded devices in non-TCP/IP networks. It is worth noting that in November 2011, IBM opened the source codes of its PAHO software for management and messaging, including based on the MQTT protocol [20].

According to many experts, the MQTT protocol has some advantages compared to the HTTP protocol: lower data transfer overhead and lower bandwidth. For its work, it does not require a constant connection between the client and the server (as in the case of HTTP). MQTT is also well adapted to work on communication channels with low bandwidth [21, 22]. A performance comparison between HTTP and MQTT is shown in Table 1.

Based on the results presented in Table 1, it can be concluded that MQTT uses 4–7 times fewer network resources than HTTP.

The MQTT protocol has the following features. A publish/subscribe messaging model is used, which provides one-to-many communication. The messaging transport is independent of the data content. The TCP/IP protocol is used to ensure connection to the core network.

When delivering messages, there are three types of service quality (Quality of Service, QoS). At most once (QoS = 0, "not more than once") means that the publisher performs a one-time message sending, but does not take any steps other than those provided by TCP/IP to ensure that the message is delivered (the message is sent to the addressee without verification of receipt). At least once (QoS = 1, "at least once"): the delivery of the message is checked, but it is allowed to deliver it more than once. Exactly once (QoS = 2, "only once") guarantees message delivery only once.

Each increase in the level of QoS leads to an additional load on the processor and the network. The choice of QoS can affect the overall scalability of messaging

Table 1 HTTP and MQTT performance comparison

Operation	HTTP	MQTT
Reading one block of data from the server	302 bytes	69 bytes
Writing one block of data to the server	320 bytes	47 bytes
Reading 100 blocks of data from the server	12,600 bytes	2445 bytes
Writing 100 blocks of data to the server	14,100 bytes	2126 bytes

systems and also creates additional overhead related to the storage of undelivered messages. Therefore, it is necessary to correctly select the appropriate QoS level for each published message. As a general rule, lower QoS values should be used unless a tighter guarantee of message delivery is required [23].

Advantages of the MQTT protocol:

- small overheads at the transport level (header of fixed size 2 bytes long);
- the exchange protocol is minimized to reduce network traffic;
- a connection control mechanism is built-in.

An important element of the whole concept in MQTT is the concept of "topic". In the exchange of messages according to the "publication/subscription" model, message recipients are called topics. The MQTT protocol uses a hierarchical topic space, that is, the topic structure should allow subscribers and publishers to specify the recipient's topic with varying degrees of precision. The length of the subject cannot exceed 32,767 characters. Subject elements are separated from each other by separators (/) and wildcard elements (#, +) that allow you to specify one or more recipients. In addition to the subject, the message transmits data (payload, data)—variable values.

Below is an example of a logically constructed message topic space for a hypothetical bus company application. Here are some meaningful examples of using this topic space:

- FastBus / City_F / Helios117 / TEMP—bus 117 temperature sensor data;
- FastBus / City_F / + / TEMP—data of all temperature sensors of all buses in City_F;
- FastBus / City_F / Helios117 / +—data of all sensors in buses 117 in City_F;
- FastBus / City_F / #—data of all sensors of all buses in City_F;
- FastBus / #—data of all sensors of all buses in all cities;
- FastBus / + / + / TEMP—data of all temperature sensors of all buses in all cities.

MQTT provides publish/subscribe messaging that allows devices to send and receive data and alarms when an event occurs (event-driven application). In the model with one publisher and many subscribers, it is possible to send information from one point to many other devices or "listeners" who are interested in receiving information.

Embedded devices can use the MQTT protocol to collect data from multiple devices with limited bandwidth and provide information to many subscribers. As a result, the system is relatively easy to configure and provides an ideal communication network protocol for cloud-based solutions in the device world.

An MQTT message has a fixed header. In addition, a variable-length header and data are added to some messages. The format of each part of the message header is described in the relevant parts of the MQTT specification. The structure of the MQTT message header is shown in Fig. 38.

<table>
<tr><td>Bit</td><td>7</td><td>6</td><td>5</td><td>4</td><td>3</td><td>2</td><td>1</td><td>0</td></tr>
<tr><td>Byte 1</td><td colspan="4">Message type</td><td>DUP flag</td><td colspan="2">QoS level</td><td>RETAIN</td></tr>
<tr><td>Byte 2</td><td colspan="8">remainder</td></tr>
</table>

Fig. 38 MQTT message header structure

Byte 1 contains the message type and flag fields (DUP, QoS level, and RETAIN), and byte 2 (at least one byte) is the remaining length field. All data values use big-endian format: high-order bytes come before lower-order bytes. The message type is represented as an unsigned 4-bit value. Acceptable type values are listed in Table 2.

The remaining bits of the first byte contain the DUP, QoS, and RETAIN fields. The DUP flag is set when the client or server attempts to redeliver a PUBLISH, PUBREL, SUBSCRIBE, or UNSUBSCRIBE message. This may occur for messages where the QoS value is greater than zero and no acknowledgment is required.

When the DUP bit is set, the variable header includes the Message ID. The recipient should consider this flag as information that the message may have been previously received. The QoS flag indicates the level of guarantees for PUBLISH message delivery. The RETAIN flag is used only for PUBLISH messages. When the client sends a PUBLISH to the server, if the RETAIN flag is set to "1", the server must hold the message after it has been delivered to the current subscribers. Saved messages should persist when the server is restarted. The developed information and measurement system is based on the compatible use of the MQTT protocol and the RESTful

Table 2 Message types in the MQTT protocol

Designation	Meaning	Description
Reserved	0	Reserved
CONNECT	1	The client's request for a connection
CONNACK	2	Connection confirmation
PUBLISH	3	Post a message
PUBACK	4	Publish confirmation
PUBREC	5	Publication received (guaranteed delivery, part 1)
PUBREL	6	Publish release (guaranteed delivery, part 2)
PUBCOMP	7	Publication completed (guaranteed delivery, part 1)
SUBSCRIBE	8	Client subscription request
SUBACK	9	Subscription confirmation
UNSUBSCRIBE	10	Customer unsubscribe request
UNSUBACK	11	Unsubscribe confirmation
PINGREQ	12	PING request
PINGRESP	13	PING response
DISCONNECT	14	The client has disconnected
Reserved	15	Reserved

web service for data exchange, which made it possible to use the advantages of both of these technologies.

2.2 Development of Hardware and Software of Intelligent Sensors Based on the uIP Library

The hardware of the STIM module of the smart sensor demonstration model is implemented based on the CP2201EK microcontroller module (Silicon Laboratories). The main system controller is a 50 MIPS Silicon Laboratories C8051F340 [24]. The Ethernet controller providing network connectivity is the Silicon Laboratories CP2201 [25]. The temperature measurements are taken from the C8051F340's on-chip temperature sensor and the ambient light reading is taken from the light sensor. The appearance of the module is presented in Fig. 39.

For the implementation of TCP/IP protocols, a stack with open source codes—uIP—has been adapted.

The uIP library provides an implementation of the TCP/IP protocol stack and is designed for 8- and 16-bit microcontrollers. It provides the necessary protocols for communication via the Internet with a very small size of code and memory—the size of the uIP code is several kilobytes of RAM and several hundred bytes of code memory. The uIP stack is open-source software, developed in the C programming language. Documentation and source code can be freely used and distributed for commercial and non-commercial use. The stack has been ported to a wide variety of 8-bit microcontrollers and is used in a large number of products and projects [26].

The uIP stack has the following features: well-documented and well-commented source code; small code size; low memory requirements that are configured during compilation; support for ARP, SLIP, IP, UDP, ICMP, and TCP protocols; a set of applications is included—web server, e-mail client (SMTP client), Telnet server, DNS; an arbitrary number of simultaneously active TCP connections is supported, the maximum number of which is configured during compilation; any number of passive

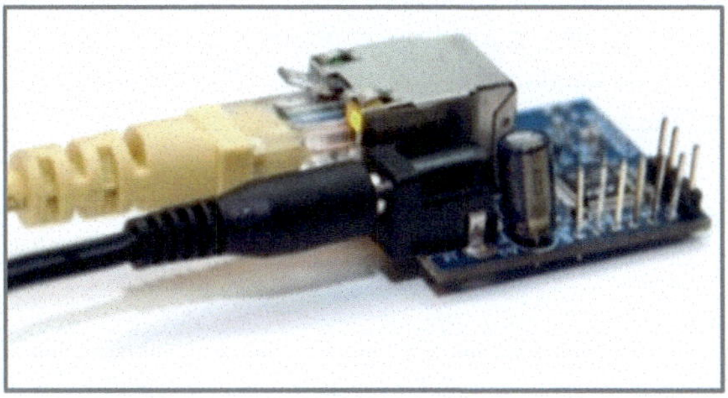

Fig. 39 Appearance of the CP2201EK module

TCP listeners (servers), the maximum number of which is configured during compilation; the stack is free for commercial and non-commercial use; RFC-compliant implementation of TCP and IP protocols is supported.

The software of the device is implemented based on uIP stack version 0.9. This choice is because the uIP stack comes with open source, so if you need to perform specific actions that are not provided by the developers of the stack, it is possible to add them by directly editing the code. In addition, quite a lot of applications have been written for uIP, which can be used in the development of device software.

For the compatible operation of the Ethernet controller CP2201 and uIP 0.9, a library was developed that ensured the transmission of messages from the uIP stack to the physical layer of CP2201. This library is located in the files cp2201.c and cp2201.h. The header file cp2201.h contains the definitions of the registers of the Ethernet controller, as well as the necessary data structures. The cp2201.c file consists of three functions—InitCP2201(), which is responsible for initializing the controller, CP2201_transmit(), which is responsible for data transmission, and CP2201_receive(), which is responsible for receiving data. In addition, since the interrupts from CP2201 come to the INT0 input of the microcontroller, the ExtInt0_ISR() interrupt processing function has been added in addition to the specified functions. The service function _wait_ms(count) is also added, which is designed to create delays for the Ethernet controller to save the configured parameters.

2.3 *Development of Software for Setting Network Parameters of Intelligent Sensors Using Flash Memory*

To be able to write and read network parameters from Flash memory, a library was developed, located in Flash.c and Flash.h files. To store these data, a code memory area of 175 bytes was allocated, in which the parameters are sequentially written. A data memory of the same size was also allocated, into which the data are transferred after reading them. The data memory was divided into blocks according to their functional purpose, while a structural data type was created for each of these blocks. The definition of these data types is located in the Flash.h file.

The LAN_params type contains four fields—a bit that indicates whether DHCP should be used or not, as well as the IP addresses of the device and the main gateway, and the subnet mask (Fig. 40).

Fig. 40 Type LAN_params

```
typedef struct LAN_params {
 u8_t dhcpc_flag;
 UNI32 uip_hostaddr;
 UNI32 uip_draddr;
 UNI32 uip_netmask;
};
```

The DNS_params type contains the IP address of the main and secondary DNS servers, as well as the name of the device and its domain name (Fig. 41).

The user_params type is used to store the user's login and password (Fig. 42).

The security_params type contains two fields. One field indicates that the data read is valid, it should always contain the value specified by the two-byte constant FLASH_VALID, and the second contains a checksum for the data. These fields are used to check the correctness of the data block after reading (Fig. 43).

The mqtt_params type contains four fields and is designed to configure data transfer using the MQTT protocol. The first two fields indicate the IP address and MQTT port of the server. The last two fields indicate the name of the device and the name of the device set to which it belongs (Fig. 44).

The domain name and device name are loaded, if their support is specified in the settings, and the received settings are written to the uIP stack variables (Fig. 45).

Fig. 41 Type DNS_params

```
typedef struct DNS_params {
 UNI32 DNS_PrimaryNum;
 UNI32 DNS_SecondaryNum;
 u8_t DNS_DeviceName[16];
 u8_t DNS_DomainName[40];
};
```

Fig. 42 Type user_params

```
typedef struct user_params {
 u8_t username[16];
 u8_t password[16];
};
```

Fig. 43 Type security_params

```
typedef struct security_params {
 u16_t valid_flash;
 u16_t checksum;
};
```

Fig. 44 Type mqtt_params

```
typedef struct mqtt_params {
 UNI32 server_ip;
 u16_t seprver_port;
 u8_t device_name[16];
 u8_t device_set_name[16];
};
```

```
#if (UIP_CONF_DEFAULT_DEVICE_NAME == TRUE)
        strcpy(tmp_DNS_params.DNS_DeviceName, (char *)&defaultDeviceName);
#else
        memset(tmp_DNS_params.DNS_DeviceName, 0,
sizeof(tmp_DNS_params.DNS_DeviceName));
#endif
#if (UIP_CONF_DEFAULT_DOMAIN_NAME == TRUE)
        strcpy(tmp_DNS_params.DNS_DomainName, &defaultDomainName);
#else
        memset(tmp_DNS_params.DNS_DomainName, 0,
sizeof(tmp_DNS_params.DNS_DomainName));
#endif
  }
  uip_sethostaddr(tmp_LAN_params.uip_hostaddr.i);
  uip_setdraddr(tmp_LAN_params.uip_draddr.i);
  uip_setnetmask(tmp_LAN_params.uip_netmask.i);
#endif /* UIP_FIXEDADDR */
}
```

Fig. 45 Configuring network settings

```
  void FLASH_Read_mqtt_Params(void) {
        if (tmp_security_params.valid_flash != FLASH_VALID) {
              tmp_mqtt_params.server_ip.b[0] = DEFAULT_MQTT_SERVER_IP_0;
              tmp_mqtt_params.server_ip.b[1] = DEFAULT_MQTT_SERVER_IP_1;
              tmp_mqtt_params.server_ip.b[2] = DEFAULT_MQTT_SERVER_IP_2;
              tmp_mqtt_params.server_ip.b[3] = DEFAULT_MQTT_SERVER_IP_3;

              tmp_mqtt_params.seprver_port = DEFAULT_MQTT_SERVER_PORT;
              strcpy(tmp_mqtt_params.device_name,
                                    (char *)&DEFAULT_MQTT_DEVICE);
              strcpy(tmp_mqtt_params.device_set_name,
                                    (char *)&DEFAULT_MQTT_DEVICE_SET);
        }
        mqtt_set_server_ip(tmp_mqtt_params.server_ip.i);
        mqtt_set_server_port(tmp_mqtt_params.seprver_port);
        mqtt_set_device_name(tmp_mqtt_params.device_name);
        mqtt_set_device_set_name(tmp_mqtt_params.device_set_name);
  }
```

Fig. 46 FLASH_Read_mqtt_Params function

The FLASH_Read_mqtt_Params function is intended for reading settings for sending data using the MQTT protocol. If the validity flag is not equal to FLASH_VALID, then the default settings must be loaded. After validating the data, functions are called to set the values required by the MQTT client (Fig. 46).

2.4 Development of DHCP Protocol Support Software

To work in modern networks, the microcontroller device must support the DHCP protocol. DHCP (Dynamic Host Configuration Protocol) is a network protocol that

```
if ((tmp_security_params.valid_flash == FLASH_VALID) &&
(tmp_LAN_params.dhcpc_flag > 0)) {
        dhcpc_init();
 }
```

Fig. 47 Starting the dhcpc client

Fig. 48 Callback function upon receipt of a package

```
void uip_udp_appcall(void)
{
 switch(uip_udp_conn->lport)
 {
       case HTONS(IANAportbootpc):
             dhcpc_appcall();
       break;
}
```

allows computers to automatically obtain an IP address and other parameters needed to operate on a TCP/IP network. DHCP support was based on the dhcpc client, which comes as an add-on for the uIP stack. To run dhcpc, the following actions were added to the main program file (Fig. 47).

Here it is checked whether the DHCP flag and the read settings validity flag are set. In addition, a callback function was specified when receiving a packet on the corresponding port (Fig. 48).

After this configuration, you can trace the main stages of DHCP operation. A DHCP client that needs to obtain network settings sends a DHCPDISCOVER broadcast packet in search of a server. The packet contains the hardware address of the requesting client. The appearance of the DHCPDISCOVER package is shown in Fig. 49.

After sending a DHCPDISCOVER packet, one or more DHCP servers examine the request and send back a DHCPOFFER packet containing the offered.

IP address and time of its use. The appearance of the DHCPOFFER package is shown in Fig. 50.

The client selects an address from the received DHCPOFFER packets. Following this, the client sends a DHCPREQUEST packet with the address of the selected server. The appearance of the DHCPOFFER package is shown in Fig. 51.

The selected server sends an acknowledgment (DHCPACK) and the negotiation process is completed. The DHCPACK packet contains the given address and its lease time. The appearance of the DHCPOFFER package is shown in Fig. 52.

After receiving the DHCPACK packet, the IP address is transferred to the uIR stack. Additionally, when handling interrupts from timer 0 every two seconds, the DHCPC_Timeout flag is set to allow the client to extend the address lease time.

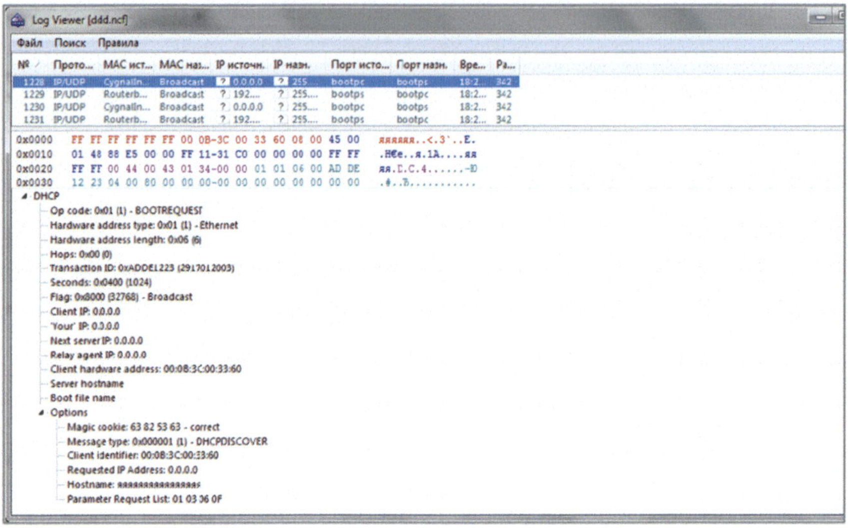

Fig. 49 View of the client's DHCPDISCOVER broadcast packet

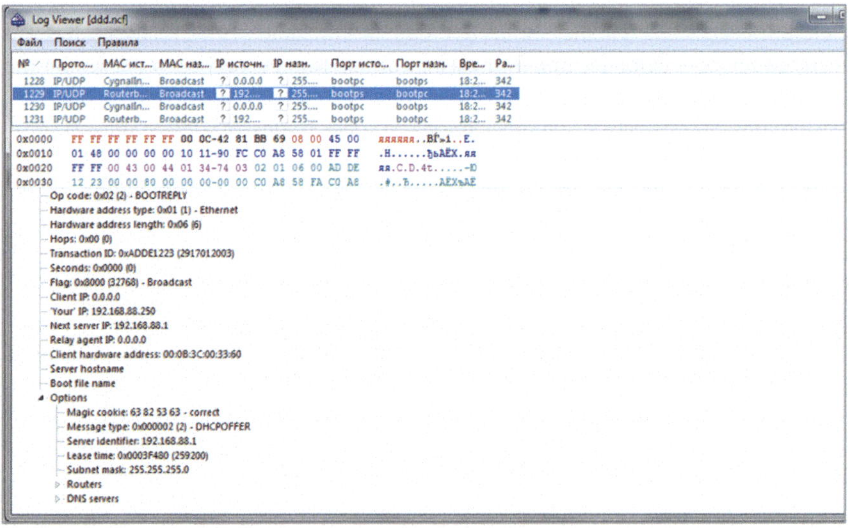

Fig. 50 View of the DHCPOFFER package from the server

2.5 *Development of Data Exchange Software Using the MQTT Protocol*

Software to support data exchange with the microcontroller module of the intelligent sensor using the MQTT protocol was developed based on the libemqtt client library.

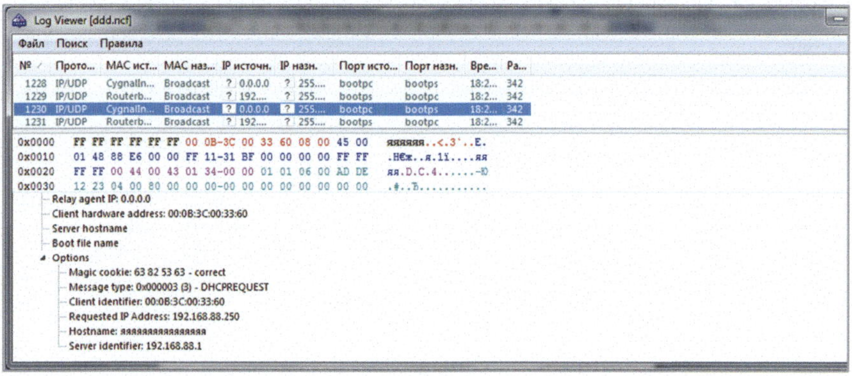

Fig. 51 View of the DHCPREQUEST packet from the client

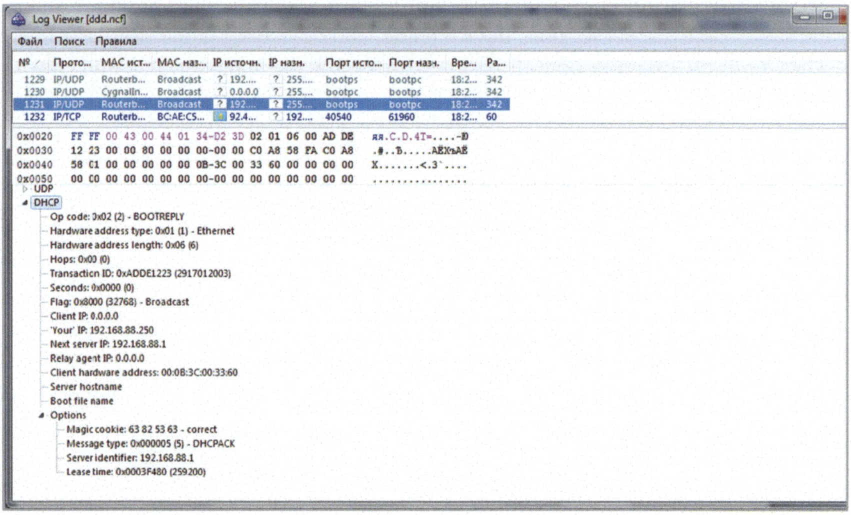

Fig. 52 View of the DHCPACK packet from the server

The library has been adapted to support the C51 compiler (Keil) and data transfer via the uIP stack. The choice is because libemqtt was created in the C language for embedded systems and is open source. To adapt this library to the uIP stack, a library was created, located in the files mqtt-client.h and mqtt-client.c.

The mqtt-client.h file contains constants that define the current state of data transfer using the MQTT protocol, default values for configuration, function prototypes, and a data structure that defines the general parameters of the current connection to the server (Fig. 53).

The mqtt-client.c file defines global constants used for data transfer using the MQTT protocol, such as the server address and port, the device name, the current

Fig. 53 The mqtt_state structure

```
typedef struct {
 u8_t state;
 u8_t *dataptr;
 u8_t *topicptr;
 u16_t *message_id;
 u8_t newdata;
 struct uip_conn *conn;
 u8_t connected;
} xdata mqtt_state;
```

transfer status, and the broker identifier of the type mqtt_broker_handle_t, which is necessary for saving the connection parameters and configuring the libemqtt library (Fig. 54).

To set settings from Flash memory, setter functions were created to set the values of the server address, server port, device name, and device set name (Fig. 55).

The init_mqtt function first allocates memory for the state variable and tries to make a TCP connection with the MQTT server. If the connection was successful,

```
mqtt_state *state;
struct mqtt_broker_handle_t *broker;
u8_t mqtt_topic_buffer[MQTT_TOPIC_BUFFER_SIZE];
u16_t mqtt_server_ipaddr[2];
u16_t mqtt_server_port;
u8_t *mqtt_device_name;
u8_t *mqtt_device_set_name;
u8_t mqtt_topic_name_buffer[MQTT_TOPIC_NAME];
```

Fig. 54 Constants for organizing data transfer using the MQTT protocol

```
void mqtt_set_server_ip(u16_t *srv_ip) {
 mqtt_server_ipaddr[0] = srv_ip[0];
 mqtt_server_ipaddr[1] = srv_ip[1];
}

void mqtt_set_server_port(u16_t server_port) {
 mqtt_server_port = server_port;
}

void mqtt_set_device_name(u8_t *device_name) {
 mqtt_device_name = (u8_t *)malloc(strlen(device_name) + 2);
 strcpy(mqtt_device_name, device_name);
}
void mqtt_set_device_set_name(u8_t *device_set_name)   {
 mqtt_device_set_name = (u8_t *)malloc(strlen(device_set_name) + 2);
 strcpy(mqtt_device_set_name, device_set_name);
}
```

Fig. 55 Functions for network settings

memory is allocated under the broker variable. Next, the value of the device name is formed by writing the name of the device settings and the name of the device with a hyphen. Then the broker variable is initialized and the initial connection state values are set (Fig. 56).

The client software implementation is located in the mqtt_send function. In it, the pointer to the data received in the uIP is assigned to the appdata pointer. Next, the need to resend data is checked by calling the uip_rexmit macro. Depending on the current state of the exchange session, a corresponding packet is sent (Fig. 57).

```
u8_t init_mqtt(){
 state = (mqtt_state*)malloc(sizeof(mqtt_state));
 state->conn = uip_connect(mqtt_server_ipaddr, HTONS(mqtt_server_port));

 if(state->conn == NULL) {
       return NULL;
 }

 broker = (mqtt_broker_handle_t*)malloc(sizeof(mqtt_broker_handle_t));
 sprintf(mqtt_topic_name_buffer, "%s-%s", mqtt_device_set_name,
                                                          mqtt_device_name);
 mqtt_init(broker, mqtt_topic_name_buffer);
 mqtt_set_alive(broker,20);
 state->state = MQTT_STATE_WAITING;
 state->message_id = 0;
 state->newdata = FALSE;
 state->connected = MQTT_NOT_CONN;
 return TRUE;
}
```

Fig. 56 The init_mqtt function

```
void mqtt_send(){
 u8_t *appdata;
 u16_t appdataSize;
 u8_t received_type;
 appdata = uip_appdata;

 if(uip_rexmit()) {

        if(state->state == MQTT_STATE_DATA_SENT || state->state ==
                                                  MQTT_STATE_ACK_RECEIVED) {
              mqtt_send_message();
              return;
        } else if(state->state == MQTT_STATE_PING_SENT) {
              mqtt_send_ping();
              return;
        } else if(state->connected == MQTT_CONN_PROCESS) {
              mqtt_connect(broker);
              return;
        }
 }
```

Fig. 57 Sending a packet from an MQTT client

```
if(uip_acked()) {
      if(state->state == MQTT_STATE_PING_SENT) {
            state->state = MQTT_STATE_WAITING;
      } else if(state->state == MQTT_STATE_DATA_SENT) {
            state->state = MQTT_STATE_ACK_RECEIVED;
      }
}
```

Fig. 58 MQTT client state change

If the received message is a confirmation of package delivery, we change the current state of the client. This type of message is checked using the uip_acked macro (Fig. 58).

Sending messages occur when uIP polls the client for new packets. This is checked using the uip_poll macro. Depending on the state of the client, a message may be sent. If there is no connection to the MQTT server, a connection packet is sent. If the connection was made, the previous message was delivered successfully and the new message is added to the queue for sending—the message that is in the state variable is sent. If the connection has taken place and the ping request has been added to the queue for sending messages, we send the ping request (Fig. 59).

Checking for the presence of received data is performed using the uip_newdata macro. Further, with the help of the libemqtt library, the received packet is processed and the MQTT type of the message is determined. If confirmation of sending the previous message is received—we change the client's status. If the connection has taken place, we write the first part of the topic (the name of the set of devices and the name of the device through "/") in the appropriate buffer and change the state of the client's connection (Fig. 60).

```
if(uip_poll()) {
      if(state->connected == MQTT_NOT_CONN) {
            if(mqtt_connect(broker) == 1) {
                  state->connected = MQTT_CONN_PROCESS;
            }
      }

      if(state->connected == MQTT_CONNECTED) {
            if((state->state == MQTT_STATE_PUBACK_RECEIVED ||
                  state->state == MQTT_STATE_WAITING) && state->newdata) {
                  mqtt_send_message();
            }

            if(state->state == MQTT_STATE_PING_REQ) {
                  mqtt_send_ping();
            }
      }
}
```

Fig. 59 Sending messages

```
if(uip_newdata()) {
        received_type = MQTTParseMessageType(appdata);
                case MQTT_MSG_PUBACK:
                        state->state = MQTT_STATE_PUBACK_RECEIVED;
                        break;
                case MQTT_MSG_CONNACK:
                        state->connected = MQTT_CONNECTED;
                        sprintf(mqtt_topic_name_buffer, "%s/%s",
                                        mqtt_device_set_name, mqtt_device_name);
                        break;
                default:
                        break;
        }
}

return;
}
```

Fig. 60 Processing of the received package

```
void mqtt_send_message() {
 int result;
 sprintf(mqtt_topic_buffer, "%s/%s", mqtt_topic_name_buffer,
                                                   state->topicptr);
 result = mqtt_publish_with_qos(broker, mqtt_topic_buffer,
            state->dataptr, MQTT_RETAIN, MQTT_QOS, state->message_id);
 if(result == 1) {
        state->state = MQTT_STATE_DATA_SENT;
 }
}
```

Fig. 61 The mqtt_send_message function

The mqtt_send_message function used in mqtt_send first writes the full topic with the specified measurement channel to a variable and sends the packet using the corresponding libemqtt library function. If the package was sent successfully, the client's status changes (Fig. 61).

The mqtt_send_ping function is similar to the mqtt_send_message function, only it sends a ping request to the server (Fig. 62).

The mqtt_quene_ping function checks whether the current state can send a ping request, and if it is possible, it changes the state of the client to send a corresponding request at the next uIP poll (Fig. 63).

Fig. 62 The mqtt_send_ping function

```
void mqtt_send_ping() {
 int result = mqtt_ping(broker);
 if(result == 1) {
        state->state = MQTT_STATE_PING_SENT;
 }
}
```

```
u8_t mqtt_quene_ping() {
    if((state->state == MQTT_STATE_PUBACK_RECEIVED || state->state ==
MQTT_STATE_WAITING)) {
        state->state = MQTT_STATE_PING_REQ;
        return TRUE;
    }
    return FALSE;
}
```

Fig. 63 The mqtt_quene_ping function

```
u8_t mqtt_quene_for_sending(u8_t *dataptr, u8_t *topicptr) {
    if(state->state == MQTT_STATE_PUBACK_RECEIVED || state->state ==
MQTT_STATE_WAITING) {
        state->dataptr = dataptr;
        state->topicptr = topicptr;
        state->newdata = TRUE;
        return TRUE;
    }

    if(state->state == MQTT_STATE_ACK_RECEIVED) {
        state->newdata = TRUE;
    }
    return FALSE;
}
```

Fig. 64 Function mqtt_quene_for_sending

The function mqtt_quene_for_sending checks whether it is possible to send a message with measurement data in the current state and records the corresponding changes to the state of the client to send the message at the next uIP poll (Fig. 64).

After starting the microcontroller, initializing it, and initializing the uIP, the init_mqtt function is called. In addition, a mqtt_send call was added to the uip_appcall function, which is periodically called by the uIP stack to run client code, when receiving data from the corresponding port (Fig. 65).

Thus, after the init_mqtt call and the first mqtt_send call, the device establishes an MQTT connection to the server. The connection message packet is shown in Fig. 66.

After connecting, the server sends a connection confirmation message, after which the client's status changes to "connected". The connection confirmation message is shown in Fig. 67.

Receiving data from the microcontroller's ADC is performed using an interrupt from timer 3. The received data is converted into the corresponding physical value and recorded in the temperature and illuminance variables. After measuring the temperature, the ADC switches to another port to measure the illumination by setting the appropriate value in the AMX0P register (Fig. 68).

During preinterruption processing, temperature and light values are alternately sent from timer 0 every 50 ms.

After the next call to mqtt_send, the message is sent using uIP. An example of a message packet with measurement data is shown in Fig. 69.

```
void uip_appcall(void)
{

 switch(uip_conn->lport)
 {
        case HTONS(IANAportwww_http):
        case HTONS(IANAporthttp_altC):
               httpd_appcall();
        break;
        default:
               switch(uip_conn->rport)
               {
                      case 1883:
                             mqtt_send();
                             goto uip_appcall_end;
                      default:
                             uip_abort();
                             break;
               }
               uip_abort();
               break;

 }
 uip_appcall_end: ;
}
```

Fig. 65 Function uip_appcall

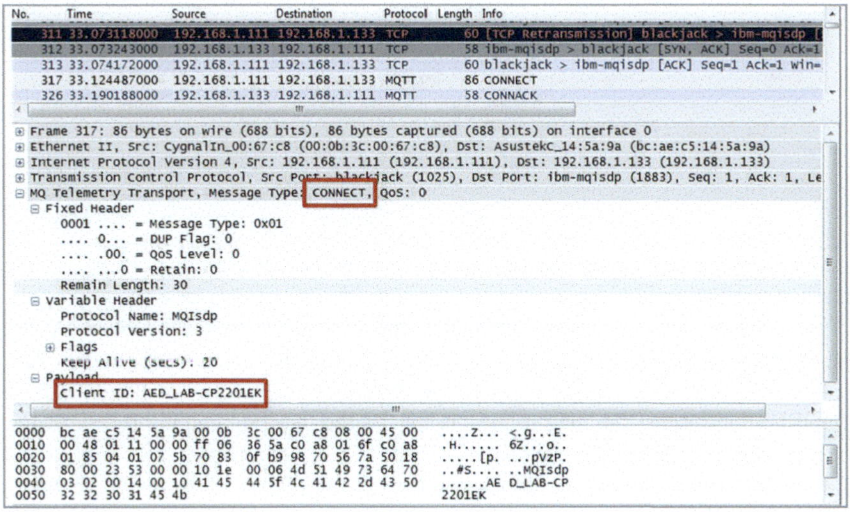

Fig. 66 Server connection message (message type—connect)

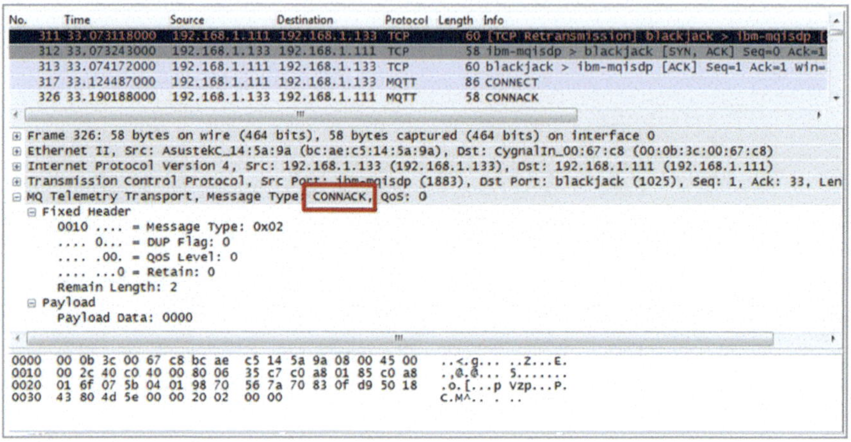

Fig. 67 Connection confirmation message

Fig. 68 Function ADC_int

```
void ADC_int() interrupt ADOINT_VECTOR using 1
{
 if(AMXOP == 0x1E) {
       if(ADCO > 1) {
             temperature = ADCO*8.536 - 2710
       }
       AMXOP = 0x02;
 } else {
if(ADCO > 1) {
 illuminance = ADCO*18.5;
}
       AMXOP = 0x1E;
 }
 adcData = ADCO;
 ADOINT = 0;
}
```

The server's response with information about the successful delivery of the message is given in Fig. 70.

The MQTT protocol uses PINGREQ (ping request) and PINGRESP (ping response) messages to check the connection. To maintain the connection, the client sends a PINGREQ message every 10 s, to which the server responds with a PINGRESP message. The ping request is sent by the interrupt handler from timer 0. An example of a ping request is shown in Fig. 71.

The ping response from the server is shown in Fig. 72.

The WMQTT utility from the IA-92 package from the IVM company was used to check data transfer using the MQTT protocol. An example of the obtained temperature value is shown in Fig. 73.

An example of the obtained illuminance value is shown in Fig. 74.

The log of MQTT messages from the mosquito server is shown in Fig. 75.

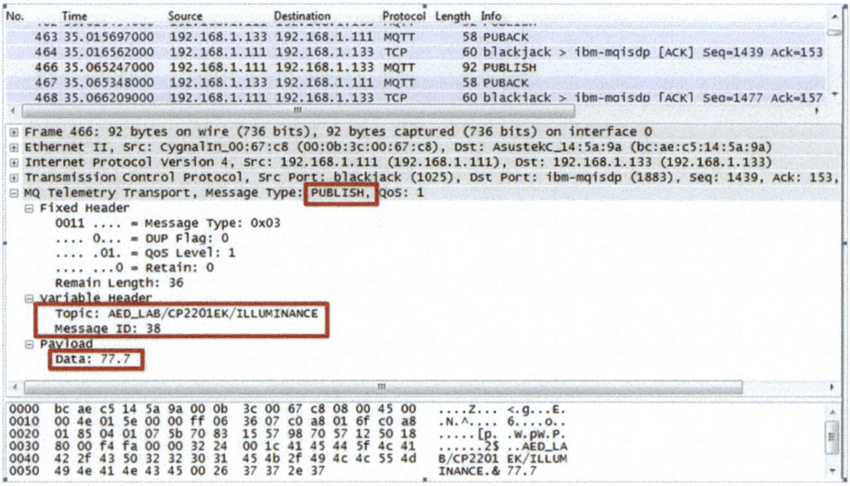

Fig. 69 Example of a message packet with measurement data

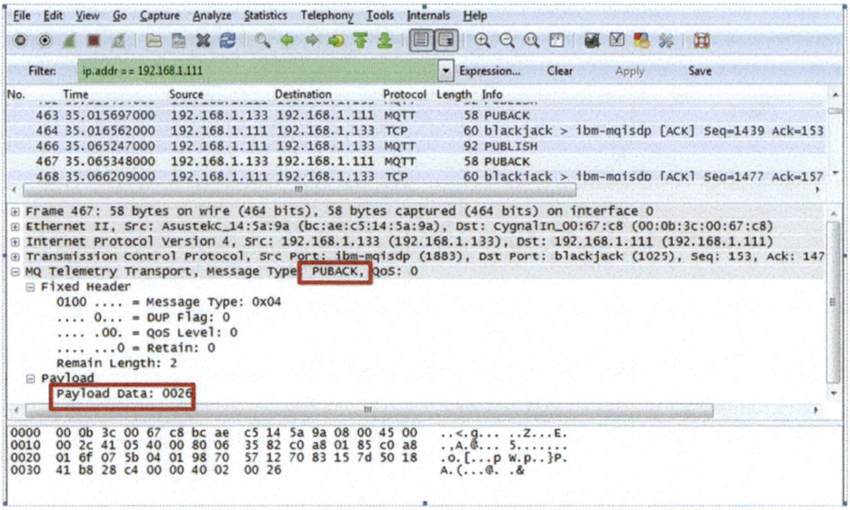

Fig. 70 Server response with information about successful message delivery

In the magazine in Fig. 75, you can clearly see the alternate transmission of temperature and illumination values, as well as receiving a ping request and sending a ping response.

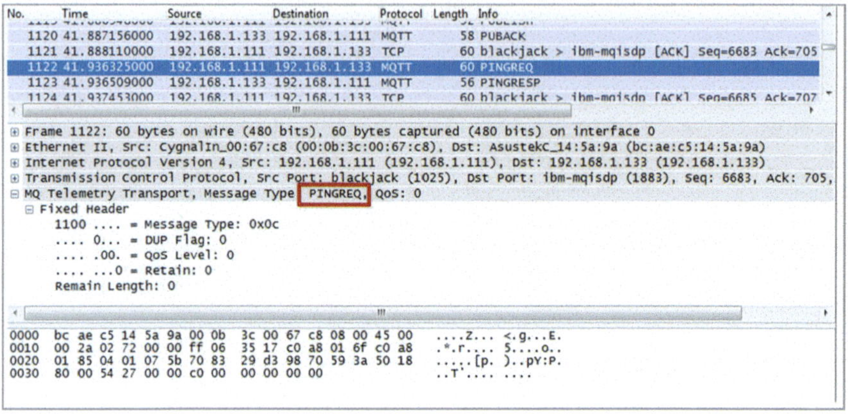

Fig. 71 An example of a ping request

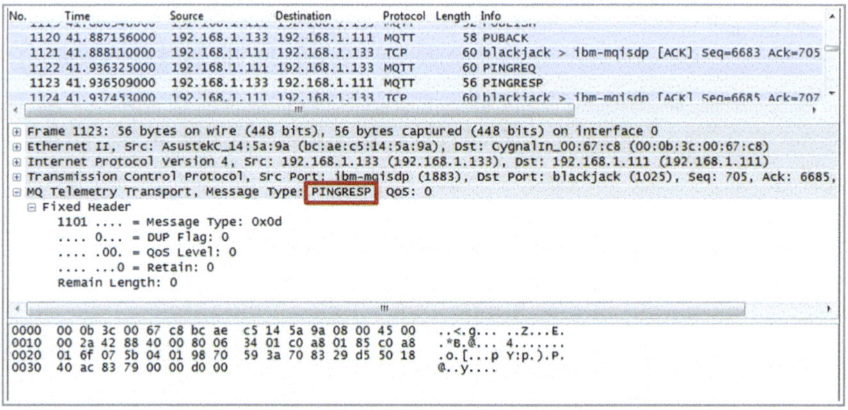

Fig. 72 Ping response from the server

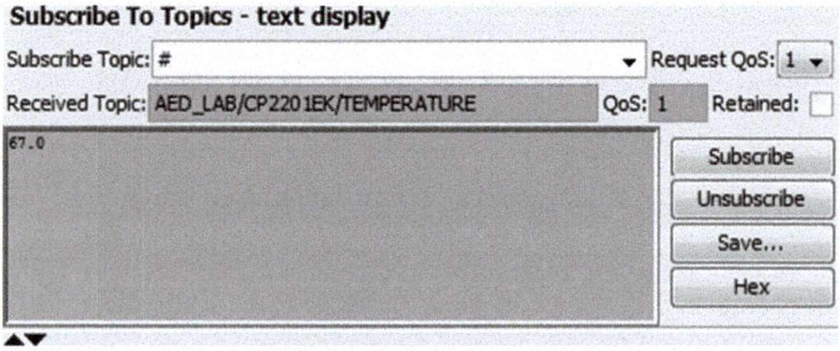

Fig. 73 Example of temperature value transfer

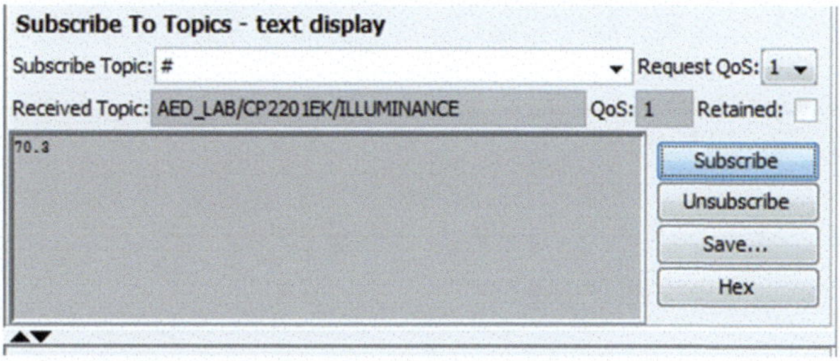

Fig. 74 An example of the transmission of the illumination value

```
1397424040: Received PUBLISH from AED_LAB-CP2201 (d0, q1, r0, m379, 'AED_LAB/CP2201/ILLUMINANCE', ... (5 bytes))
1397424040: Sending PUBACK to AED_LAB-CP2201 (Mid: 379)
1397424040: Received PUBLISH from AED_LAB-CP2201 (d0, q1, r0, m380, 'AED_LAB/CP2201/TEMPERATURE', ... (4 bytes))
1397424040: Sending PUBACK to AED_LAB-CP2201 (Mid: 380)
1397424040: Received PUBLISH from AED_LAB-CP2201 (d0, q1, r0, m381, 'AED_LAB/CP2201/ILLUMINANCE', ... (5 bytes))
1397424040: Sending PUBACK to AED_LAB-CP2201 (Mid: 381)
1397424040: Received PUBLISH from AED_LAB-CP2201 (d0, q1, r0, m382, 'AED_LAB/CP2201/TEMPERATURE', ... (4 bytes))
1397424040: Sending PUBACK to AED_LAB-CP2201 (Mid: 382)
1397424040: Received PINGREQ from AED_LAB-CP2201
1397424040: Sending PINGRESP to AED_LAB-CP2201
1397424040: Received PUBLISH from AED_LAB-CP2201 (d0, q1, r0, m383, 'AED_LAB/CP2201/TEMPERATURE', ... (4 bytes))
1397424040: Sending PUBACK to AED_LAB-CP2201 (Mid: 383)
1397424040: Received PUBLISH from AED_LAB-CP2201 (d0, q1, r0, m384, 'AED_LAB/CP2201/ILLUMINANCE', ... (5 bytes))
1397424040: Sending PUBACK to AED_LAB-CP2201 (Mid: 384)
1397424040: Received PUBLISH from AED_LAB-CP2201 (d0, q1, r0, m385, 'AED_LAB/CP2201/TEMPERATURE', ... (4 bytes))
1397424040: Sending PUBACK to AED_LAB-CP2201 (Mid: 385)
1397424040: Received PUBLISH from AED_LAB-CP2201 (d0, q1, r0, m386, 'AED_LAB/CP2201/ILLUMINANCE', ... (5 bytes))
1397424040: Sending PUBACK to AED_LAB-CP2201 (Mid: 386)
1397424040: Received PUBLISH from AED_LAB-CP2201 (d0, q1, r0, m387, 'AED_LAB/CP2201/TEMPERATURE', ... (4 bytes))
1397424040: Sending PUBACK to AED_LAB-CP2201 (Mid: 387)
1397424040: Received PUBLISH from AED_LAB-CP2201 (d0, q1, r0, m388, 'AED_LAB/CP2201/ILLUMINANCE', ... (5 bytes))
1397424040: Sending PUBACK to AED_LAB-CP2201 (Mid: 388)
1397424040: Received PUBLISH from AED_LAB-CP2201 (d0, q1, r0, m389, 'AED_LAB/CP2201/TEMPERATURE', ... (4 bytes))
1397424040: Sending PUBACK to AED_LAB-CP2201 (Mid: 389)
1397424040: Received PUBLISH from AED_LAB-CP2201 (d0, q1, r0, m390, 'AED_LAB/CP2201/ILLUMINANCE', ... (5 bytes))
1397424040: Sending PUBACK to AED_LAB-CP2201 (Mid: 390)
1397424041: Received PUBLISH from AED_LAB-CP2201 (d0, q1, r0, m391, 'AED_LAB/CP2201/TEMPERATURE', ... (4 bytes))
1397424041: Sending PUBACK to AED_LAB-CP2201 (Mid: 391)
```

Fig. 75 MQTT message transmission log from the mosquito server

2.6 Development of Software for Remote Configuration of Intelligent Sensors

The httpd library, which is part of the uIP stack, was used to develop the configuration web page of the device. This library has been supplemented with functions for working with requests of the POST method, as well as CGI functions for outputting dynamic data.

The CGI function for outputting the TCP/IP parameters edit form outputs parts of the form sequentially as HTML with the values entered using the sprintf function added. After a certain block of fields is output and delivery confirmation is received, the i variable points to the next HTML block (Fig. 76).

Implementation of fields for entering the IP address (Fig. 77).

Implementation of fields for entering the primary DNS address (Fig. 78).

```
static u8_t LAN_settings(u8_t next){
 static int i = 0;

 if (next || (i == 0)) {
      switch (i++) {
```

Fig. 76 Generation of the configuration page

```
     case 1:
          uip_send(uip_appdata, sprintf((char *)uip_appdata,
"<Table width=\"500\"><TR><td width=\"200\" valign=\"middle\" "
"align=\"left\">IP Address:</td>"
"<td><input size=\"3\" name=\"T20\" value=\"%i\" /><B>.</B>"
"<input size=\"3\" name=\"T21\" value=\"%i\" /><B>.</B>"
"<input size=\"3\" name=\"T22\" value=\"%i\" /><B>.</B>"
"<input size=\"3\" name=\"T23\" value=\"%i\" /></td></tr></table>",
                    (htons(uip_hostaddr[0]) >> 8),
                    (htons(uip_hostaddr[0]) & 0xff),
                    (htons(uip_hostaddr[1]) >> 8),
                    (htons(uip_hostaddr[1]) & 0xff)));
     break;
```

Fig. 77 Fields for entering the IP address

```
     case 4:
          uip_send(uip_appdata, sprintf((char *)uip_appdata,
"<Table width=\"500\"><TR><td width=\"200\" "
"valign=\"middle\" align=\"left\">Primary DNS IP Address:</td>"
"<td><input size=\"3\" name=\"T50\" value=\"%i\" /><B>.</B>"
"<input size=\"3\" name=\"T51\" value=\"%i\" /><B>.</B>"
"<input size=\"3\" name=\"T52\" value=\"%i\" /><B>.</B>"
"<input size=\"3\" name=\"T53\" value=\"%i\" /></td></tr></table>",
                    (htons(tmp_DNS_params.DNS_PrimaryNum.i[0]) >> 8),
                    (htons(tmp_DNS_params.DNS_PrimaryNum.i[0]) & 0xff),
                    (htons(tmp_DNS_params.DNS_PrimaryNum.i[1]) >> 8),
                    (htons(tmp_DNS_params.DNS_PrimaryNum.i[1]) & 0xff)
                    ));
     break;
```

Fig. 78 Fields for entering the Primary DNS address

Implementation of fields for entering the IP address and port of the MQTT server (Fig. 79).

Implementation of fields for entering the name of a set of devices and a device for data transfer using the MQTT protocol (Fig. 80).

The HTML page for the remote configuration of the intelligent sensor is shown in Fig. 81.

```
case 7:
       uip_send(uip_appdata, sprintf((char *)uip_appdata,
"<b><font color=\"#0000FF\" size=\"4\">MQTT Parameters</font></b>"
"<Table width=\"600\"><TR><td width=\"200\" valign=\"middle\"
                              align=\"left\">Server IP Address / Port:</td>"
"<td><input size=\"3\" name=\"T80\" value=\"%i\" /><B>.</B>"
"<input size=\"3\" name=\"T81\" value=\"%i\" /><B>.</B>"
"<input size=\"3\" name=\"T82\" value=\"%i\" /><B>.</B>"
"<input size=\"3\" name=\"T83\" value=\"%i\" /><B>:</B>"
"<input size=\"5\" name=\"T84\" value=\"%i\" /></td></tr></table>",
               (htons(tmp_mqtt_params.server_ip.i[0]) >> 8),
               (htons(tmp_mqtt_params.server_ip.i[0]) & 0xff),
               (htons(tmp_mqtt_params.server_ip.i[1]) >> 8),
               (htons(tmp_mqtt_params.server_ip.i[1]) & 0xff),
                htons(tmp_mqtt_params.seprver_port)
               ));
       break;
```

Fig. 79 Fields for entering the IP address and port of the MQTT server

```
case 8:     // MQTT Parameters 2
       uip_send(uip_appdata, sprintf((char *)uip_appdata,
"<Table width=\"625\"><TR><td width=\"266\" valign=\"middle\"
                                   align=\"left\">Device Set Name:</td>"
"<TD><input type=\"text\" size=\"17\" maxlength=\"16\" name=\"T85\"
                                   value=%s></TD></TR>\r\n"
"<TR><td width=\"266\" valign=\"middle\" align=\"left\">Device
                                   Name:</th>"
"<TD><input type=\"text\" size=\"17\" maxlength=\"16\" name=\"T86\"
                                   value=%s></TD></TR></Table>\r\n",
       tmp_mqtt_params.device_set_name, tmp_mqtt_params.device_name));
       break;
       default:
              i = 0;
              return 1;
       break;
       }
       return 0;
 }
 return 0;
}
```

Fig. 80 Fields for entering device names for the MQTT protocol

Data is saved using the process_LAN_form function, which determines its type by the first character of the HTML field name and performs the corresponding function (Fig. 82).

The process_LAN_button function performs the functions of cleaning, resetting, and updating the Flash memory. Other functions make changes to temporary variables, which will then be written to Flash memory using the FLASH_Update function. Shown is the part of the process_LAN_token function that makes changes to the MQTT server settings (Fig. 83).

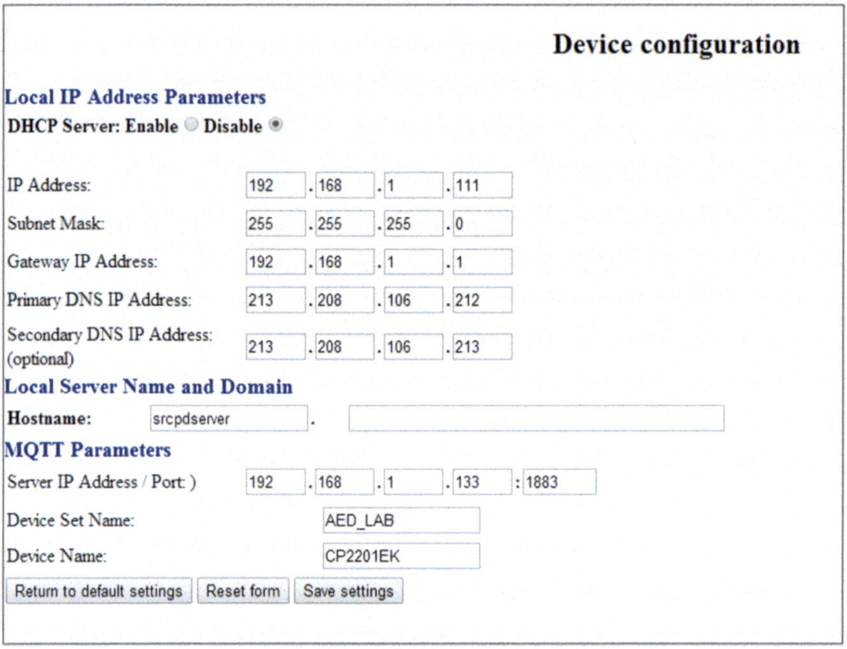

Fig. 81 Smart sensor configuration page

```
int process_LAN_form(struct formParams *post) {
 switch (post->postType)  {
       case ISO_B:
              return process_LAN_button(post);
       break;
       case ISO_C:
              return process_LAN_checkbox(post);
       break;
       case ISO_R:
              return process_LAN_radiobutton(post);
       break;
       case ISO_T:
              return process_LAN_token(post);
       break;


       default :
              return FALSE;
       break;
 }
}
```

Fig. 82 Function process_LAN_form

```
int process_LAN_token(struct formParams *post) {
 int i;
 switch (post->postNum) // What is the button number?
 {
       case 80:    // MQTT Server byte 0
             tmp_mqtt_params.server_ip.b[0] = (u8_t) atoi(post->postptr);
             return TRUE;
       break;
       case 81:    // MQTT Server byte 0
             tmp_mqtt_params.server_ip.b[1] = (u8_t) atoi(post->postptr);
             return TRUE;
       break;
       case 82:    // MQTT Server byte 0
             tmp_mqtt_params.server_ip.b[2] = (u8_t) atoi(post->postptr);
             return TRUE;
       break;
       case 83:    // MQTT Server byte 0
             tmp_mqtt_params.server_ip.b[3] = (u8_t) atoi(post->postptr);
             return TRUE;
       break;
       case 84:    // MQTT Server port
             tmp_mqtt_params.seprver_port = (u16_t) atoi(post->postptr);
             return TRUE;
       break;
       default :
             return FALSE;
       break;
 }
}
```

Fig. 83 Process_LAN_token function

Its type is determined by the name of the field and its value is entered in the corresponding variable. Thus, after processing the information of all fields, the values in the memory will be updated and in the process of processing the button click event will be saved on the server.

3 Development of Components of Network Microcontroller Data Collection Systems Based on RESTful Web Services

3.1 Peculiarities of Using RESTful Web Services in Network Data Collection Systems

Web services are distributed application components accessible over a network. They are used to integrate computer programs written in different programming languages that run on different platforms. Web services do not depend on the programming language and the platform on which the program is executed, as there is an agreement on general web service standards [27]. Web service developers use several programming models. These models can be divided into the following two categories.

Model based on SOAP/WSDL. In standard models, the web service itself and its interfaces are provided using WSDL (XML document type) documents. Messaging is done via SOAP, another type of XML document [28].

Model based on REST. Representational State Transfer is a fairly new way of creating and interacting with web services. When using REST, each resource has a URI and is managed through operations with various HTTP headers [29].

On the server side that provides a SOAP-based web service, utilities create a WSDL file that is accessible over the network. Clients using the web service make a programmatic request based on the information in the WSDL. Messages are exchanged in the SOAP format. The set of operations that can be performed based on SOAP is much wider than the capabilities of REST technology, especially in the area of security [30].

SOAP-based web services are used both for large applications that use complex operations and for applications that require special security, reliability, or other standards-supported capabilities. They are also used when a transport protocol other than HTTP must be used. Many of Amazon's web services, particularly those related to commercial transactions and web services used by banks and government agencies, are based on the SOAP model [31–33].

REST-based ("RESTful") web services are a collection of web resources identified by their Uniform Resource Identifier (URI). Each document and each process appears as a web resource with a unique URI. These web resources can be controlled using the actions defined in the HTTP header. SOAP and WSDL standards are not used, instead messages can be exchanged in any format—XML, JSON, HTML, etc. In many cases, the client can be a standard browser.

The data transfer protocol in REST is HTTP. Its four methods are available: GET, PUT, POST, and DELETE. Queries can be tabbed and responses can be cached. A network administrator can monitor the performance of a RESTful service by viewing the HTTP headers.

REST is a technology for creating those applications that do not require special security technologies other than those available in the HTTP infrastructure, and which are implemented in the HTTP protocol. REST services implement fairly complex functionality while using a narrower data transfer channel. RESTful web services are used by companies such as Flickr, Google Maps, and Amazon.

The REST-style architecture consists of clients and servers. Clients initiate requests to servers, servers process requests, and return the necessary responses. Requests and responses are created based on resource URIs.

The REST exchange model is described in the context of HTTP but is not limited to this protocol. RESTful architectures can be based on other application-level protocols if they already implement a sufficiently large and unified vocabulary for stateful applications. RESTful applications extend the use of existing interfaces and other built-in capabilities provided by the chosen network protocol and make it easy to add new application-specific capabilities to it.

The REST architectural style defines the following constraints on the architecture, leaving the implementation of individual components free.

Client–server technology. Clients are separated from the server by a single interface. This separation of responsibilities means, for example, that clients are not responsible for the data store that is internal to each server, resulting in an easier process of migrating client code. Servers are not responsible for user interface or user state, so servers can be simpler and more scalable. Servers and clients can be developed and replaced independently as long as the exchange interface does not change.

Lack of status. Client–server interaction is characterized by the lack of saving the client session on the server side between requests. Every request from any client contains all the information needed to service it, and any session state is stored client-side.

Caching. Clients can cache responses. Therefore, the response must explicitly or implicitly indicate whether it can be cached to prevent the client from using old or inappropriate data when sending subsequent requests. Well-managed caching partially or eliminates some of the client–server interactions, increasing scalability and performance.

Multi-level system. The client cannot clearly determine whether it is connecting directly to the server or an intermediary on the connection path. A server broker can improve system scalability by providing load balancing and providing a shared cache. An intermediary may also be a requirement to comply with security policies.

Code on demand (optional). Servers can temporarily extend or customize the functionality of the client by passing the logic to be executed. Examples of this include compiled components such as Java applets and client-side scripts such as JavaScript.

The only additional limitation of the REST architecture is code on demand. If a service violates any other constraints, it cannot be unambiguously called RESTful.

Adhering to these constraints, and therefore conforming to the REST architectural style, will allow a distributed system to have the required properties, such as performance, scalability, simplicity, modifiability, visibility, mobility, and reliability.

The key goals of REST include scalability of component interactions; community of interfaces; independent implementation of components; middleware components that reduce latency, strengthen security, and encapsulate legacy systems.

An important concept in REST is the availability of resources (sources of specific information), each of which is defined by a link with a global identifier. To manipulate these resources, network components exchange a set of commands through a standardized interface (for example, HTTP) and a representation of these resources (actual information). Web service resources are represented by a URL string.

RESTful web service (another name is RESTful web API) is a web service implemented using HTTP and REST principles. It is a set of resources with three specific aspects. The base URL for the web service, for example, http://example.com/resources/. The content type for the data is supported by the web service. This is often JSON, XML, or YAML, but any other valid content type for the web can be used. A set of operations supported by the web service using the main mechanisms of the protocol (POST, GET, PUT, or DELETE).

URL is an abbreviation for Uniform Resource Locator. According to this name, the URL should serve as a reference to the resource. Moreover, all URL addresses are not only a link to a resource, but also URI identifiers, or Uniform Resource Identifier.

No two resources can be hosted from the same URL, so the URL can be considered a means of identifying the resource. Oftentimes, URLs don't point to anything or identify anything—they point to requirements. Instead of identifying a resource, they indicate the performance of some action.

For example, an example URL processed by the displaySpittle() method of the DisplaySpittleController is http://localhost:8080/Spitter/displaySpittle.htm?id=123.

The example URL does not refer to or identify any resource. It requires the server to render a Spittle object. The only part of the URL that identifies anything is the id query parameter. The basic part of this URL is the action that expresses the request. This indicates that this URL is against the REST architecture.

In contrast to URLs that conflict with the REST architecture, URLs that conform to it demonstrate that the HTTP protocol is intended specifically for transferring resources.

The following shows what the previous URL might look like after making it conform to the REST architecture—http://localhost:8080/Spitter/spittles/123.

This example shows how a URL can refer to a resource. Specifically, it refers to a resource that represents a Spittle object. Actions that will be performed on this request depending on the type of HTTP request.

The URL from the example not only refers to the resource but also uniquely identifies it. It is equal parts URL and URI. The URL itself is used to fully identify the resource, not the request parameters. Although parameters are still considered a valid way to pass information to the server, they are only instructions for the server to help it render the resource. Query parameters should not be used to identify a resource.

RESTful URLs have a hierarchical organization. When reading from left to right, more general concepts are replaced by more specialized ones. In this example, the URL contains several levels, each of which defines a resource:

- http://localhost:8080—defines the domain name and port;
- http://localhost:8080/Spitter—defines the root address of the web service (this URL is more specialized—it defines the application running on the server);
- http://localhost:8080/Spitter/spittles/123—the most specialized URL that defines a specific Spittle resource.

The most important feature of RESTful URLs is that their paths are parameterized. RESTful URLs carry input data as request parameters, while RESTful URLs carry the same data inside the URL path.

The application can interact with resources, having information only about the ID of the resource and the required action. It does not need information about whether there is a cache, proxy, gateway, firewall, tunnel, or anything else between it and the server. The application, however, must have information about what data format is being returned. This is typically an HTML, XML, or JSON document, although it can be an image, text, or any other content [27].

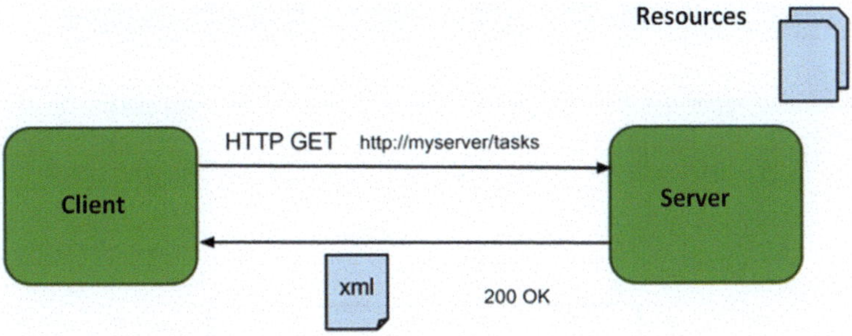

Fig. 84 Data exchange process when using a RESTful web service

The data exchange process when using a RESTful web service is shown in Fig. 84.

A key role in a RESTful web service is played by HTTP methods that represent the operations that the service can perform. HTTP Method (English HTTP Method) is a sequence of any characters that indicates the main operation on the resource. Usually, the method is a short English word written in capital letters [28].

The GET method is used to request the content of the specified resource. GET can also be used to start any process. In this case, information about the progress of the process is included in the response body.

The client may pass request execution parameters in the URI of the target resource after the "?" character:

GET /path/resource?Param1 = value1&param2 = value2 HTTP/1.1

The POST method is used to transfer data from the user to a given resource. For example, in blogs, visitors can usually enter their comments to entries in an HTML form, after which they are transmitted to the server by the POST method and it places them on the page. In this case, the transferred data is included in the body of the request.

If the result of the request has a code of 200 (Ok), a message about the result of the request should be included in the body of the response. If a resource was created, the server should return a 201 (Created) response with the URI of the new resource in the Location header.

Message—the server's response to the execution of the POST method is not cached.

The PUT method is used to download the content of the resource to the URI specified in the request. If the resource did not exist for the specified URI, the server creates it and returns the status 201 (Created). If the resource has been changed, the server returns 200 (Ok) or 204 (No Content). The server MUST NOT ignore invalid headers of the Content-* data type sent by the client along with the message. If any of these headers cannot be recognized or are invalid under the current conditions, then a 501 (Not Implemented) error code must be returned. The fundamental difference between the POST and PUT methods lies in the understanding of resource URI assignments. The POST method assumes that the specified URI will process the content submitted by the client. Using PUT, the client assumes that the downloaded

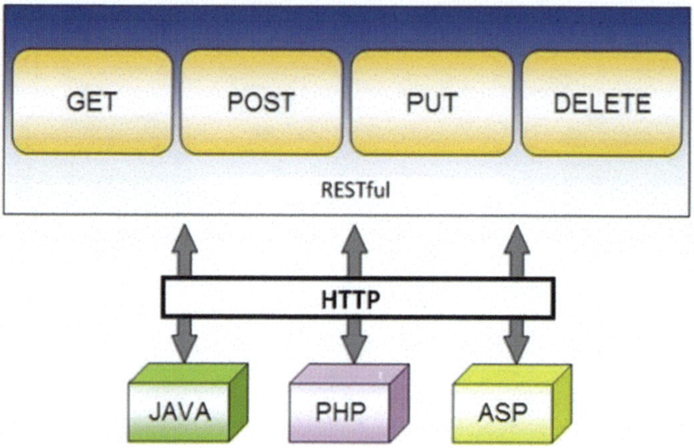

Fig. 85 Exchange of customer data with a RESTful web service

content is already located at this resource URI. The DELETE method deletes the specified resource.

Customer data exchange with a RESTful web service is presented in Fig. 85.

3.2 Development of Software for Supporting RESTful Web Services Based on Java Microservers

The developed implementation of the STIM and NCAP modules of intelligent sensors according to the IEEE 1451 standard (Fig. 86) is based on the use of web services technology for both the STIM module and the NCAP module.

This approach is due to the wide popularity of web services in general and the spread of RESTful architectures in particular. The REST architecture has fairly

Fig. 86 Appearance of the demonstration data collection system

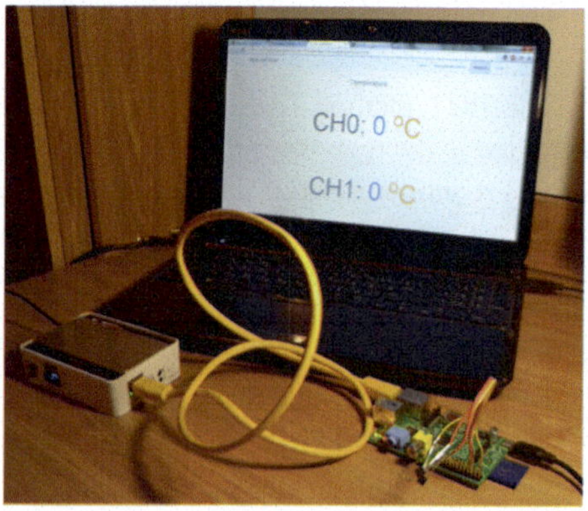

simple implementation requirements, which gives it advantages over SOAP. This helps speed up software development. Support for the REST specification is available in many programming languages. In Java, there are several approaches to implementing a RESTful architecture, one of which is based on the JAX-RS API. This API, in turn, has several implementations. Among all implementations of this API, the Jersey framework is considered the most popular.

The Jersey framework has such advantages as ease of use; ease of configuration; availability of the implementation of the component life cycle manager and DI container; easy integration with third-party implementations of DI containers.

The structure of the developed data collection system according to the IEEE 1451 standard based on REST web services from lower-level devices to the user level is shown in Fig. 87.

As an interface for data exchange between STIM and NCAP modules, the Ethernet physical level standard was chosen, and as a data transmission medium—twisted pair. Since Raspberry Pi and Cubieboard have built-in support for the Ethernet interface, the configuration for data exchange between TIM and NCAP via Ethernet is reduced

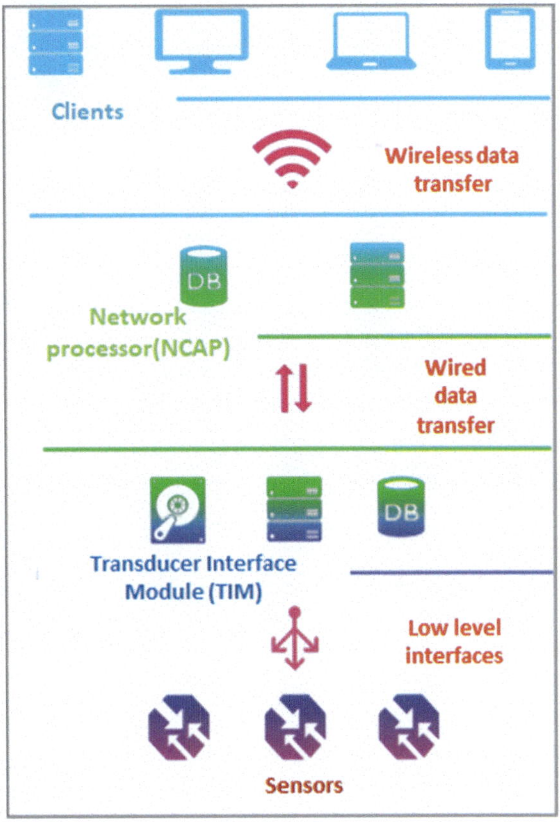

Fig. 87 Structure of implementation of IEEE-1451 based REST web services

to the corresponding configurations of the interface file located in the /etc./network directory of each module.

3.3 Development of RESTful Web Service Support Software for the STIM Module According to the IEEE 1451 Standard Based on a Raspberry Pi Single-Board Computer

This development assumes that electronic documentation of sensors will be stored in the file system in the form of XML files. Meta TEDS (electronic documentation for the TIM module) must be present in one instance for each TIM. The structure of the implemented XML Meta TEDS for Raspberry Pi is shown in Fig. 88.

Example of Meta TEDS for Raspberry Pi (Fig. 89).

Fig. 88 XML structure of meta TEDS document

```xml
<?xml version="1.0" encoding="UTF-8" standalone="yes"?>
<metaTeds id="4821">
    <dataStructure>
        <calibrationKey>1</calibrationKey>
        <standartFamily>1</standartFamily>
        <majorVersion>2</majorVersion>
        <implementedChannels>2</implementedChannels>
    </dataStructure>
    <timing>
        <worstUpdateTime>2.00E-5</worstUpdateTime>
        <worstReadeTime>8.00E-5</worstReadeTime>
        <worstWriteTime>0</worstWriteTime>
        <worstSamplingTime>2.00E-4</worstSamplingTime>
        <ioResponseTime>5.00E-4</ioResponseTime>
    </timing>
    <identification>
        <uuid>550e8400-e29b-41d4-a716-446655440000</uuid>
        <manufacturer>Raspberry Pi Foundation</manufacturer>
        <model>Raspberry Pi B</model>
        <revision>15</revision>
        <serialNumber>2ABCB-RPI21</serialNumber>
        <description>The Raspberry Pi is a credit card-sized single-board
computer developed in the UK by the Raspberry Pi Foundation.</description>
    </identification>
</metaTeds>
```

Fig. 89 Meta TEDS example for Raspberry Pi

A TEDS channel is electronic documentation corresponding to one specific sensor. The structure of the implemented XML Channel TEDS for Raspberry Pi is shown in Fig. 90.

The Channel TEDS implementation for the DS18B20 temperature sensor looks like this (Fig. 91).

Calibration TEDS—electronic documentation containing calibration data for a specific corresponding sensor channel. The implementation of Calibration TEDS for the DS18B20 temperature sensor is shown in Fig. 92.

TEDS Calibration example, for DS18B20 (Fig. 93).

Calibration TEDS contains a set of data elements. In turn, such elements consist of an offset and a calibration factor.

The implementation of the TIM module as a RESTful WEB service (Fig. 94) provides an opportunity to reduce the dependence of the TIM module on the NCAP module and simplify data structures for exchanging information between modules.

To test the developed WEB service, an application for the Chrome browser was used—the Postman REST client. It performs formatting for JSON and XML text formats if the content of the response corresponds to the MIME type (Fig. 95).

An example of an XML response received from the server is shown in Fig. 96.

An example of the coded response received from the server is shown in Fig. 97.

The package structure of the developed software includes five main packages: *.service—contains classes responsible for the temperature measurement process, and serialization of the class to a byte array that complies with the IEEE-1451 standard regarding the TEDS record format; *.listener—contains a class responsible for

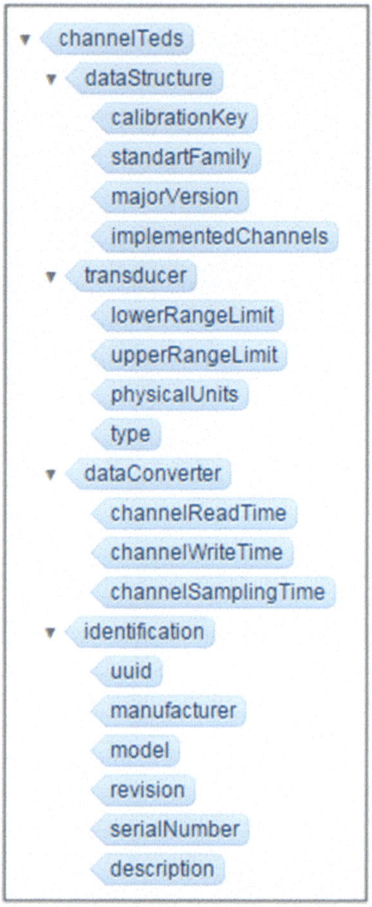

Fig. 90 XML structure of channel TEDS document

loading the necessary modules to the kernel of the Linux operating system; *.entity—contains entity classes that correspond to the subject area of this STIM module; *.rest—contains the so-called resource classes that are responsible for processing the request and forming a response to the corresponding request; *.util—contains auxiliary classes for implementing the operation of the STIM module.

The *.listener package contains one class—KernelModuleLoader, which is responsible for loading two necessary modules to the kernel of the Linux operating system: the w1-gpio module, which implements data transfer via the 1-Wire interface through input/output ports; the w1-therm module, which is a driver for the DS18B20 temperature sensor.

This class is an implementation of the ServletContextListener interface. This interface declares two methods contextInitialized() and contextDestroyed(). This implementation uses the contextInitialized() method (Fig. 98).

It calls the method directly responsible for loading modules (Fig. 99).

```
<?xml version="1.0" encoding="UTF-8" standalone="yes"?>
<channelTeds>
    <dataStructure>
        <calibrationKey>0</calibrationKey>
        <standartFamily>1</standartFamily>
        <majorVersion>2</majorVersion>
    </dataStructure>
    <transducer>
        <lowerRangeLimit>-55</lowerRangeLimit>
        <upperRangeLimit>125</upperRangeLimit>
        <physicalUnits>00f0f0f0f0f0f0f2f0f0</physicalUnits>
        <type>0</type> <!--Sensor-->
    </transducer>
    <dataConverter>
        <channelReadTime>1.50E-5</channelReadTime>
        <channelWriteTime>1.50E-5</channelWriteTime>
        <channelSamplingTime>1.90E-4</channelSamplingTime>
    </dataConverter>
    <identification>
        <uuid>8a0e84ee-e29b-41d4-28-0000039e958b</uuid>
        <manufacturer>MAXIM Integrated</manufacturer>
        <model>DS18B20</model>
        <revision>24</revision>
        <serialNumber>39e958b</serialNumber>
        <description>The DS18B20 Digital Thermometer provides 9 to 12-bit
(configurable) temperature readings which indicate the temperature of the
device.</description>
    </identification>
</channelTeds>
```

Fig. 91 Channel TEDS for DS18B20 sensor

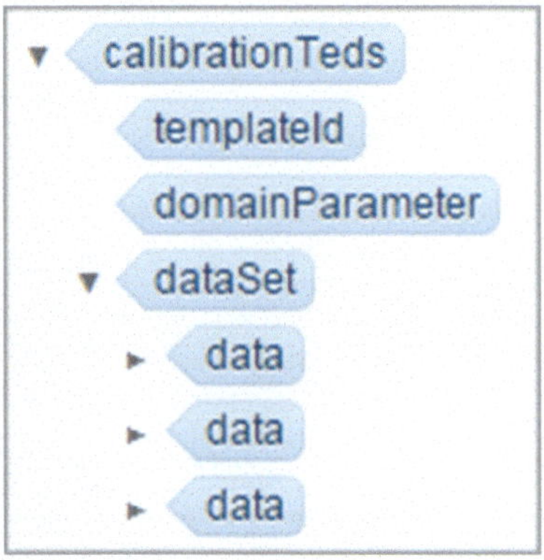

Fig. 92 XML structure of TEDS Calibration document

To increase the versatility of the software, the loading commands for each of the modules are written in the configuration file—command.properties, which is located in the archive with the software (Fig. 100).

The *.service package implements the main logic of the developed library and consists of the following sub-packages: *.service.measurement—contains classes responsible for reading DS18B20 sensor readings; *.service.converter—contains classes responsible for serializing entity classes to byte format according to the IEEE-1451 standard; *.service.teds—contains classes, the main purpose of which is to transform an XML file describing TEDS into its corresponding entity class;

Fig. 93 TEDS Calibration example for the DS18B20 sensor

```xml
<calibrationTeds>
    <templateId>40</templateId>
    <domainParameter>1</domainParameter>
    <dataSet>
        <data>
            <offset>152.5</offset>
            <factor>0.001</factor>
        </data>
        <data>
            <offset>152.5</offset>
            <factor>0.001</factor>
        </data>
  <data>
            <offset>167.3</offset>
            <factor>0.001</factor>
        </data>
    </dataSet>
</calibrationTeds>
```

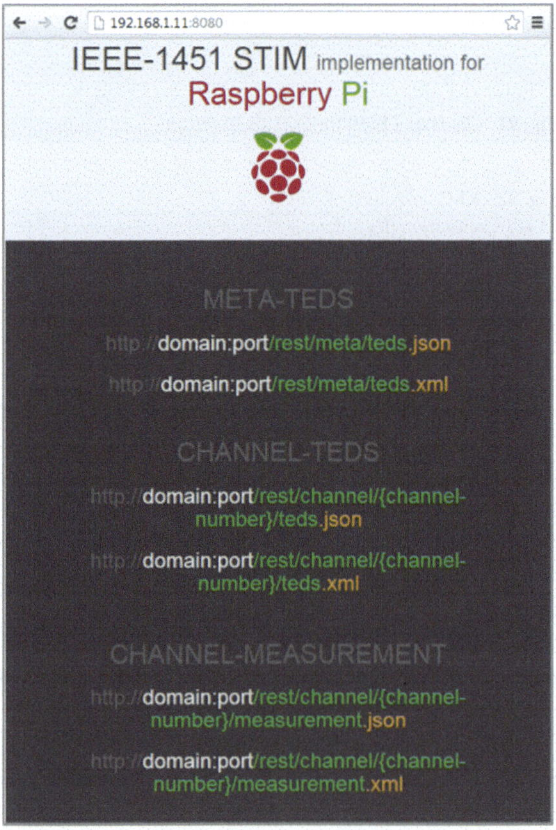

Fig. 94 Appearance of the main page of the STIM module with URLs for available resources

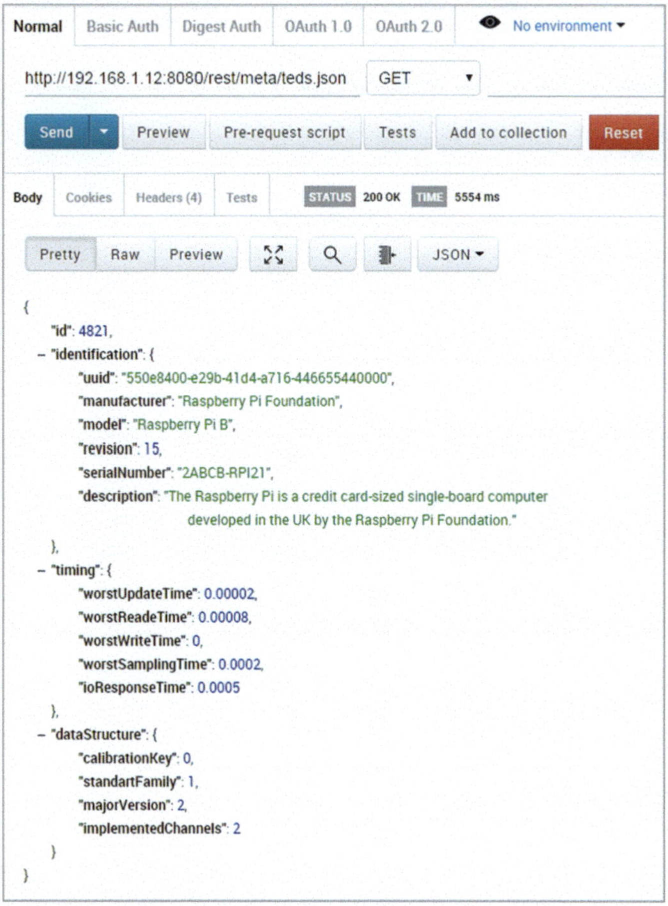

Fig. 95 Meta teds response from the server in JSON format

*.service.encoder—contains classes that are a specific implementation of converting XML files; *.service.util—contains auxiliary classes to implement the operation of the *.service package.

The ByteConverter interface contains a set of methods that perform the conversion of entity classes that represent implemented TEDS into a byte array according to the IEEE-1451 standard (Fig. 101).

This interface has one implementation—StimByteConverter (Fig. 102).

In the *.service.encoder package, the main class is ByteEncoder, which has three methods.

The encode() method converts the class-entity accepted as an argument into a byte array by successively calling the conversion function for each of the fields available in the class, which are obtained using the reflection mechanism from the class object: The standartTypeToCharacters() method accepts a class field as an argument. The

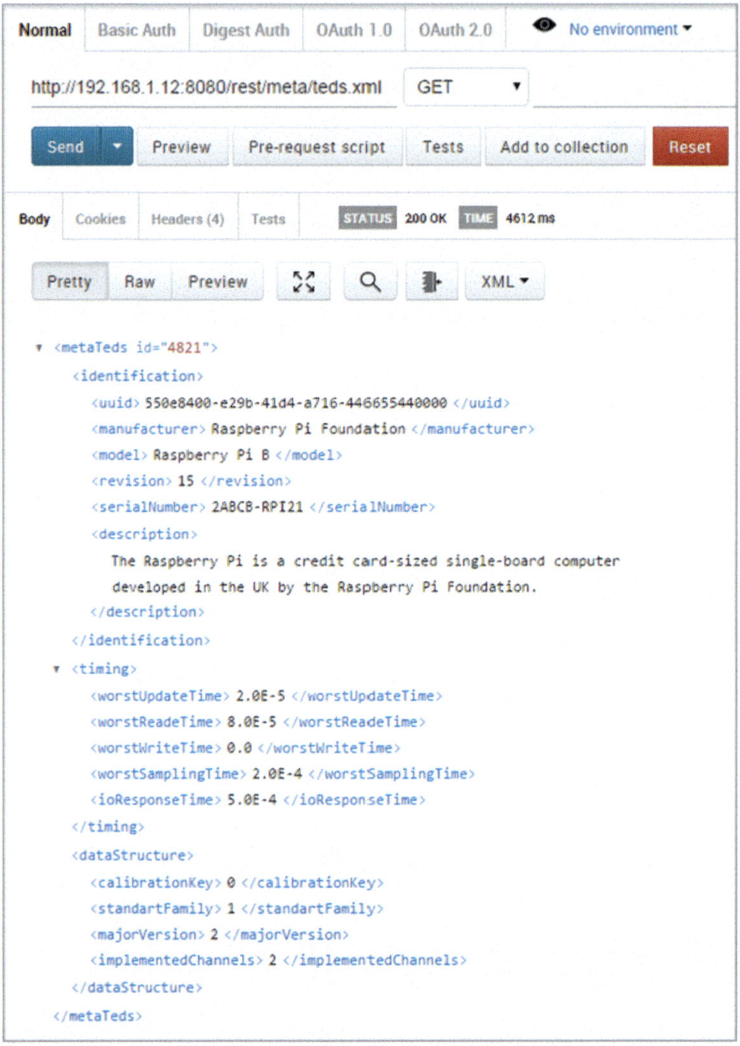

Fig. 96 Meta teds response from the server in XML format

mechanism of the method is as follows—the function checks the type of the field. If the field belongs to one of the primitive types, it turns it into a byte array, in the other case, an exception is created, indicating that the field type is incompatible with this implementation.

The *.service.measurement package consists of classes responsible for reading sensor readings. This package contains one interface (Fig. 103).

The class that implements this interface—Ds18b20Measurement. This is an implementation for the DS18B20 temperature sensor. The reading algorithm consists of 3 parts: read the file with the measurement result from the directory /sys/bus/w1/

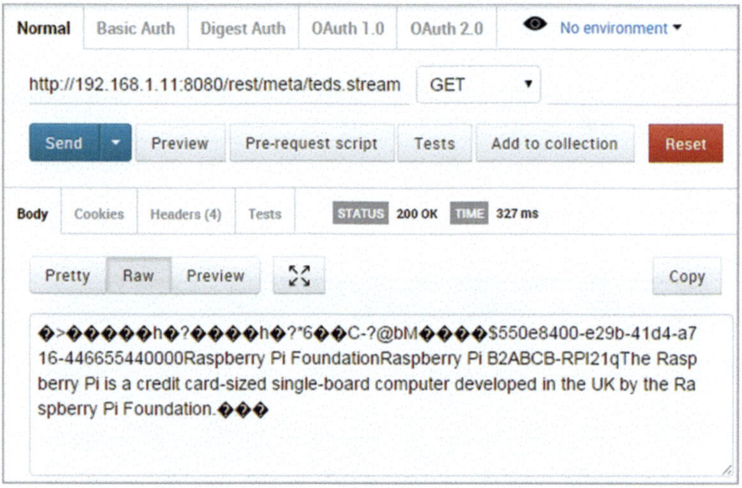

Fig. 97 Byte meta teds response from the server

```
@Override
public void contextInitialized(ServletContextEvent sce) {
        try {
            loadModules();
        } catch (IOException e) {
            e.printStackTrace();
    }
}
```

Fig. 98 The contextInitialized method

```
        private void loadModules() throws IOException {
         Properties properties =
getPropertiesByFileName(Path.COMMAND_PROPERTIES.getPath());
                CommandExecutor executor = new CommandExecutor();
           executor.execute(properties
        .getProperty(Command.MODPROBE_GPIO.getProperty()));
                executor.execute(properties
        .getProperty(Command.MODPROBE_THERM.getProperty()));
        }
```

Fig. 99 The loadModules method

Fig. 100 The command.properties file

```
## Commands ##
modprobe.gpio=sudo modprobe w1-gpio
modprobe.therm=sudo modprobe w1-therm
```

```
public interface ByteConverter {
    byte[] convertMetaTeds(MetaTeds metaTeds);
    byte[] convertChannelTeds(ChannelTeds channelTeds);
    byte[] convertCalibrationTeds(CalibrationTeds calibrationTeds);
    byte[] convertMeasurement(Measurement measurement);
}
```

Fig. 101 ByteConverter interface

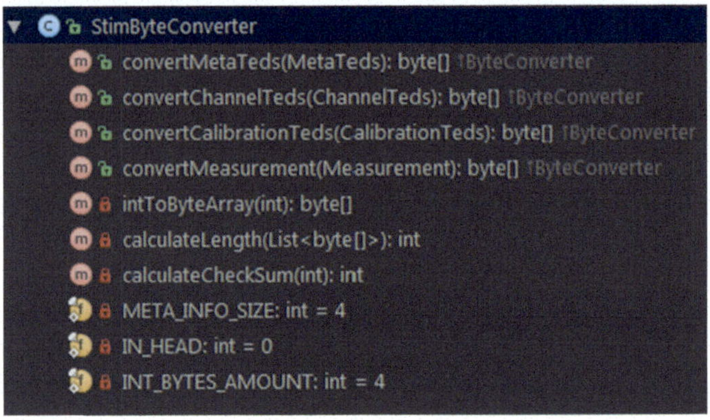

Fig. 102 StimByteConverter class structure

```
public interface MeasurementService {
    Measurement measure(int channel);
}
```

Fig. 103 MeasurementService interface

devices/{id}/w1_slave, where id is the serial number of the sensor; decode the read file (the result of this stage is an object of type String, which represents the measurement result, which in turn is a floating-point number; reducing the type of the result to the double data type.

The measureTemp(int channelNumber) method looks like this (Fig. 104).

The *.entity package consists of entity classes, which are object-oriented representations of domain models. The package structure is shown in Fig. 105.

The format of each TEDS may be partially repeated. To reduce duplication of code, the structure of implemented TEDS was decomposed. Thus, six classes have been developed to describe TEDS: data—is a display of calibration points or coefficients; dataConverter—displays time delays for specific data; dataStructure—is a display of meta-information about a specific TEDS type; identification—displays meta-information for the physical device described by TEDS data; timing—is a

```
        public double measureTemp(int channelNumber) {
                BufferedReader reader = null;
                String destinationLine = null;
                String channelId =
propertyHandler.getProperty(Stim.CHANNEL_ID.getProperty() + channelNumber);
                try {
                    reader = new BufferedReader(new
FileReader(buildPath(channelId)));
                    reader.readLine();
                    destinationLine = reader.readLine();
                } catch (IOException e) {
                    e.printStackTrace();
                } finally {
                    try {
                        if (reader != null)
                            reader.close();
                    } catch (IOException e) {
                        e.printStackTrace();
                    }
                }

                if (destinationLine != null)
                    return parseLineToDouble(destinationLine);
                return 0;
        }
```

Fig. 104 The measureTemp method

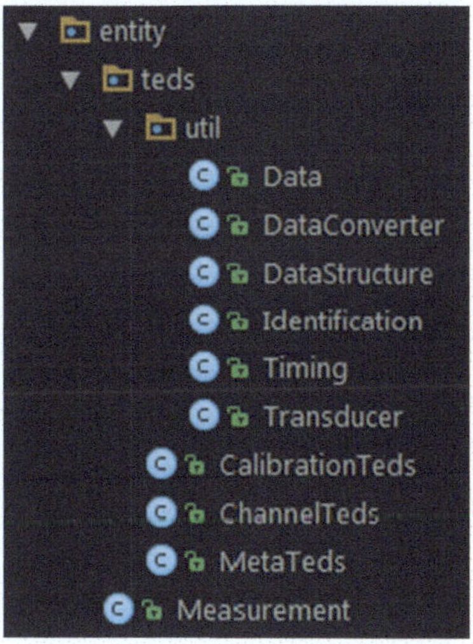

Fig. 105 Structure of the *.entity package

reflection of the range of permissible time delay for data; transducer—displays meta-information about the sensor represented by TEDS data.

An example of the implementation of the Timing class (Fig. 106).

Also, in the *.entity package, a class has been developed that represents the measurement entity—Measurement (Fig. 107).

```
@XmlAccessorType(XmlAccessType.NONE)
@XmlRootElement
public class Timing implements EncodableMarker{

    @XmlElement
    private double worstUpdateTime;

    @XmlElement
    private double worstReadeTime;

    @XmlElement
    private double worstWriteTime;

    @XmlElement
    private double worstSamplingTime;

    @XmlElement
    private double ioResponseTime;
...
}
```

Fig. 106 The timing class

```
@XmlAccessorType(XmlAccessType.NONE)
@XmlRootElement(name = "measurement")
public class Measurement implements EncodableMarker{

    @XmlElement
    private int channel;

    @XmlElement
    private double value;

    @XmlElement
    private long timestamp;
...
}
```

Fig. 107 Measurement class

The *.rest package is responsible for REST API configuration and resource classes. The main configuration class of the Jersey framework is the RestConfig class (Fig. 108).

This class in the constructor calls the register() method, which registers the necessary classes, which include resource classes and service classes. Also, in the constructor, classes are bound to implement control inversion using the AbstractBinder class.

An example of a method from the MetaResource resource class that implements the request processing logic (Fig. 109).

```
public RestConfig() {
        register(new AbstractBinder() {
            @Override
            protected void configure() {
                bind(XmlTedsService.class)
                        .to(TedsService.class)
                        .in(Singleton.class);
                bind(StimPropertyHandler.class)
                        .to(StimPropertyHandler.class)
                        .in(Singleton.class);
                bind(Ds18b20Measurement.class)
                        .to(MeasurementService.class)
                        .in(Singleton.class);
                bind(StimByteConverter.class)
                        .to(ByteConverter.class)
                        .in(Singleton.class);
            }
        });
        register(MetaResource.class);
        register(ChannelResource.class);
}
```

Fig. 108 RestConfig class

```
@GET
@Path("/teds.xml")
@Produces(MediaType.APPLICATION_XML)
public Response getXmlMetaTeds() {
return Response
       .ok()
              .entity(tedsService.getMetaTeds())
              .build();
}
```

Fig. 109 getXmlMetaTeds method

In this case, to reduce the dependency of the MetaResource class on the class that implements the TedsService interface, the control inversion mechanism is used (Fig. 110).

To increase configurability and flexibility, part of the parameters is written to the configuration file—stim.properties (Fig. 111).

Fig. 110 The tedsService field

```
@Inject
private TedsService tedsService;
```

```
## General ##
stim.meta.teds=meta-teds.xml
stim.teds.path=/home/pi/stim/teds/

## Channels ##
channel.amount=2
channel.teds.prefix=channel-teds-
channel.teds.extension=.xml
channel.0=28-0000039e958b
channel.1=28-0000039ed12b

## Calibration ##
calibration.teds.prefix=calibration-teds-
calibration.teds.extension=.xml

## Measurement path ##
measurement.path.prefix=/sys/bus/w1/devices/
measurement.path.suffix=/w1_slave
```

Fig. 111 The stim.properties file

Such parameters include the path to the directory in which TEDS are stored; the name of the Meta Teds file; identifier of the physical device of each of the channels and the total number of channels; Channel and Calibration Teds file name patterns in the file system; the path to the channel driver.

3.4 Development of RESTful Web Service Support Software for the NCAP Module Based on Cubieboard Java Microserver

This implementation uses a single-board Cubieboard computer as the basis for developing an NCAP module according to the IEEE 1451 standard. An operating system based on the Linux kernel—Cubian is used as an operating system. The programming language is Java.

The Apache Tomcat 8 servlet container was chosen as the server. The JAX-RS specification with its JAX-RS API was chosen to implement the RESTful architecture. The Jersey framework is chosen as the provider's API. Apache Tomcat is selected as the server.

To implement the WEB client, a browser is selected, to which static HTML pages are sent. Of all the currently available versions of HTML and CSS, the fifth and third versions are selected, respectively. To send asynchronous JavaScript requests (AJAX) to the server, the Jquery2 library is used. The Chart.js library (Fig. 112),

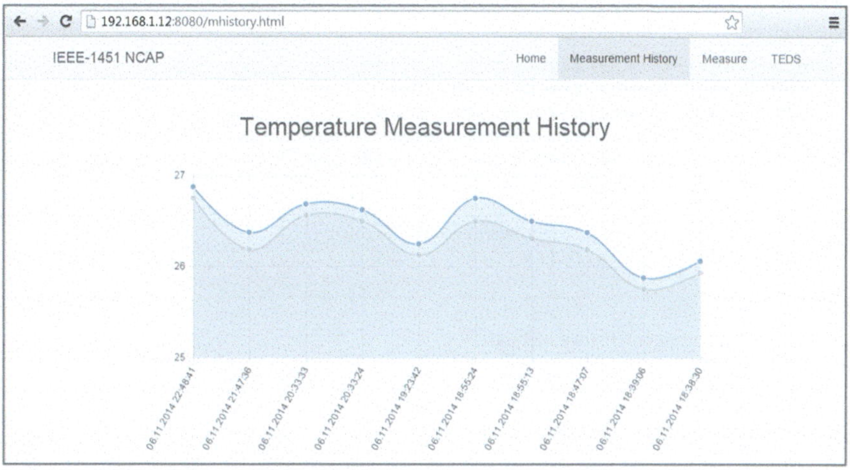

Fig. 112 Graphic display of the last 10 measurement results

which uses HTML5 canvas elements, was used to construct graphs of temperature changes for each of the TIM channels.

A MySQL relational database is used to store information from TIM. The Hibernate framework was used to communicate with the database on the server side.

The developed software for the NCAP module consists of the following Java packages: *.ncap.rest—the package contains classes responsible for the operation of the RESTful API; *.ncap.util—contains auxiliary classes, which by their functionality cannot be attributed to any of the other packages; *.ncap.service—contains classes and interfaces responsible for implementing the logic of executing functions and commands defined within the IEEE 1451 standard; *.ncap.entity—contains entity classes representing the subject area of the IEEE 1451 standard; *.ncap.dao—contains classes and interfaces responsible for the interaction of the developed application with the database.

The *.ncap.rest package consists of classes that implement the RESTful API: *.rest.resources—resource classes that process client requests, delegating request processing to the appropriate services, form and return responses received from services; *.rest.config—contains a class responsible for configuring the Jersey framework.

The configuration class RestConfig has a structure similar to the corresponding class in the TIM module, which was described above.

Classes from the *.rest.resources package have the same structure. Example of implementation of the getMeasurementHistory() method in the MeasurementResource class (Fig. 113).

The *.ncap.util package consists of the PropertyHandler class, which is responsible for reading properties from the ncap.properties configuration file (Fig. 114).

```
@GET
@Path("/measurement/history")
@Produces(MediaType.APPLICATION_JSON)
public Response getMeasurementHistory() {
MeasurementHistory measurementHistory = measurementService
                .getLastMeasurements(MEASUREMENT_HISTORY_LENGTH);
        return Response
                .ok()
                .entity(measurementHistory)
                .build();
}
```

Fig. 113 getMeasurementHistory method

```
    public class PropertyHandler {

        private static final String NCAP_PROPERTIES = "ncap.properties";
        private Properties properties = new Properties();

        public PropertyHandler() throws IOException {
            InputStream fileStream = Thread.currentThread()
.getContextClassLoader().getResourceAsStream(NCAP_PROPERTIES);
            properties.load(fileStream);
        }

        public String getProperty(String propertyName) {
            return properties.getProperty(propertyName);
        }
    }
```

Fig. 114 PropertyHandler class

This approach allows for flexible configuration of the developed NCAP module. The structure of the ncap.properties configuration file looks like this (Fig. 115).

This file defines the following configuration data: protocol type; IP address of the STIM module; the port on which the RESTful web server of the STIM module is located; URIs of available TEDS.

The *.ncap.service package consists of the following Java packages: *.service.url—contains classes responsible for constructing the URL to the corresponding TIM resource; *.service.client—contains classes that implement the client for the TIM RESTful web server; *.service.measurement—contains service classes responsible for processing measurement results; *.service.accuracy—contains classes for correcting measurement results.

The RaspberryJsonStimClient class from the *.service.client package implements the client for the TIM module. The function to get Meta TEDS looks like this (Fig. 116).

The *.ncap.dao package contains classes that ensure the interaction of the developed server with the database server. The HibernateUtil class is responsible for

```
## General Stim info ##
stim.protocol=http://
stim.ip=200.10.10.11
stim.port=:8080

## Meta TEDS uri ##
stim.uri.meta.teds=/rest/meta/teds


## Channel uri ##
stim.uri.channel.teds.prefix=/rest/channel/
stim.uri.channel.teds.suffix=/teds
stim.uri.calibration.teds.suffix=/calibration/teds
stim.uri.measurement.suffix=/measurement
```

Fig. 115 The ncap.properties file

```
    @Override
        public MetaTeds getMetaTeds() {
            MetaTeds metaTeds = null;
            WebTarget webTarget =
client.target(urlBuilder.buildMetaTedsUrl(DATA_FORMAT.JSON));
            String entity = generateResponse(webTarget);

            try {
                metaTeds = objectMapper.readValue(entity, MetaTeds.class);
                System.out.println(metaTeds.toString());
            } catch (IOException e) {
                e.printStackTrace();
            }
            return metaTeds;
    }
```

Fig. 116 getMetaTeds method

configuring the Hibernate framework and creating a SessionFactory based on the configuration XML file (Fig. 117).

The MeasurementDAO class has two methods that interact with the database. The save(…) method is for writing one measurement result to the database (Fig. 118).

And the getLastMeasurements(…) method—for reading from the database the specified number of the last measurement results (Fig. 119).

The *.ncap.entity package contains entity classes that display the TEDS types available in this implementation in an object-oriented representation. Also, this package contains two entities for representing the result of measurements, Measurement, and MeasurementHistory, respectively. An example of the implementation of the Measurement class (Fig. 120).

The appearance of the NCAP page with temperature sensor readings is shown in Fig. 121.

```
      public class HibernateUtil {
          private static final String XML_CONFIG_FILE_PATH = "/hibernate.cfg.xml";

          private static final SessionFactory sessionFactory =
configureSessionFactory();
          private static ServiceRegistry serviceRegistry;

          private static SessionFactory configureSessionFactory() {
              Configuration config = new Configuration();
              config.configure(XML_CONFIG_FILE_PATH);

              serviceRegistry = new
StandardServiceRegistryBuilder().applySettings(config.getProperties()).build();

              return config.buildSessionFactory(serviceRegistry);
          }
          public static SessionFactory getSessionFactory() {
              return sessionFactory;
          }
      }
```

Fig. 117 HibernateUtil class

```
public void save(Measurement measurement) {
 try {
      session.beginTransaction();
            session.save(measurement);
            session.getTransaction().commit();
        } catch (HibernateException e) {
            e.printStackTrace();
        }
}
```

Fig. 118 The save method

```
public List<Measurement> getLastMeasurements(int amount, int channel) {
        List<Measurement> measurementList = null;

        if (!session.isOpen())
            session = HibernateUtil.getSessionFactory().openSession();

        try {
            session.beginTransaction();
            Query query = session.createQuery(QUERY);
            query.setMaxResults(amount);
            query.setParameter(CHANNEL_PARAMETR, channel);
            measurementList = (List<Measurement>) query.list();
            session.getTransaction().commit();
        } catch (HibernateException e) {
            e.printStackTrace();
        }

     return measurementList;
}
```

Fig. 119 The getLastMeasurements method

```
@XmlAccessorType(XmlAccessType.NONE)
@XmlRootElement(name = "measurement")
@Entity
@Table(name = "measurements")
public class Measurement {

    @XmlElement
    @Id
    @GeneratedValue(strategy = GenerationType.IDENTITY)
    private int id;

    @XmlElement
    private int channel;

    @XmlElement
    private double value;

    @XmlElement
    private long timestamp;

    public Measurement() {}

    public Measurement(int channel, double value, long timestamp) {
        this.channel = channel;
        this.value = value;
        this.timestamp = timestamp;
    }
...
}
```

Fig. 120 Measurement class

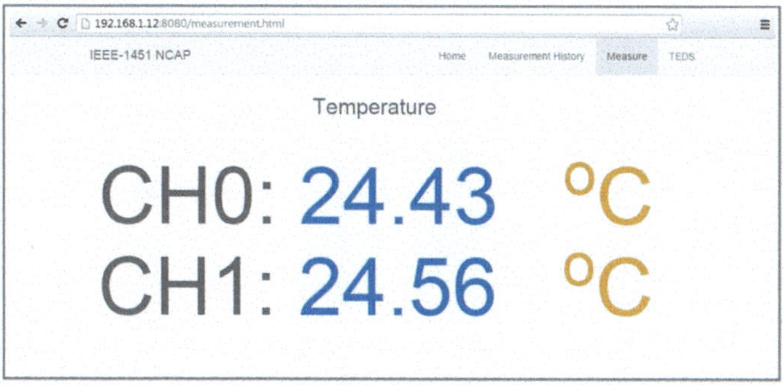

Fig. 121 The appearance of the NCAP page with temperature sensor readings

4 Development of a Network Data Collection System Based on MQTT Protocol and RESTful Web Service

4.1 Development of the Structure of the Interaction of the Components of the Network Data Collection System

The information and measurement system is based on a server with support for the MQTT protocol and a RESTful web service. Each device can send messages to the server using one of these protocols. Devices that support sending data to a RESTful web service can send requests to multiple devices sequentially. Device settings (such as web service address, MQTT server, and device name) must be done through its web interface.

The diagram of the network data collection system is presented in Fig. 122.

The full name of the device consists of the name of the device, as well as the name of a set of devices, separated by a slash. For example "AED_LAB/MULTIMETER". Each device can measure one or more physical quantities. To indicate a physical quantity measured by the device, the name of that quantity is added to the name of the device through a slash, for example "AED_LAB/MULTIMETER/VOLTAGE". If the instrument has several channels for measuring a value, the channel name is written with a hyphen after the name of the value, for example "AED_LAB/MULTIMETER/VOLTAGE-CH1". All characters in the name of a set of devices, a device, a physical quantity, and a channel, except for the Latin alphabet, underscores, and numbers must be encoded using percent-encoding, for example, %2B for the " +" sign. The value of all quantities must be written in the SI system. Magnitude values can be

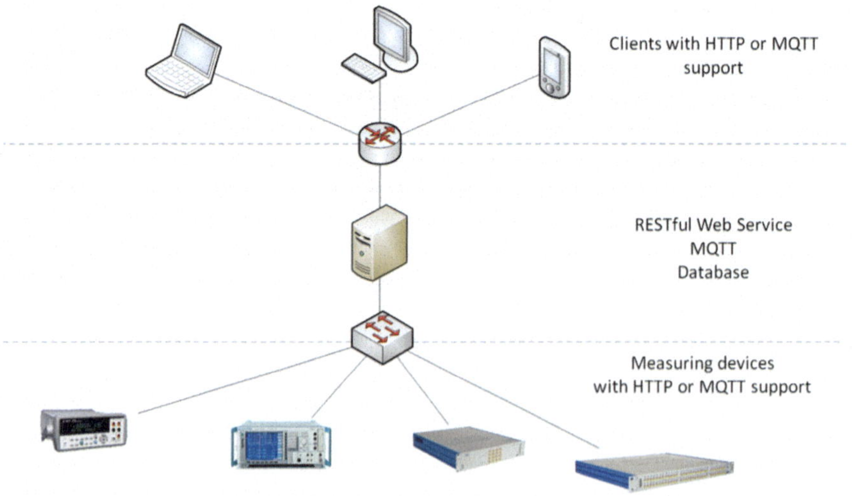

Fig. 122 Scheme of the network data collection system

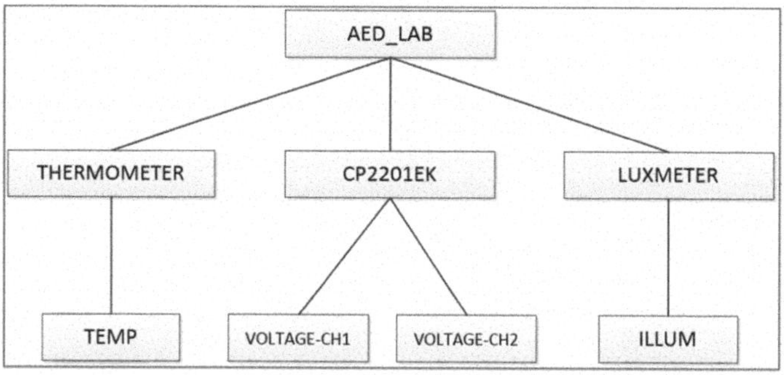

Fig. 123 MQTT protocol topic structure

passed in exponential representation, for example, 1.048576E + 06, 2.043E06, and 0.6514e-06.

Based on the specified naming rules, MQTT topics are presented in the form of a hierarchy Instrument Set -> Instrument -> Channel. A graphical representation of this hierarchy is shown in Fig. 123.

4.2 *Development of a RESTful Web Service Interface of a Network Data Collection System*

Data can be sent to the server in the form of XML format, or using the form-urlencoded format. Data is added using the HTTP POST method. The request must be sent to each of the web services specified in the device configuration.

In the case of data transfer in XML, the request will have the following form (Fig. 124).

It is necessary to pay attention to the empty line between the request header and its body. In the example above, {ws_root} points to the root directory of the

```
POST {ws_root}/AED_LAB/TERMOMETER HTTP/1.1
Host: {ws_host}
Cache-Control: no-cache
Content-Type: application/xml

<measure>
     <data sensor="TEMP-CH1">273.445</data>
     <data sensor="TEMP-CH2">272.342</data>
     <data sensor="TEMP-CH3">274.120</data>
</measure>
```

Fig. 124 Request type for sending XML data to the server

```
POST /measurementWS/AED_LAB/TERMOMETER HTTP/1.1
Host: 192.168.33.1:8080
Cache-Control: no-cache
Content-Type: application/xml

<measure>
     <data sensor="TEMP-CH1">273.445</data>
     <data sensor="TEMP-CH2">272.342</data>
     <data sensor="TEMP-CH3">274.120</data>
</measure>
```

Fig. 125 Request to send XML data to the server

```
POST {ws_root}/AED_LAB/TERMOMETER HTTP/1.1
Host: {ws_host}
Cache-Control: no-cache
Content-Type: application/x-www-form-urlencoded

TEMP-CH1=273.445&TEMP-CH2=272.342&TEMP-CH3=274.120
```

Fig. 126 Query view for form-urlencoded format

web service, and {ws_host} points to the server address. For example, the web service is located at 192.168.33.1:8080/measurementWS, then the request will be sent to 192.168.33.1:8080, {ws_root} will be /measurementWS, and {ws_host} will be 192.168.33.1:8080. Based on this, the request specified in the example for the web service at the address 192.168.33.1:8080/measurementWS will look like this (Fig. 125).

In the case of data transmission using the form-urlencoded format, the request will have the following form (Fig. 126).

In case the devices are synchronized with the server, the data block in XML will have the form <data sensor ="TEMP-CH1" time ="2013–11-01T12:00:01.112+02:00">273.445< /data > . n this case, when processing data, the time of measurement will be taken into account, and not the time of receiving data by the server. In the case of data transfer in form-urlencoded format, the data will look like this:

TEMP-CH1 = 273.445%402,013–11-01T12%3A00%3A01.112%2B02%3A00&
TEMP-CH2 = 273.445%402,013–11-01T12%3A00%3A01.231%2B02%3A00&
TEMP-CH3 = 274.120273.445%402,013–11- 01T12%3A00%3A01.230%2B02%3A00.

That is, the data for the first channel has the form: 273.445@2013–11-01T12:00:01.112 + 02:00, but before sending, a percentage encoding procedure is performed on it to exclude reserved characters and it becomes 273.445%402,013–11-01T12% 3A00%3A01.112%2B02%3A00.

```
GET {ws_root}/AED_LAB/TERMOMETER?count=25 HTTP/1.1
Host: {ws_host}
Cache-Control: no-cache
Accept: application/xml
```

Fig. 127 The view of the request to receive data

As a result of sending a request, the web service returns the execution result with the following codes: 200 OK, in case of successful execution, 400 Bad Request, in case of an error processing the transmitted request.

The HTTP GET method is used to receive measurement results from the web service. The results can be obtained for all channels of the corresponding device, or only for the selected channel. Data can be output in text format and XML format. For example, to receive data in XML format, the following query is used (Fig. 127).

As a result, a response in the following format will be sent (Fig. 128).

The count parameter passed in the request is optional, its default value is 1. If the request is executed for a specific channel of the device, for example {ws_root}/AED_LAB/TERMOMETER/TEMP-CH2, a similar XML will be output, but with only one block sensor. Example (Fig. 129),

In the data block, the time parameter is entered, which indicates the time at which the server received data from the device, or the measurement time if the devices are synchronized. Time is represented according to the ISO 8601 standard in the following format: YYYY-MM-DDThh:mm:ss.sTZD, where YYYY is the year (4 digits), MM is the month (2 digits, starting with 01 for January, DD is the day (2 digits, from 01 to 31), Hh—hours (3 digits, from 00 to 23), mm—minutes (2 digits, from 00 to 59), ss—minutes (2 digits, from 00 to 59), s—milliseconds (1–3 digits, from 0 to 999), TZD—time zone (+hh:mm or -hh:mm).

```
HTTP/1.1 200 OK
Content-Type: application/xml
<measure instrument="/AED_LAB/TERMOMETER">
 <sensor name="TEMP-CH1">
     <data time="2013-11-01T12:00:00.210+02:00">273.445</data>
     <data time="2013-11-01T12:00:00.716+02:00">272.412</data>
     <data time="2013-11-01T12:00:01.112+02:00">273.122</data>
     <data time="2013-11-01T12:00:01.956+02:00">271.789</data>
  ...
 </sensor>
 <sensor name="TEMP-CH2">
       <data time="2013-11-01T12:00:00.210+02:00">273.445</data>
  ...
 </sensor>
...
</measure>
```

Fig. 128 View of the response from the server

```
Request

GET {ws_root}/AED_LAB/TERMOMETER/TEMP-CH2 HTTP/1.1
Host: {ws_host}
Cache-Control: no-cache
Accept: application/xml


Response

HTTP/1.1 200 OK
Content-Type: application/xml

<measure instrument="/AED_LAB/TERMOMETER">
  <sensor name="TEMP-CH2">
        <data time="2013-11-01T12:00:00.210+02:00">273.445</data>
  </sensor>
</measure>
```

Fig. 129 Request and response for one channel

To receive the Instrument Set, the list of instruments, and their channels in XML format, you need to perform the following request (Fig. 130).

Thus, it is possible to query channels for a specific device (by querying, for example, {ws_root}/instruments/AED_LAB/TERMOMETER), or devices and their channels for a set of instruments (by querying, for example, {ws_root}/instruments/AED_LAB).

To register a device or its channel in the web service list, it is enough to send data from this channel once via the web service or MQTT.

When transmitting data using the MQTT protocol, values must be sent on a topic that is the full name of the channel, such as AED_LAB/MULTIMETER/VOLTAGE. The requirements for data transmitted are the same as for data transmitted through a web service.

When using MQTT, one message is sent per channel. The data is in the form of either number, for example, 234.12, or numbers and measurement dates, for example, 273.122@2013–11-01T12:00:01.112 + 02:00 In this case, percentage encoding of symbols is optional.

In the case of an unexpected disconnection, the message "LOST CONNECTION" should be sent to clients. The Quality of Service (QoS) value of this message must be 1 (deliver at least once) or 2 (deliver only once).

In the case where the device adds the measurement date to the transmitted value, the QoS shall be 1, in the absence of this date, the QoS shall be 2.

The appearance of the demonstration data collection system is shown in Fig. 131.

Fig. 130 Request and response for a list of devices

```
Request

GET {ws_root}/instruments HTTP/1.1
Host: {ws_host}
Cache-Control: no-cache
Accept: application/xml


Response

HTTP/1.1 200 OK
Content-Type: application/xml

<instruments>
  <set name="AED_LAB">
    <instrument name="TERMOMETER" >
       <channel name="TEMP-CH1" />
       <channel name="TEMP-CH2" />
       <channel name="TEMP-CH3" />
    </instrument>


    <instrument name="MULTIMETER" >
       <channel name="VOLTAGE" />
       <channel name="RESISTANCE" />
       <channel name="CURRENT" />
       ...
    </instrument>
    ...
  </set>
  ...
</instruments>
```

4.3 *Software Development of a Network Data Collection System Based on the MQTT Protocol and a RESTful Web Service*

The JAX-RS technology, which is part of Java EE, was used to implement the web service. It facilitates the development process and allows you to abstract from low-level details of information transfer. In addition, there is an opportunity to output data in XML format, passing to the output of the library objects describing the data contained in XML. To access the data, the Hibernate library was used, which converts records in the database into objects. The open server mosquito was selected to implement data transfer capabilities using the MQTT protocol. The server is free software, requires few resources, and has high speed. The PostgreSQL system was used as a DBMS. The choice is due to its high performance under parallel loading.

The diagram of the database for storing measurement information is presented in Fig. 132.

Fig. 131 The appearance of the demonstration data collection system

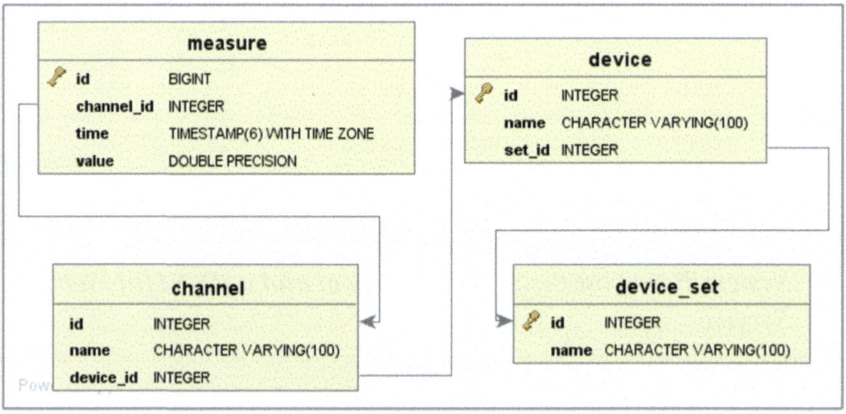

Fig. 132 Diagram of a database for storing measurement information

According to this scheme of tables, Java objects describing them, so-called entities, were created. A simplified entity class for the device_set table describing a set of devices looks like this (Fig. 133).

A simplified entity class for the device table describing a specific device looks like this (Fig. 134).

A simplified entity class for the channel table describing the full channel name looks like this (Fig. 135).

```
public class DeviceSet implements Serializable {
 private static final long serialVersionUID = 1L;

 @Id
 @SequenceGenerator(name="SET_SEQ_GEN", sequenceName="device_set_id_seq")
 @GeneratedValue(strategy=GenerationType.SEQUENCE, generator="SET_SEQ_GEN")
 private Integer id;

 private String name;

 //bi-directional many-to-one association to Device
 @OneToMany(mappedBy="deviceSet")
 private List<Device> devices;
}
```

Fig. 133 DeviceSet class

```
public class Device implements Serializable {
 private static final long serialVersionUID = 1L;

 @Id
 @SequenceGenerator(name="DEVICE_SEQ_GEN", sequenceName="device_id_seq")
 @GeneratedValue(strategy=GenerationType.SEQUENCE,
   generator="DEVICE_SEQ_GEN")
 private Integer id;

 private String name;

 //bi-directional many-to-one association to Channel
 @OneToMany(mappedBy="device")
 private List<Channel> channels;

 //bi-directional many-to-one association to DeviceSet
 @ManyToOne
 @JoinColumn(name="set_id")
 private DeviceSet deviceSet;
}
```

Fig. 134 Device class

```
public class Channel implements Serializable {
 private static final long serialVersionUID = 1L;

 @Id
 @SequenceGenerator(name="CHANNEL_SEQ_GEN", sequenceName="channel_id_seq")
 @GeneratedValue(strategy=GenerationType.SEQUENCE,
 generator="CHANNEL_SEQ_GEN")
 private Integer id;

 private String name;

 //bi-directional many-to-one association to Device
 @ManyToOne
 private Device device;

 //bi-directional many-to-one association to Measure
 @OneToMany(mappedBy="channel")
 private List<Measure> measures;
}
```

Fig. 135 Channel class

A simplified entity class for the measuring table, which describes a specific measurement result, looks like this (Fig. 136).

It is worth noting that the use of entity objects makes it convenient to work with a one-to-many relationship. For example, the Measure class has a reference to the Channel class, and the Channel class has a reference to the Measure class collection (that is, a reference to all measurement results that have been stored for a given channel).

The CommonDAO class is used to retrieve data using the HQL query language. The method of obtaining a channel object by its full name looks like this (Fig. 137).

The parameters are the session and transaction received by the client class after connecting to the database and the MeasureUtil.ChannelName class, which contains information about the full name of the channel—a set of devices, the device, and the channel itself. The method uses the query execution tool built into Hibernate. The request specified in the annotations of the entity object is received from the session,

```
public class Measure implements Serializable {
 private static final long serialVersionUID = 1L;

 @Id
 @SequenceGenerator(name="MEASURE_SEQ_GEN", sequenceName="measure_id_seq")
 @GeneratedValue(strategy=GenerationType.SEQUENCE,
       generator="MEASURE_SEQ_GEN")
 private Long id;

 private Timestamp time;

 private double value;

 //bi-directional many-to-one association to Channel
 @ManyToOne
 private Channel channel;
}
```

Fig. 136 The Measure class

```
public static Channel getChannelByFullName(Session session,
      Transaction transaction, MeasureUtil.ChannelName fullChannelName) {
       Query query =
                 session.getNamedQuery("Channel.findByNameDeviceAndSet");
       query.setString("channel_name", fullChannelName.getChannel());
       query.setString("device_name", fullChannelName.getDevice());
       query.setString("set_name", fullChannelName.getDeviceSet());

       List<Object> list = query.list();
       if (!list.isEmpty()) {
            return (Channel) list.get(0);
       } else {
            return null;
       }
}
```

Fig. 137 getChannelByFullName method

```
public static Device getDeviceByFullName(Session session,
      Transaction transaction, MeasureUtil.ChannelName fullChannelName) {
       Query query = session.getNamedQuery("Device.findByNameAndSet");
       query.setString("set_name", fullChannelName.getDeviceSet());
       query.setString("device_name", fullChannelName.getDevice());

       List<Object> list = query.list();;
       if (!list.isEmpty()) {
            return (Device) list.get(0);
       } else {
            return null;
       }
}
```

Fig. 138 The getDeviceByFullName method

```
public static DeviceSet getDeviceSetByFullName(Session session,
      Transaction transaction, MeasureUtil.ChannelName fullChannelName) {
       Query query = session.getNamedQuery("DeviceSet.findByName");
       query.setString("set_name", fullChannelName.getDeviceSet());

       List<Object> list = query.list();
       if (!list.isEmpty()) {
            return (DeviceSet) list.get(0);
       } else {
            return null;
       }
}
```

Fig. 139 The getDeviceSetByFullName method

and the values of the corresponding parameters of this request are set. Queries always return a list of results, but a query for a channel object by its full name can return at most one result due to database restrictions, so only the first element of the list is returned, or a null value.

The method of obtaining a device object by its name looks like this (Fig. 138).

The method of obtaining a set of devices by its name looks like this (Fig. 139).

The getDeviceByFullName and getDeviceSetByFullName methods are similar to the getChannelByFullName methods and differ only in the entity objects and the number of transmitted parameters.

The following method has been developed to obtain all measurement results for a given channel or device with a time limit for the results to be displayed (Fig. 140).

The method first obtains a list of channel entities for the specified name, sets parameter values for the query, and returns a list of measurement entity objects.

These methods are used in Enterprise Java Bean (EJB), which implements the business logic of the application. To implement the program, three EJBs were created, which are used for data exchange. The DataBean interface that describes the EJB used to retrieve measurement data looks like this (Fig. 141).

The InstrumentsBean interface describing the EJB used to retrieve all measurement channels looks like this (Fig. 142).

```
public static List<Measure> getMeasuresWithTimeLimit(
                  Session session, Transaction transaction,
                  MeasureUtil.ChannelName fullChannelName,
                  Date dateFrom, Date dateTo) {
      List<Channel> channels = getChannelsByFullName(session,
                        transaction, fullChannelName);
      Query query = session.getNamedQuery(
                        "Measure.findAllByChannelsAndTimeLimit");
      query.setParameterList("channels", channels);
      query.setTimestamp("date_from", dateFrom);
      query.setTimestamp("date_to", dateTo);
      return (List<Measure>)query.list();
}
```

Fig. 140 The getMeasuresWithTimeLimit method

```
@Local
public interface DataBean {
 public List<Measure> getMeasureData(String channelName, int count);
 public List<Measure> getMeasureData(String channelName,
                                     Calendar timeFrom, Calendar timeTo);
}
```

Fig. 141 DataBean interface

```
@Local
public interface InstrumetsBean {
 public List<Channel> getChannels(String fullChannelName);
}
```

Fig. 142 InstrumentsBean interface

The MeasureDataSaverBean interface that describes the EJB used to save measurement information looks like this (Fig. 143).

Implementations of these interfaces have appropriate names—DataBeanImpl, InstrumetsBeanImpl, and MeasureDataSaverBeanImpl. Saving data to the database is performed by creating an entity object of the appropriate type and calling the save

```
@Local
public interface MeasureDataSaverBean {
 public void save(String fullChannelName, double data, Calendar time);
 public void save(String fullChannelName, double data, Calendar time,
                                                   boolean fromMqtt);
}
```

Fig. 143 MeasureDataSaverBean interface

```
Session session = HibernateUtil.getSessionFactory().getCurrentSession();
Transaction transaction = session.beginTransaction();
MeasureUtil.ChannelName fullChannelName =
 MeasureUtil.getChannelNameContainer(channelName);
Channel channel = getOrCreateChannelByFullName(session, transaction,
 fullChannelName);
Measure msr = new Measure();
msr.setChannel(channel);
msr.setTime(new Timestamp(time.getTimeInMillis()));
msr.setValue(data);
session.save(msr);
transaction.commit();
mqttEngine.publish(channelName, data, time);
```

Fig. 144 Saving measurement results

function for the transaction. For example, the commands for saving measurement results look like this (Fig. 144).

The xenqtt library was used to receive and send data using the MQTT protocol. An EJB called MqttEngine was created to implement a separate background process that waits for a message and sends it if necessary. This EJB is a Singleton object, that is, it can exist only in one instance on the server. After the server starts, the thread in which this EJB is executed is started. The implementation of the connection with the MQTT server looks like this (Fig. 145).

The implementation of the method of sending a message using the MQTT protocol looks like this (Fig. 146).

This method is executed in the MeasureDataSaverBeanImpl after saving the data. To receive messages, an internal observer object has been created, which updates data in the database using MeasureDataSaverBean when data is received. The publishReceived observer method is executed when receiving data using the MQTT protocol (Fig. 147).

After receiving the data, a confirmation is sent that the data has been received. After that, the message is divided into the measurement value and time. If the measurement

```
@PostConstruct
  private void startup() {
        client = new AsyncMqttClient(MQTT_CONNECTION_STRING, listener,
                                                    MQTT_THREAD_POOL_SIZE);
        client.connect(MQTT_IDENTIFIER, false);
        try {
              connectLatch.await();
        } catch (InterruptedException e) {}

        List<Subscription> subscriptions = new ArrayList<Subscription>();
        subscriptions.add(new Subscription(MQTT_TOPIC_SUBSCRIBED,
                                                          QoS.AT_LEAST_ONCE));
        client.subscribe(subscriptions);
        connected = true;
        log.info("MqttEngine started and connected to brocker.");
 }
```

Fig. 145 Startup method

```
public void publish(String topic, double data, Calendar time){
      if (connected && !client.isClosed()) {
            log.info("Publishing info to " + topic);
            String timeStr = DatatypeConverter.printDateTime(time);
            String message = data + DATE_SEPARATOR + timeStr;
            messagePool.add(topic + message);
            client.publish(new PublishMessage(topic,
                                          QoS.AT_LEAST_ONCE, message));
      }
}
```

Fig. 146 The publish method

```
public void publishReceived(MqttClient client, PublishMessage message) {
  message.ack();
  String payload = message.getPayloadString();
  String topic = message.getTopic();
  String messageIdentifier = topic + payload;
  String messageIdentifier = topic + payload;
  if(messagePool.contains(messageIdentifier)) {
        messagePool.remove(messageIdentifier);
        return;
  }
  String[] splittedValue = payload.split(DATE_SEPARATOR);

  double value = Double.parseDouble(splittedValue[0]);
  Calendar date;
  if(splittedValue.length > 1) {
        date = DatatypeConverter.parseDateTime(splittedValue[1]);
  } else {
        date = new GregorianCalendar();
  }
  saverBean.save(topic, value, date, true);
}
```

Fig. 147 The publishReceived method

time is not specified, the current time is used by calling the new GregorianCalendar(). After the message is processed, a MeasureDataSaverBean is used to store the data in the database, making it available via a RESTful web service. In this way, the synchronization between the web service and the MQTT protocol is performed. In addition, to eliminate circular dependencies between MeasureDataSaverBean and MqttEngine, a message pool is used, which allows not duplicating a request if it was sent from MeasureDataSaverBean and received again because MqttEngine is subscribed to all MQTT topics, including those to which it sends.

The UML diagram of the EJB level is presented in Fig. 148.

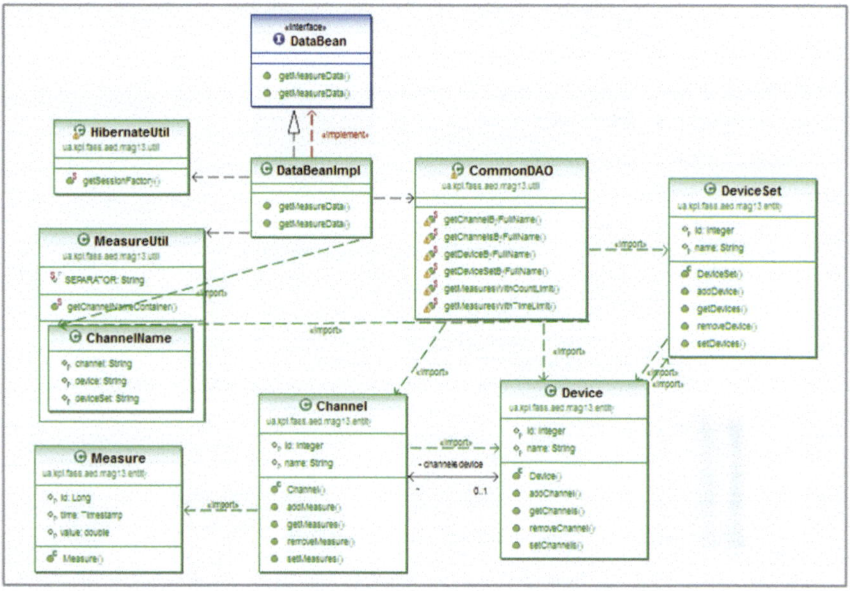

Fig. 148 UML diagram of the EJB level

4.4 Development of RESTful Web Service Implementation Software

The JAX-RS specification was used to implement the web service [34]. The annotations defined in this specification have been used to create classes that handle requests to a web service. To save data, the MeasureDataPublishService class has been created, which has the following form (Fig. 149).

The @Path annotation is used to specify the access address of a resource. The @PathParam annotation will receive the value from the corresponding parts of the URL and write the result to the class fields. The @Consumes annotation specifies the data type that the class accepts. Two methods are implemented—for receiving data in XML format and form-urlencoded format. When receiving data in XML format using JAX-RS, an object of the Measure class will be created, which, together with its descendants, will contain information from the request received from an XML document according to the JAXB specification.

The JAXB Measure class looks like this (Fig. 150).

This class uses annotations to describe how to input XML should be processed, which method fields correspond to which XML attributes, and which classes correspond to which XML elements. Based on the specified annotations, XML is converted into objects using the JAXB library.

```
@Path("/{instrument_set}/{instrument}")
public class MeasureDataPublishService {

 @PathParam("instrument_set")
 private String instrumentSet;

 @PathParam("instrument")
 private String instrument;

 @POST
 @Consumes(value={"text/xml", "application/json"})
 @Path("/")
 public void publish(Measure measureData) {

       PostRequestHandler handler = new PostRequestHandler(
             new XmlRequestParser(measureData,
                   instrumentSet + MeasureUtil.SEPARATOR + instrument)
             );
       handler.handleRequest();
 }


 @POST
 @Consumes("application/x-www-form-urlencoded")
 @Path("/")
 public void publish(MultivaluedMap<String, String> formData) {
       PostRequestHandler handler = new PostRequestHandler(
                   new FormRequestParser(formData, instrumentSet +
                                     MeasureUtil.SEPARATOR + instrument)
             );
       handler.handleRequest();
 }
}
```

Fig. 149 MeasureDataPublishService class

The XmlRequestParser and FormRequestParser classes are responsible for processing the received data for their type and transforming them into a Request-Container container object that contains request data for one measurement result in a unified form. Depending on how the data was obtained, a corresponding class is created. Its data is processed using the PostRequestHandler class (Fig. 151).

During the processing of the result, for each received value, saving is performed using EJB MeasureDataSaverBean.

The class that implements the service for obtaining measurement results looks like this (Fig. 152).

This class also uses the @Path and @PathParam annotations to get information about which instrument to get measurement data for. In addition, @QueryParam annotations were used to obtain query parameters, @DefaultValue to specify default values for query parameters, as well as @CalendarFormat annotation, which was created to convert a date range into the appropriate object.

The DataRequestHandler class is used to retrieve measurement values from the database using the business logic class and return the result (Fig. 153).

Depending on the transmitted request parameters, it is determined which of the EJB DataBean methods to use, and the result of processing the received data is

```
@XmlAccessorType(XmlAccessType.FIELD)
@XmlType(name = "", propOrder = {
    "data"
})
@XmlRootElement(name = "measure")
public class Measure {

    protected List<Measure.Data> data;

    @XmlAccessorType(XmlAccessType.FIELD)
    @XmlType(name = "", propOrder = {
        "value"
    })
    public static class Data {

        @XmlValue
        protected double value;
        @XmlAttribute(name = "sensor", required = true)
        protected String sensor;
        @XmlAttribute(name = "time")
        @XmlSchemaType(name = "dateTime")
        protected XMLGregorianCalendar time;
    }
}
```

Fig. 150 JAXB class Measure

```
public class PostRequestHandler {

  private MeasureDataSaverBean saverBean = (MeasureDataSaverBean)
                        EjbHelper.loadBean(MeasureDataSaverBean.class);

  private Iterator<RequestContainer> parser;

  public PostRequestHandler(Iterator<RequestContainer> parser) {
        this.parser = parser;
  }

  public void handleRequest(){
        while(parser.hasNext()){
              RequestContainer request = parser.next();
              saverBean.save(request.getFullChannelName(),
                                request.getData(), request.getTime());
        }
  }

}
```

Fig. 151 PostRequestHandler class

returned using the corresponding implementation of the DataResponseBulder interface. This interface is implemented by two classes—StringDataResponseBulder and XmlDataResponseBulder to represent data in text format and XML format, respectively.

```
@Path("/{instrument_set}/{instrument}")
public class DataService {
        @PathParam("instrument_set")
    private String instrumentSet;
        @PathParam("instrument")
    private String instrument;
        @DefaultValue("1")
        @QueryParam("count")
    private int count;
        @QueryParam("timefrom")
        @CalendarFormat
    private Calendar timeFrom;
        @QueryParam("timeto")
        @CalendarFormat
    private Calendar timeTo;
        @GET
        @Produces(value={"text/xml", "application/json"})
        @Path("/")
    public Object returnXml() {
        return new DataRequestHandler(new XmlDataResponseBulder())
                .handleRequest(
                        instrumentSet + MeasureUtil.SEPARATOR + instrument,
                        count,
                        timeFrom,
                        timeTo);
    }
        @GET
        @Produces(value={"text/plain"})
        @Path("/")
    public Object returnString() {
        return new DataRequestHandler(new StringDataResponseBulder())
                .handleRequest(
                        instrumentSet + MeasureUtil.SEPARATOR + instrument,
                        count,
                        timeFrom,
                        timeTo);
    }
}
```

Fig. 152 DataService class

Also created an InstrumentsService web service class to retrieve device list data. Its structure and principle of operation are similar to the DataService class.

The UML diagram of the implementation of the web service is presented in Fig. 154.

The container for JEE and web services was the JBoss AS open-source software server, which has open source, fully supports the JEE standard, has high-quality documentation, and a large developer community.

The Postman REST Client application was used to test the web service.

The result of sending a request to save measurement data to the server is shown in Fig. 155.

The result of receiving test measurement data from the server in text format is shown in Fig. 156.

```java
public class DataRequestHandler {

 private DataBean dataBean = (DataBean)EjbHelper.loadBean(DataBean.class);
 private DataResponseBulder builder;

 public DataRequestHandler(DataResponseBulder builder) {
       this.builder = builder;
 }

 public Object handleRequest(String fullChannelName, int count,
                                  Calendar timeFrom, Calendar timeTo){
       List<Measure> measureData;

       if(timeFrom != null) {
             if(timeTo == null) {
                   timeTo = new GregorianCalendar();
             }
             measureData = dataBean.getMeasureData(fullChannelName,
                                                    timeFrom, timeTo);
       } else {
             measureData = dataBean.getMeasureData(fullChannelName, count);
       }

       return builder.build(measureData);
 }

}
```

Fig. 153 DataRequestHandler class

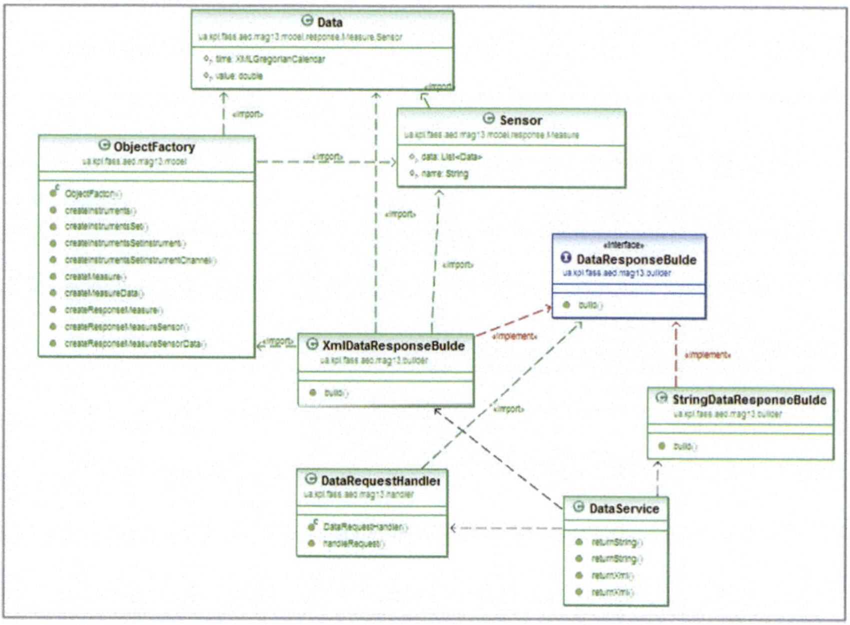

Fig. 154 UML diagram of web service implementation

Fig. 155 The result of sending test measurement data to the server

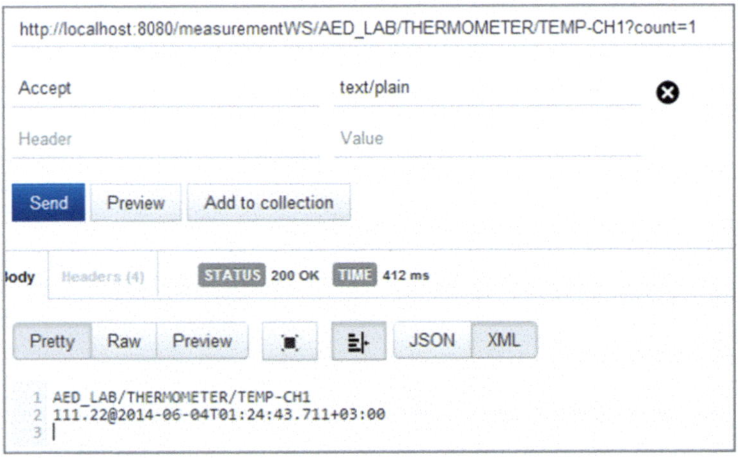

Fig. 156 Obtaining test measurement data in text format

The result of receiving data from the measuring device in XML format is given in Fig. 157.

The server's response to a request for a list of devices in text format is shown on Fig. 158.

Fig. 157 The result of receiving data in XML format

Fig. 158 The result of executing a device list request

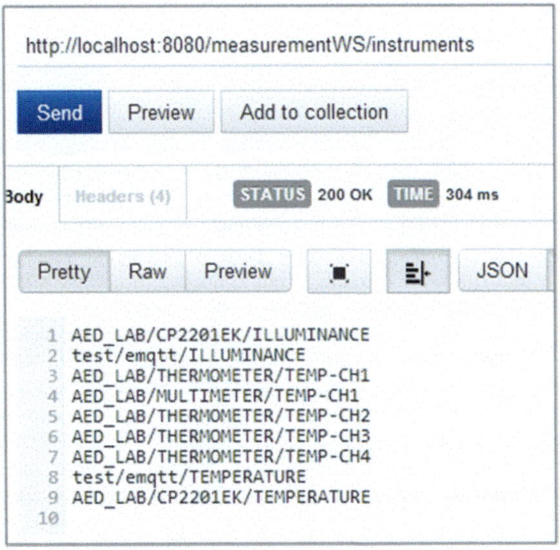

References

1. Chong, C.-Y., Kumar, S.P.: Sensor networks: evolution, opportunities, and challenges. Proc. IEEE **91**(8), 1247–1256 (2003). https://doi.org/10.1109/JPROC.2003.814918
2. IEEE 1451.0-2007; IEEE Standard for a Smart Transducer Interface for Sensors and Actuators—Common Functions, Communication Protocols, and Transducer Electronic Data Sheet (TEDS) Formats, pp. 1–335. IEEE, Piscataway Township, NJ, USA (2007)
3. IEEE Std 1451.1-1999, Standard for a Smart Transducer Interface for Sensors and Actuators - Network Capable Application Processor (NCAP) Information Model, Institute of Electrical and Electronics Engineers, Inc., PiscataNetwork way, New Jersey 08855, 25 June 1999 (1999)
4. IEEE Std 1451.2-1997, Standard for a Smart Transducer Interface for Sensors and Actuators - Transducer to Microprocessor Communication Protocols and Transducer Electronic Data Sheet (TEDS) Formats, Institute of Electrical and Electronics Engineers, Inc., Piscataway, New Jersey 08855, 26 September 1997 (1997)
5. Oliveira, D., Pinheiro, J., Neto, L., Pinto, V.H., Gonçalves, G.: A plug-and-play solution for smart transducers in industrial applications based on IEEE 1451 and IEC 61499 standards. Sensors **22**, 7694 (2022). https://doi.org/10.3390/s22197694
6. Barrero, F., Toral, S.L., Vargas, M., Becerra, J.: Networked electronic equipments using the IEEE 1451 standard—VisioWay: a case study in the ITS area. Int. J. Distrib. Sens. Netw. **2012**, Article ID 467124, 12 pp. https://doi.org/10.1155/2012/467124
7. Javed, A.R., Shahzad, F., Rehman, S.U., Bin Zikria, Y., Razzak, I., Jalil, Z., Xu, G.: Future smart cities: requirements, emerging technologies, applications, challenges, and future aspects. Cities **129**, 103794 (2022)
8. Wang, H., Zhou, Y.: Research and implementation on networked smart sensor based on IEEE1451 standard. Open Electr. & Electron. Eng. J. **8**, 492–500 (2014)
9. Lee, K.: IEEE 1451: a standard in support of smart transducer networking. In: IEEE Instrumentation and Measurement Technology Conference, 1–4 May 2000, Baltimore, MD USA (2000)
10. Schmalzel, J., Figueroa F., Morris, J., Mandayam, S., Polikar, R.: An architecture for intelligent systems based on smart sensors. IEEE Trans. Instrum. Meas. **54**(4), 1612–1616 (2005)
11. Viegas, V., Postolache, O., Dias Pereira, J.M.: Transducer electronic data sheets: anywhere, anytime, anyway. Electronics **8**, 1345 (2019). https://doi.org/10.3390/electronics8111345
12. Popelka, J., Paces, P.: Performance of smart sensors standards for aerospace applications. PRZEGLĄD ELEKTROTECHNICZNY (Electr. Rev.), pp. 229–232 (2012). ISSN 0033-2097, R. 88 NR 1a/2012
13. TY - RPRT TI - Second workshop on wireless sensing: proceedings AU - Gilsinn, James AU - Lee, Kang AU - Schneeman, Rick AU - Huang, Hui-Min PY - 2002 PB - National Institute of Standards and Technology CY - Gaithersburg, MD SN - NIST IR 6930. https://doi.org/10.6028/NIST.IR.6930, ER. https://nvlpubs.nist.gov/nistpubs/Legacy/IR/nistir6930.pdf
14. Mishra, B., Kertesz, A.: The use of MQTT in M2M and IoT systems: a survey. IEEE Access **8**, 201071–201086 (2020). https://doi.org/10.1109/ACCESS.2020.3035849
15. Dizdareviă, J., Carpio, F., Jukan, A., Masip-Bruin, X.: A survey of communication protocols for Internet of Things and related challenges of fog and cloud computing integration. ACM Comput. Surv. **51**(6), 1–29 (2019). https://doi.org/10.1145/3292674
16. Manowska, A., Wycisk, A., Nowrot, A., Pielot, J.: The Use of the MQTT protocol in measurement, monitoring and control systems as part of the implementation of energy management systems. Electronics **12**, 17 (2023). https://doi.org/10.3390/electronics12010017
17. Bhatt, C., Dey, N., Ashour, A.: Internet of Things and Big Data Technologies for Next Generation Healthcare. Springer (2017). https://doi.org/10.1007/978-3-319-49736-5
18. Hunkeler, U., Truong, H.L., Stanford-Clark, A.: MQTT-S—A publish/subscribe protocol for wireless sensor networks. In: Communication Systems Software and Middleware and Workshops (2008)

19. Luzuriaga, J.E., Cano, J.C., Calafate, C., Manzoni, P., Perez, M., Boronat, P.: Handling mobility in IoT applications using the MQTT protocol. Internet Technol. Appl. (ITA) (2015). https://doi.org/10.1109/ITechA.2015.7317403
20. Prada, M.A., Reguera, P., Alonso, S., Moran, A., Fuertes, J.J., Domínguez, M.: Communication with resource-constrained devices through MQTT for control education. IFAC-PapersOnLine **49**(6), 150–155 (2016). https://doi.org/10.1016/j.ifacol.2016.07.169
21. Yokotani, T., Sasaki, Y.: Comparison with HTTP and MQTT on required network resources for IoT. In: Proceedings of International Conference on Control, Electronics, Renewable Energy and Communications (ICCEREC), September 2016, pp. 1–6 (2016). https://doi.org/10.1109/iccerec.2016.7814989
22. Naik, N.: Choice of effective messaging protocols for IoT systems: MQTT, CoAP, AMQP and HTTP. In: Proceedings of IEEE International Systems Engineering Symposium (ISSE), October 2017, pp. 1–7 (2017). https://doi.org/10.1109/syseng.2017.8088251
23. Jaffey, T.: MQTT and CoAP, IoT protocols (2020). Accessed 23 Aug 2020. https://eclipse.org/615community/eclipsenewsletter/2014/febru-ary/article2.php
24. C8051F34x Data Sheet, 277pp. https://www.silabs.com/documents/public/data-sheets/C8051F34x.pdf
25. CP2200/CP2201 Single-Chip Ethernet Controller, 107pp. https://www.silabs.com/documents/public/data-sheets/CP2200.pdf
26. Barnett, D., Massa, A.J.: Inside the uIP Stack. Dr Dobbs J. (2005). Accessed 2 Feb 2015. https://www.drdobbs.com/inside-the-uip-stack/184405971
27. Belqasmi, F., et al.: RESTful web services for service provisioning in next generation networks: a survey. IEEE Commun. Mag. **49**(12), 66–73 (2011)
28. Mumbaikar, S., Padiya, P.: Web services based on SOAP and REST principles. Int. J. Sci. Res. Publ. **3**(5) (2013). ISSN 2250-3153
29. Deljouyi, A., Ramsin, R.: MDD4REST: model-driven methodology for developing RESTful web services. In: Proceedings of the 10th International Conference on Model-Driven Engineering and Software Development (MODELSWARD 2022), pp. 93–104 (2022). https://doi.org/10.5220/0011006300003119, ISBN: 978-989-758-550-0; ISSN: 2184-4348
30. Hernandez-Mendez, A., Scholz, N., Matthes, F.: A model-driven approach for generating RESTful web services in single-page applications. In: International Conference on Model-Driven Engineering and Software Development, pp. 480–487 (2018)
31. Ong, S.P., et al.: The materials Application Programming Interface (API): a simple, flexible and efficient API for materials data based on REpresentational State Transfer (REST) principles. Comput. Mater. Sci. **97**, 209–215 (2015)
32. Terzić, B., Dimitrieski, V., Kordić, S., Milosavljević, G., Luković, I.: Development and evaluation of MicroBuilder: a model-driven tool for the specification of REST microservice software architectures. Enterp. Inf. Syst. **12**(8–9), 1034–1057 (2018)
33. Zolotas, C., Diamantopoulos, T., Chatzidimitriou, K.C., Symeonidis, A.L.: From requirements to source code: a model-driven engineering approach for RESTful web services. Autom. Softw. Eng. **24**(4), 791–838 (2017)
34. Hadley, M., Sandoz, P.: JAX-RS: java API for RESTful web services. Java Specif. Req. (JSR) **311**, 2009 (2009)

Model of Information Signals Formation in the Diagnostics of Composite Products

Anastasiia Shcherban, Volodymyr Eremenko, Valentyn Mokiichuk, and Artur Zaporozhets

Abstract Products made of composite materials, unlike products made of metals, are formed from primary raw materials simultaneously with the formation of the materials themselves. In this case, the process of diagnosing of these materials is characterized by a large influence of random factors caused by changes in the properties of composites that arise as a result of the complexity of manufacturing processes, a large number of types of possible defects that do not lend themselves to a formalized description, the imperfection of diagnostic methods and defectoscopic equipment, and other factors. The reliability of the diagnostics in this case is determined not only by the physical methods used to obtain information about the technical condition of the product, but also by the mathematical models that are the basis of the diagnostic methods and the methods of processing the received information for the purpose of forming the spaces of diagnostic signs and making diagnostic decisions. In the tasks of diagnosing products made of composite materials, under the condition of a limited number of reference samples used for setting up diagnostic systems, the presence of an adequate model, according to which the synthesis of information signals characteristic of objects with various types of defects or degrees of damage occurs, is of great importance, because allows to solve several problems at the same time. First, the existence of such a model allows to create a set of information signals that correspond to possible defects of composites, and therefore can be used to train and adjust computerized diagnostic systems without physically manufacturing reference samples. Secondly, the developed model of the information signal can be used to select the limit value of the sensitivity of the computerized diagnostic system, evaluate its efficiency and adjust the main parameters, to determine the reliability of diagnosis and classification, etc.

A. Shcherban · V. Eremenko · V. Mokiichuk
National Technical University of Ukraine "Igor Sikorsky Kyiv Polytechnic Institute", Kyiv, Ukraine

A. Zaporozhets (✉)
General Energy Institute of NAS of Ukraine, Kyiv, Ukraine
e-mail: a.o.zaporozhets@nas.gov.ua

Green Technology Research Center, Yuan Ze University, Taoyuan, Taiwan

V. Eremenko and A. Zaporozhets (eds.), *Advanced Information-Measuring Technologies and Systems I*, Studies in Systems, Decision and Control 439,
https://doi.org/10.1007/978-3-031-40718-5_5

Keywords Hilbert linear process · Low-frequency acoustic methods · Benchmarkless diagnostics · Composite materials · Diagnosis of defects · Disturbance field

1 Structure of Diagnostic Information Processes

The theoretical basis for the creation of computerized diagnostic systems is based on the use of known and obtained new results:

- the study of physical processes that occur in composite materials during mechanical loads and the action of macro- and microstructure disturbance fields and the creation of new models of physical fields and information signals obtained using acoustic control methods;
- development of methods for processing information signals to identify diagnostic signs most sensitive to product defects;
- the development of statistical methods of processing input data taking into account the peculiarities of their distribution laws for the construction of vectors of diagnostic signs and decisive rules of diagnosis;
- development of modern hardware and software tools for diagnostic systems, methods, and techniques of their practical application.

The structure of information processes for diagnosing products made of composites was proposed (Fig. 1), which made it possible to specify the main directions of research, information processing methods, and technical solutions that can be used in the construction of computerized diagnostic systems [1, 2].

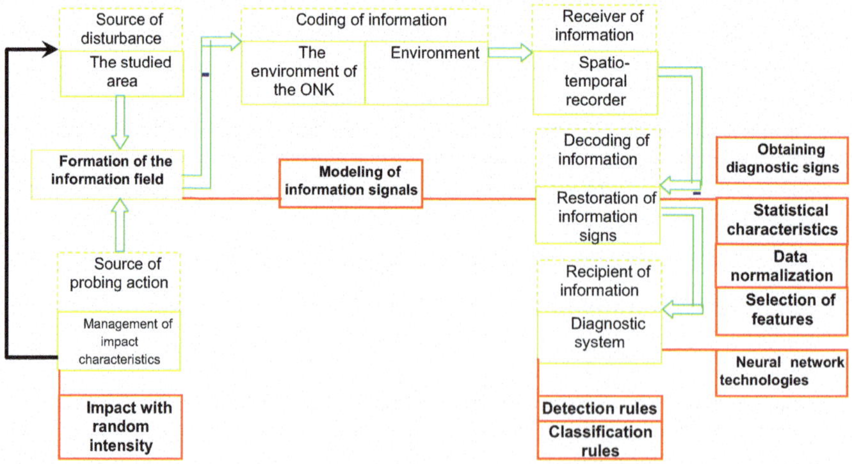

Fig. 1 The structure of information processes for diagnosing composite products

The source of information during diagnosis is the investigated area of the product, in which the probing action created by the corresponding transducer forms an information field [3]. The characteristics of this field, on the one hand, reflect the physical and mechanical properties of the studied area and encode information signals in a certain way, and on the other hand, they are determined by the physical method of diagnosis, which is the basis of the system. The properties of the field determine the specificity of further operations of coding, reception, and processing of information [4].

In general, a computerized diagnostic system should provide decoding of information, reception of information signals, processing and selection of diagnostic signs. Methods of decoding and processing information signals are largely determined by the characteristics and features of the signals themselves and can be improved in the process of designing diagnostic systems.

The obtained diagnostic signs are by their nature random values, so their statistical processing allows to significantly increase immunity and reliability [5].

The information processes of diagnosis also include the construction of feature vectors, which are based on the determination of their diagnostic value and largely conditions the algorithms of decision-making rules. These processes generally determine the structure of the computerized diagnostic system, its hardware, information processing algorithms, software architecture, as well as diagnostic methods and technologies.

Due to the large nomenclature of composite materials and their defects, as well as the difficulty of manufacturing standard samples of all types of possible structures with all possible defects (it is necessary to take into account that the composite material itself is formed simultaneously with the product and most technologies do not allow obtaining separate material samples), an important task is the implementation of the principles of standard-free diagnosis of composite materials, which imposes additional conditions on the information processes of diagnosis [6]. The principle of standard-free diagnostics is that the initial set of diagnostic systems is carried out without the use of standard samples of composite materials with standardized defect characteristics.

Standard-free diagnostics is possible in two variants:

1. there are samples of the studied composite materials that do not contain defects, and the system is adjusted with the help of reference samples, and with the help of classifiers that implement non-reference defectoscopy methods, samples of diagnostic features from the reference sample are compared with samples of features from the sample under study and, according to certain criteria, it is accepted the decision.
2. in the absence of defect-free samples of the investigated materials, a sample of diagnostic features from the defect-free zone of the investigated product is taken as a training sample, as the most probable, for example, by the 3-point method.

Standard-free diagnostics is based on the determination of the differences in the diagnostic feature in one zone of the product under study relative to another (other zones). To build decision-making rules in this case, statistical criteria for testing hypotheses are used, which are used to assess the statistical significance of differences between compared samples of informative parameters. Therefore, for the construction and application of unstandartized decision rules, the accumulation of statistical material and its processing is necessary.

2 Mathematical Model of the Information-Signal Field

One of the most important objects in solving the problem of diagnosing products from composite materials is physical and mathematical models of the information-signal field that is formed in the process of diagnosing composites. Although such models are virtual copies of real processes, they play a decisive role in choosing methods for processing information signals and determining diagnostic signs, decision-making methods, and the final result of the diagnosis.

The theoretical model of the source information obtained from the object of diagnosis includes a set of knowledge, assumptions, hypotheses, initial and boundary conditions, which homomorphically reflect the main properties and characteristics in space and time of the source information and is built in the form of a coherent logically sustained structure, formulated using necessary objects, terms, and symbols to solve a given class of problems. Theoretical models can be divided into physical and mathematical models. Each of these models complements the other and makes it possible to make the diagnostic process more effective, providing the specified indicators of probability, productivity, etc.

In the process of building a physical model, first of all, are displayed the physical properties of the research object, its elements, and their interaction, taking into account physical law. The development of a physical model for diagnosing composite materials is significantly complicated by their significant heterogeneity, dispersion of properties between layers, large number of types of defects, and various physical mechanisms of their influence on structural characteristics.

The mathematical model to a greater extent formalizes the process of mapping the properties of the object into a model, which allows, on the one hand, depending on the statement of the diagnostic problem, to build different models for one object of research, and on the other hand, to use one model for the study of different objects [7].

In the process of diagnosing composites by low-frequency acoustic methods, a physical field of mechanical disturbances of the macro- and microstructure of the studied area of the composite is formed under the influence of pulsed mechanical loading (Fig. 2), which is carried out by the primary transducer. This is the impact of the impact energy of a fight in diagnosis by the method of free oscillations and the method of low-speed impact, harmonic damping force in diagnosis by the pulse impedance method [8, 9].

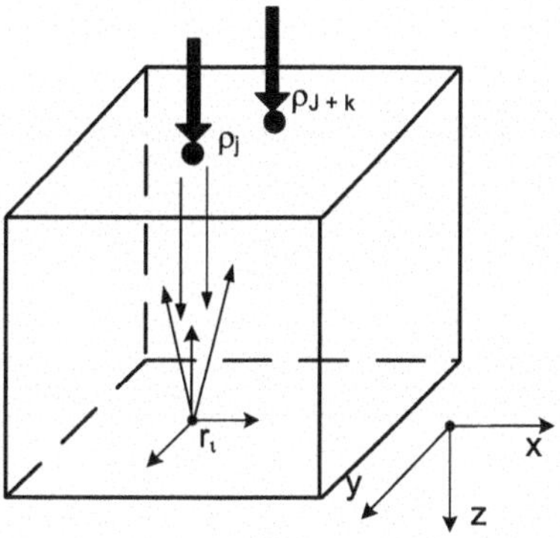

Fig. 2 Formation of the field of mechanical disturbances

The mechanism of field formation is determined by three main physical phenomena that characterize the material under study—this is propagation, scattering, and absorption of energy, they generate processes of elementary disturbances of particles of the material structure.

The field of mechanical disturbances is formed by the joint influence of many disparate micro- and macro-particles involved in the processes of propagation, dispersion, and absorption of mechanical excitation energy, among which defects in the structure of the composite play a significant role. In general, it is impossible to determine the entire set of sources of elementary disturbances and classify them by types, intensity, features, and significance, which does not allow to specify the field in each case. This significantly complicates the study of the properties of information signals [10].

It is possible to distinguish the main mechanisms of the formation of the field of mechanical disturbances:

- volumetric, due to the redistribution of energy on the macro- and microparticles of the structure, which fill the entire volume of the composite as a whole;
- in the layer determined by the special physical and mechanical properties of each layer of the composite;
- marginal, due to the redistribution of energy at the boundary of the layers.

These physical mechanisms of field formation determine the main characteristics of acoustic low-frequency diagnostic methods. The impedance method is based on the disturbance field formed in the surface layer and at its interface with other layers, the free oscillation method, as an integral method, is built on the volumetric mechanism of field formation, the low-speed shock method uses both the volumetric mechanism and mechanism of field formation in a separate layer. This classification is convenient for the mathematical description of field formation processes since the resulting field can be represented as a linear combination of several formation mechanisms.

The complex mechanisms of formation of the resulting disturbance field, the great variety of heterogeneous particles, defects, properties of layers forming local fields, and various combinations of these fields do not allow to create deterministic model that takes into account all these components. Therefore, a probabilistic approach to the development of the such model is proposed. It is based on the adoption of a hypothesis about the distribution of particles in a layer, the sizes of particles and layers, and their physical characteristics, and allows not to analyze processes of scattering and absorption of energy on each particle and in each layer, but to conduct research statistical characteristics of information signals and build the entire diagnostic complex on their basis [11, 12].

A constructive mathematical model of the researched information field of signals is proposed based on the use of methods of statistical mechanics, and the theory of random processes and fields. This model includes an impulse transient function that reflects the physical nature of the propagation of mechanical disturbances in each local area of the composite, takes into account both the temporal and spectral, spatial features of this propagation and the probabilistic character of the disturbances in the general case is described by the properties of the generating noise. On the other hand, such a model makes it possible to form realizations of informational signals using computer simulation.

Taking into account the need to build a model for practical use, certain conditions must be met: the examined zone of the composite in which elementary disturbances propagate is a linear medium, for which the property of superposition occurs when summing up the elementary mechanical pulses of disturbance of the zone of the macrostructure of the composite; these pulses are independent random variables of the form:

$$\alpha_{ij}(\mathbf{r}_i, \boldsymbol{\rho}_j; \tau_k, t), \ t \in T, \tag{1}$$

where i $\boldsymbol{\rho}_j = (x_j, y_j, z_j)$—respectively, the spatial coordinates of the point of elementary disturbance and the place of observation (location of the sensor); τ_k and t—moments of load action time and current observation time.

Suppose that the elements that form the structure of the composite in a certain layer are distributed discretely, and the elementary pulses of disturbances have random amplitudes $\alpha_{i,j}$ and moments of time of occurrence τ_k.

Each event (applying a mechanical impulse to the studied area) has some finite intensity A, causes elementary perturbations, and, taking into account the mechanisms of absorption and scattering, we obtain:

$$A = \int_0^\infty \alpha^2(t)dt < \infty.$$

Thus, the formed field can be attributed to the class of random Hilbert fields with finite energy.

If we imagine that the duration of the excitation pulses (mechanical load) is insignificant, and the duty cycle is sufficiently large, which allows us to neglect their overlap, then the total field of elementary disturbance pulses, which arises under the action of mechanical excitation, forms a random Hilbert linear field of the following form:

$$\xi(\boldsymbol{\rho}_j, t) = \sum_{=1}^{n} \sum_{k=1}^{m} {}_{\mathrm{j}}(\mathbf{r}_i, \boldsymbol{\rho}_j; \tau_k, t). \tag{2}$$

A number of elementary signals n, that create a field at a point of measurement at a moment in time t is also a random variable. If we assume that the number of particles in the elementary volume of the composite layer is constant, then for a given instant of time t there will be a defined value $< n_1(t) >$ equal to the average number of disturbance signals arriving at the receiving point per unit of time.

Making the following assumptions about the generated field:

- sources of elementary excitations in the environment of the composite are located statistically independently;
- for a certain region of the layer, the average density of macro- and microparticles that cause disturbances is assumed to be constant in space;
- we get that quantity n of elementary excitation signals arriving at the reception point in the time interval $[t - {}^{T}/_{2}; t + {}^{T}/_{2}]$ is distributed according to the Poisson law:

$$P(n) = \frac{(< n_1 > T)^n}{n!} \exp(- < n_1 > T).$$

Thus, the superposition of elementary pulses of disturbance, taking into account the physical nature of their formation and propagation, which forms the studied information-signal field at a spatial point $\boldsymbol{\rho}_j$, makes it possible to use the constructive model of the Hilbert linear field:

$$\xi(\boldsymbol{\rho}_j, t) = \sum_{=1}^{n} \sum_{k=1}^{m} \phi(\mathbf{r}_i, \boldsymbol{\rho}_j; \tau_k, t)\eta(\boldsymbol{r}_i, \tau_k), \tag{3}$$

where respectively $\phi(\mathbf{r}_i, \boldsymbol{\rho}_j, \tau_k, t)$ is the deterministic impulse transient function of a spatial filter with time-varying parameters; $\eta(\mathbf{r}_i, \tau_k)$ is a random field of Poisson white noise with discrete arguments, due to the input action of a disturbance at a point in space $\mathbf{r}_i$ and in a moment of time τ_k.

This form of representation of a random field is a partial case of canonical expansions of random functions, which are based on infinitely divisible distribution laws, partial cases of which are Gaussian, Poisson, and gamma distributions. The theory of canonical distributions makes it possible to determine the main statistical characteristics of the total field based on the known properties of the random variables that form it.

It is established that in the limiting case, when the sequence of elementary perturbation pulses grows and goes to infinity, a continuous Hilbert linear random field of the form:

$$\xi(\boldsymbol{\rho}, t) = \int_{\mathrm{R}^3} \int_0^{\infty} \phi(\mathbf{r}, \boldsymbol{\rho}, \tau, t)\eta(\mathbf{r}, \tau) d\mathbf{r} d\tau. \tag{4}$$

It is shown that the characteristics of the impulse transient function of the spatial filter with time-varying parameters $\phi(\mathbf{r}, \boldsymbol{\rho}, \tau, t)$ are determined by the physical and mechanical characteristics of each examined zone of the composite, their values depend on the degree of defectiveness (type and size of defects) of the controlled zone. The field $\eta(\mathbf{r}, \tau)$ is formed by a sequence of pulses of mechanical disturbances, the random change in the characteristics of which is due to the stochasticity of their appearance in time and space, and which can be represented by a random field of white noise. The process of excitation of the controlled zone of the composite is limited by the observation interval T finite duration of time $t \in T$, and the corresponding finite spatial domain $\mathbf{r} \in G$, i.e., the studied information-signal field is localized in time and space and is represented by a Hilbert linear random field of the form:

$$\xi(\boldsymbol{\rho}, t) = \int_{R^3} \int_0^\infty \phi(\mathbf{r}, \boldsymbol{\rho}, \tau, t)\eta(\mathbf{r}, \tau)d\mathbf{r}d\tau \cdot I(\mathbf{r}, \mathrm{t}), \tag{5}$$

where $I(r, t) = \begin{cases} 1, r \in G, t \in T \\ 0, r \notin G, t \notin T \end{cases}$—multi-parameter indicator function.

This random field model for each diagnostic task has a specific representation for given research conditions and can be piecewise uniform and piecewise stationary on finite intervals of space and time.

For the practical use of the proposed model, a transition to its digital analog was made. An image of a discrete linear random field is obtained in the form:

$$\xi_{j,k} = \sum_{u=1}^{N} \sum_{s=0}^{L} \phi_{u,v,s,m} \cdot \varsigma_{u,s} \cdot I(\mathbf{r}, t), \tag{6}$$

where $\phi_{u,v,s,m} = \phi(u\mathbf{h}, v\mathbf{p}, s\Delta\tau, m\Delta t)$; $\mathbf{h} = (\Delta x_i, \Delta y_i, \Delta z_i)$—the quantization step by spatial variables in the perturbation space; $\mathbf{p} = (\Delta x_j, \Delta y_j, \Delta z_j)$– quantization step by spatial variables in the measurement domain; $\Delta\tau$—time discretization step of mechanical load; Δt—the sampling step in time of the information signal; $\varsigma_{u,s}$—discrete readings of white noise, which describes the random component of the information signal, due to the heterogeneity of the composite material, reverberation and frictional noises and other influential factors; u, v, s, m—whole numbers.

The use of the image (6) made it possible to obtain models of realizations of random fields, the characteristics of which depend on the kernel parameters $\phi_{i,j,s,m}$ and generating field $\varsigma_{i,s}$. For fixed $\boldsymbol{\rho}$, a partial case of the model (6) is obtained, namely a linear random process ξ_m with discrete time:

$$\xi_m = \sum_{s=-S}^{S} \phi_{s,m} \cdot \varsigma_s \cdot I(\mathbf{r}, t), \tag{7}$$

which corresponds to the process of determining the characteristics of the information signal in a specific area of the product [6, 11].

3 Diagnostic Method

The developed stochastic model of the information-signal field made it possible to propose and investigate a diagnostic method based on the randomization of the intensity of the stimulating effect on the studied area [12], i.e. applying a low-speed blow to the controlled area of the product using a striker with variable impact energy. The change in energy occurs according to a random law with known statistical characteristics. The decision about the presence of a defect in the controlled zone is made based on the measurement of the pulse parameters of the force of impact interaction using statistical criteria.

If we consider the impact interaction process as linear in the area of elastic deformations, then the signal $\xi(t)$ recorded by the force sensor installed on the striker of the impact mechanism can be considered as the response of a linear system with an impulse response $\phi(\tau, t)$ to increments of the generating process $\Delta\eta(\tau_j)$. This response is described by a stochastic integral over a random function and is also a linear random process by analogy with Eq. (7), whose integral image:

$$\xi(t) = \int_{-\infty}^{\infty} \phi(\tau, t) d\eta(\tau), \quad t \in T,$$

where T—process definition area.

Kernel parameters $\phi(\tau, t)$ of the process $\xi(t)$ are determined by the mechanical properties of the controlled area, which in turn depend on the presence of defects in the product. If we consider this system to be invariant in time:

$$\phi(\tau, t) = \phi(\tau - t),$$

then its response is a linear stationary process, which is used as a stochastic model of shock interaction signals. This process must satisfy the condition of physical realizability and have a finite capacity:

$$M(\xi^2(t)) < \infty,$$

that is, the variance of process increments $\eta(\tau)$ must be finite, and the kernel of the process square integrable. These conditions are satisfied by the class of Hilbert linear random processes.

A linear link with an impulse response $\phi(\tau, t)$, which models the transformation of the generating process into an informative signal at the output of the impact force sensor, can be represented in the form of a mechanical system "mass-spring-damper" with one degree of freedom [13], which is acted upon by a disturbance force in the form of a sequence impulses $\eta(t)$. Moreover, the compliance of the spring in this model is determined by the compliance of the collision zone B, and the mass attached to the spring corresponds to the mass of the striker γ of the shock mechanism. The impulse response of this link is determined by the expression:

$$\phi(\tau) = U(\tau)\frac{\omega^2}{\Psi}e^{-\beta\tau}\sin(\Psi\tau),$$

where β—attenuation coefficient; ω—resonant frequency; $\Psi = \sqrt{\omega^2 - \beta^2}$;
$U(t) = \begin{cases} 1, t \geq 0 \\ 0, t < 0 \end{cases}$—the Heaviside function.

In turn, the ductility B of the impact zone will be determined mainly by the ductility of the material of the product at the point of impact (since the striker is made of metal, the ductility of which are several orders of magnitude lower than the ductility of the composite material under control) [13]:

$$B \approx \frac{(1 - \frac{\mu^2}{E})^{2/3}}{1.82\sqrt[3]{F_m R}},$$

where μ, E—respectively, Poisson's ratio and Young's modulus of the controlled material; R—the contact area of the striker with the product during impact; F_m—the maximum impact force value.

Under the condition of constant contact area, elastic compliance behaves like a linear spring. Therefore, the presence of a defect in the controlled product is similar to the insertion of a spring with a compliance B_d between the striker and the product, the stiffness of which decreases with an increase in the size of the defect:

$$B^1 = B + B_{\mathrm{d}}.$$

In the general case, it is not possible to calculate the ductility of the area, because the shape of the defect, the boundary conditions, and the coordinates of the shock action have not been determined. Therefore, the presence of a defect in the product leads to a change in the susceptibility of the mechanical system and its impulse response [14].

The transient characteristic can be roughly described by the appearance of impact force pulses $\phi(\tau)$ impulse response of the second-order link.

To transform the mechanical system accepted as a model into an aperiodic link, it is necessary to consider the limiting case when the actual damping C approaches for critical $C_k = 2\sqrt{B\gamma}$. Then the relative damping coefficient $\gamma = {}^{C}/_{C_k} \to 1$, and attenuation $\beta \to \omega$. In this case, the impulse response can be calculated as a limit in the form:

$$\lim_{\beta\to\omega}\phi_1(\tau) = \lim_{\beta\to\omega}\omega^2\tau e^{-\beta\tau}\frac{\sin\Psi\tau}{\Psi\tau}U(\tau) = \omega^2\tau e^{-\omega\tau}U(\tau).$$

For a real Hilbert linear process $\xi(t)$ the mathematical expectation and variance can be written as:

$$M\xi(t) = \kappa_1 \int_{-\infty}^{\infty} \phi(\tau, t)dt;$$

$$D\xi(t) = \kappa_2 \int_{-\infty}^{\infty} \phi^2(\tau, t)dt,$$

where $\kappa_1 = M\eta(t)$ and $\kappa_2 = D\eta(t)$—corresponding cumulants of the input process $\eta(t)$.

For a stationary input process $\kappa_1 = const$, $\kappa_2 = const$, then changes in the mathematical expectation and variance of the process $\xi(t)$, that are registered by the force sensor will be determined only by the parameters ω, Ψ, β transient characteristic $\phi(t)$, which depend on the mechanical properties of the controlled area of the product.

Similar considerations can be extended to moments of the third and higher orders. Thus, based on the above, it can be concluded that the stochastic characteristics of the impact process $\xi(t)$ depend on the mechanical properties of the controlled area of the product and can be used as informative parameters in the diagnosis of composite products [15].

4 Experimental Studies

The above theoretical provisions of the proposed diagnostic method confirm the experimental results obtained during the non-destructive control of cellular panels by the method of low-speed impact.

The surfaces of the samples were scanned by the primary impact force transducer, in which the impact energy on the surface varied quasi-randomly with a uniform distribution law in the range of values from 10 to 40 mJ. The samples were 5 composite panels with a cellular aggregate of the ISP-1 type and sheathing based on fiberglass T42/1-76. The thickness of the panel and cladding was 12 mm and 1.5 mm, respectively. The models of defects served as zones on which a point impact with normalized energy of 2.3 was previously inflicted by a steel ball; 2.8, 3.2, and 5.1 kJ, which caused the destruction of the sample in the contact zone (Fig. 3). For each of the 4 samples with defects and a sample without defects, realizations of shock interaction pulses were obtained, according to which the evaluation of the density of the probability distribution of the diagnostic feature—the amplitude of the shock interaction pulses (Fig. 4) and the duration of these pulses (Fig. 5) was carried out. Tables 1 and 2 show the values of statistical characteristics (mathematical expectation and root mean square deviation) of distributions of diagnostic features.

As can be seen from the examples given, the statistical characteristics of the shock interaction pulse parameters differ significantly in defect-free and defective zones and, thus, can be used as informative parameters during non-destructive testing. In addition, the given graphs indicate that in defective zones, not only the mathematical expectation and variance of the characteristic undergo significant changes, but also

Fig. 3 Sample composite panel

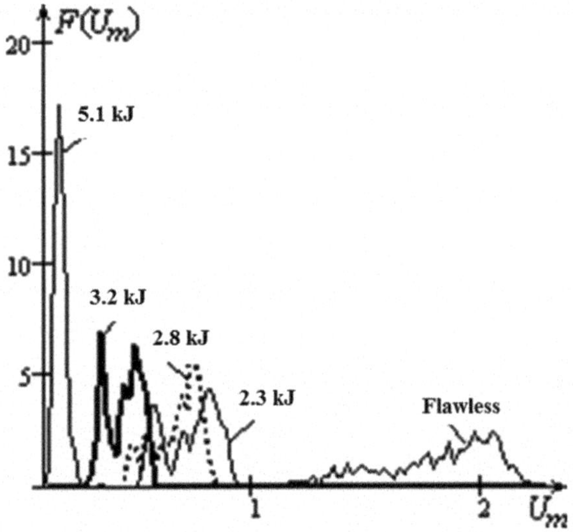

Fig. 4 Graphs of histograms of amplitude distributions of shock interaction pulses

the type of distribution law, which can serve as a source of additional characteristics and determines the need to use specific statistical procedures for processing non-Gaussian data for diagnosis. For example, the application of the Lehmann-Rosenblatt robust criterion. In the Table 3 shows the calculated values of the T-statistics of the Lehmann-Rosenblatt homogeneity criterion [6] for the values of the duration and amplitude of pulses in the defect-free zone and zones with corresponding impact damage for samples of 20 values (the limiting value of the statistic for a confidence probability of 0.95 is equal to 0.47).

Also, two experiments were carried out on the identification of defects in three groups of samples of composite materials No. 1–No. 3, for each group 100 results

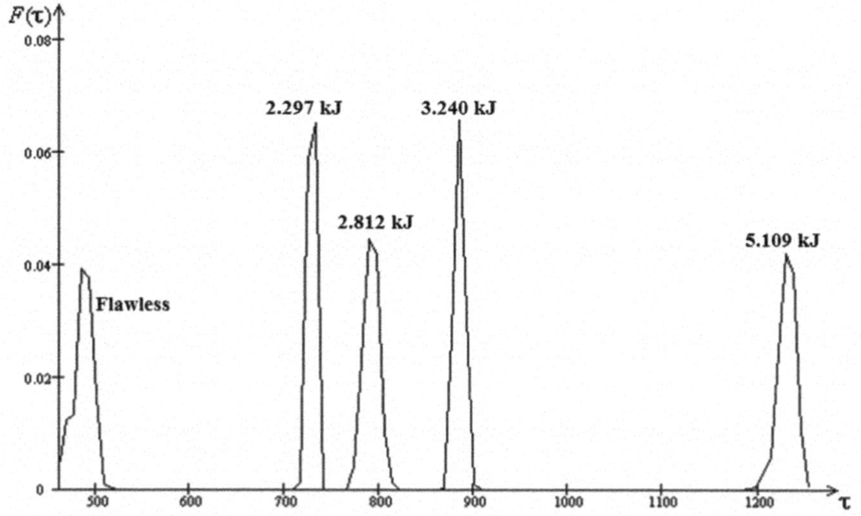

Fig. 5 Graphs of histograms of pulse duration distributions of impact interaction

Table 1 Values of statistical mathematical expectation of distributions of diagnostic features

Statistical parameters of pulse duration	Defectivity of the zone (destructive impact energy, [kJ])				
	Without def.	2.297	2.812	3.240	5.108
Mathematical expectation, [ms]	489.4	732.4	794.6	888.9	1233.8
MSD [ms]	10.3	2.9	7.5	4.9	7.1

Table 2 The values of statistical characteristics root mean square deviation of distributions of diagnostic features

Statistical parameters of pulse amplitude	Defectivity of the zone (destructive impact energy, [kJ])				
	Without def.	2.297	2.812	3.240	5.108
Mathematical expectation, [V]	1.82	0.74	0.69	0.46	0.18
MSD [V]	0.25	0.11	0.09	0.07	0.02

Table 3 The calculated values of the T-statistics of the Lehmann-Rosenblatt homogeneity criterion

The value of T-statistics when comparing parameters	Defectivity of zones (destructive impact energy [kJ])			
	2.297	2.812	3.240	5.108
Pulse durations	3.376	3.376	3.376	3.376
Amplitudes of pulses	3.376	3.376	3.376	3.376
Limit value T-statistics for $\alpha = 0.95$	0.47			

were obtained. At the first, the impact of energy did not change (classification matrix CL1), and at the second—it changed randomly (classification matrix CL2):

CL1				CL2			
Group	Number of results assigned to the group			Group	Number of results assigned to the group		
	№1	№2	№3		№1	№2	№3
№1	98	4	0	№1	100	1	0
№2	2	89	12	№2	0	98	2
№3	0	7	88	№3	0	1	98

It was established that in the case of diagnosis without changing the impact energy, the reliability value was 91.7%, and the randomization of the impact energy made it possible to increase it to 98.7%, which confirmed the feasibility of using the developed diagnosis method.

5 Conclusions

According to the results of theoretical and practical studies, the use of the proposed model allows to construct a set of information signals that correspond to possible defects of composites and therefore can be used for training and setting up computerized diagnostic systems without the physical production of reference samples. Also, the developed model of the information signal can be used to select the limit value of the sensitivity of the computerized diagnostic system, evaluate its efficiency and adjust the main parameters, to determine the reliability of diagnosis and classification.

References

1. Eremenko, V.S., Babak, V.P., Zaporozhets, A.O.: Method of reference signals creating in non-destructive testing based on low-speed impact. Tech. Electrodyn. **4**, C. 70–82 (2021)
2. Cooper, J., McGillan, K.: Probabilistic methods of analysis of signals and systems, 376p (1989)
3. Babak, V., Babak, S., Eremenko, V., Kuts, Yu., Myslovych, M., Scherbak, L., Zaporozhets, A.: Models and measures in measurements and monitoring. In: Studies in Systems, Decision and Control, vol. 360, 266p. Springer (2021)
4. Adams R.D., Cawley P.: Low-velocity impact inspection of bonded structures. In: Proceedings of the International Conference on Structural Adhesives in Engineering, pp. 139–142. Bristol (1986)
5. Campbell, N.R.: Physics: The Elements, 576p. Cambridge University Press (2013)
6. Babak S.V., Muslovych M.V., Susak R.M.: Statystycheskaia dyahnostyka elektrotekhnycheskoho oborudovanyia: Monohrafyia, 456 s. K.: Yn-t elektrodynamyky NAN Ukrain (2015)

7. Ieremenko, V.S., Kuts, Yu.V., Mokiichuk, V.M., Suslov, Ye.F.: Patent Ukrainy na vynakhid № 87864, MPK G01 N 29/00. Sposib neruinivnoho kontroliu materialiv i vyrobiv ta prystrii dlia yoho zdiisnennia.; zaiavnyk ta patentovlasnyk Natsionalnyi aviatsiinyi universytet. № u200702489; zaiava 06.03.2007; opubl. 25.09.2009. Biul. № 16
8. Cawley, P., Adams, R.D.: The mechanics of the coin-tap method of non-destructive testing. J. Sound Vib. **2**(122), 299–313 (1988)
9. Kutcheruk, V.Yu.: Classification and analysis of methods for assessing the condition of electric machines. Vymirjuval'na ta obchysljuval'na tehnika v tehnologichnyh procesah **4**, 56–62 (1999)
10. Wang, L., Zhang, Z.: Automatic detection of wind turbine blade surface cracks based on uav-taken images. IEEE Trans. Ind. Electron. **64**(9), 7293–7303 (2017)
11. Patsios, C., Wu, B., Chatzinikolaou, E., Rogers, D.J., Wade, N., Brandon, N.P.: An integrated approach for the analysis and control of grid connected energy storage systems. J. Energy Storage **5**(2), 48–61 (2016)
12. Babak, S., Babak, V., Zaporozhets, A., Sverdlova, A.: Method of statistical spline functions for solving problems of data approximation and prediction of object state. In: CEUR Workshop Proceedings, vol. 2353, pp. 810–821 (2019). http://ceur-ws.org/Vol-2353/paper64.pdf
13. Eremenko, V., Zaporozhets, A., Isaenko, V., Babikova, K.: Application of wavelet transform for determining diagnostic signs. In: CEUR Workshop Proceedings, vol. 2387, pp. 202–214 (2019). http://ceur-ws.org/Vol-2387/20190202.pdf
14. Dolinskiy, A.A.: Energozberezhennja ta ekologichni problemy energetyky. Visnyk Natl. Acad. Sci. Ukr. **2**, 24–32 (2006)
15. Sushko, M.Ya., Semeno, A.K.: Effective electrical conductivity of composite polymer electrolytes. In: 8th International Conference Physics of Liquid Matter: Modern Problems, p. 81 (2018)

Theory and Practice of Ensuring the Validity in Testing Laboratories

Valentyn Mokiichuk, Olha Samoilichenko, and Artur Zaporozhets

Abstract The features of the organization of ensuring the validity in testing laboratories are investigated. Measures to ensure the validity of the results depending on modern quality concepts are analyzed. The author shows the necessity of building a systematic model of ensuring the validity to achieve a balance between the customer's expectations regarding quality and the laboratories' quality statement and quality concept. The processes in the laboratory were analyzed in terms of the requirements of the ISO 17025 standard and the developed system model for ensuring reliability. The standards that allow applying qualitative and quantitative methods of validity assurance are considered. Examples of practical assurance measures for processes are given: equipment and personnel management facilities, deviation from the contract, externally provided products, LIMS, impartiality, sampling, risk management, management reviews, and improvement. The role of method validation to ensure validity is defined. The uncertainty of the result is considered as one of the options for generalizing the quantitative indicator of validity. Among the main disadvantages of quantitative methods are the insufficient power of classical criteria for small sample sizes and when the Gaussian distribution law is violated. The features of the implementation of methods for ensuring higher levels of validity are investigated. The algorithm of actions to improve the reliability of test results depending on the identified risks, available financial and human resources, and adopted quality goals, is shown.

V. Mokiichuk (✉)
National Technical University of Ukraine "Igor Sikorsky Kyiv Polytechnic Institute", Kyiv, Ukraine
e-mail: uncertainty@ukr.net

O. Samoilichenko
National University of Life and Environmental Sciences of Ukraine, Kyiv, Ukraine

A. Zaporozhets
General Energy Institute of NAS of Ukraine, Kyiv, Ukraine

Green Technology Research Center, Yuan Ze University, Taoyuan, Taiwan

A. Zaporozhets
e-mail: a.o.zaporozhets@nas.gov.ua

V. Eremenko and A. Zaporozhets (eds.), *Advanced Information-Measuring Technologies and Systems I*, Studies in Systems, Decision and Control 439,
https://doi.org/10.1007/978-3-031-40718-5_6

Keywords Validity of results · Testing laboratories · Risk management · Quality concepts · ISO/IEC 17025 · Uncertainty evaluation · Management of personnel · Accuracy · Equipment management · Sample size · LIMS · System approach

1 Features of Ensuring the Validity in Testing Laboratories

Ensuring quality testing in the context of the modern Total Quality Management (TQM) approach involves paying considerable attention to the issues of a risk-based approach to managing the testing process in testing laboratories. This approach has led to the publication of a new edition of the ISO 9000 and ISO 17000 series of standards with a special focus on the laboratory's ability to demonstrate the validity of the results obtained.

The term "test validity" is not defined in the International vocabulary of metrology – Basic and general concepts and associated terms (VIM), ISO 9000, and ISO 17000. The term "reliability" in the metrological sense means confidence in the measurement results. Measurements can be valid and invalid, depending on whether the probable characteristics of their deviations from the actual values of the relevant quantities are known or unknown. Measurement results whose probability is unknown are of no value and in some cases can serve as a source of misinformation [1].

In ISO/IEC 17025 [2], compliance Clauses 4 to 7 are a confirmation of the laboratory's ability to produce technically valid data and results. Clause 7.7 is devoted to methods for ensuring the validity of results.

Based on the requirements of the standard [2], the laboratory shall:

- document its procedures to the extent necessary to ensure the consistent application of its laboratory activities and the validity of the results (Clause 5.5 (c));
- ensure that the facilities and environmental conditions will not adversely affect the validity of results (Clause 6.3.1);
- monitor, control, and record environmental conditions by relevant specifications, methods, or procedures or where they influence the validity of the results (Clause 6.3.3);
  the equipment used for measurement shall be capable of achieving the measurement accuracy and/or measurement uncertainty required to provide a valid result (Clause 6.4.5);
- take practicable measures to prevent unintended adjustments of equipment from invalidating results (Clause 6.4.12);
- prevent deviations in contracts that affect the validity of test results (Clause 7.1.4);
- identify data in the reports that could affect the validity of the obtained results;
- carry out overall risk management based on the principle of proportionality to the potential impact on the validity of laboratory results (Clause 8.5.3);

- the sampling method shall address the factors to be controlled to ensure the validity of subsequent testing or calibration results (Clause 7.3.1).

The laboratory shall also have a procedure for monitoring the validity of results. This monitoring shall be planned and reviewed according to the (a)–(k) points of Clause 7.7.1, Clause 7.7.2 where available and appropriate.

From the point of view of the formal approach, the provision of objective evidence that a given item fulfills during the accreditation audit and subsequent surveillance by the conformity assessment body is sufficient to confirm competence according to [2]. However, most testing and calibration laboratories have long since moved from Quality Inspections to Business Excellence in their understanding of quality management [3]. Testing laboratories are almost always an element of larger and more complex or-organizations (as a component or stakeholder) that have not stopped at TQM, which implies the continuous improvement of customer-oriented quality as both requiring active management and involving the entire company [4].

Business Excellence is an overall way of working that balances stakeholder interest the likelihood of sustainable competitive advantages and hence long-term organizational success through operational, customer-related, financial, and marketplace performance excellence [5, 6]. Based on the Principles of Excellent Organizations [7], the conformity assessment body is guided by the PDCA principle in its testing activities [8]. This allows you to achieve Quality Management Benefits [9], reduce risks, and keep up with the competition.

Given the above, laboratories are not limited to the direct requirements of [2] but carry out an in-depth analysis of test processes to manage all subprocesses that have a direct impact on the validity of the test result.

The validity of test results is considered both in relation to certain types of tests and in terms of general approaches to the organization.

In [10] and its sequel [11], the validity of the results either by internal or external means in the field of non-automatic weighing instruments in applications in the pharmaceutical/medical field is considered. In particular, the implementation of the requirements of the standard [2] Clauses 7.7.1. and 7.7.2 as intermediate checks and participation in proficiency testing is shown. For each measurement, the instrument is presented with an acceptance criterion.

The validity of the results of hydrophone calibration by the mutual comparison calibration of hydrophone magnitude sensitivity is discussed in [12]. In the article [13] ensuring the validity of the results of calibrations and tests by interlaboratory check is given. Statistical methods help the need to identify the drift of the obtained data and use the analysis of variance.

When assessing validity, special attention should be paid to the size of the samples studied.

Validity issues related to the use of statistical criteria are quite relevant and have been discussed in many papers. The statistical criteria regulated by ISO 5725, ISO/TR 10017, and other standards are not able to provide the necessary validity with the limited amount of data that testing laboratories usually work with. This leads to an uncontrolled decrease in the validity of the data obtained, and laboratories make erroneous decisions.

For example, [14] shows that the power of the Cochrane criterion, which is used in various statistical methods of validity monitoring, significantly depends on the sample size. A methodology for determining the power of the criterion for different sample sizes is proposed. Study [15] is devoted to the determination of power on small sample sizes of other criteria – Fisher criterion and Student criterion. For their statistical power estimates, the coefficients of the approximating equation with uncertainty are obtained using regression dependencies, which allow us to determine the validity of the data obtained with limited sample sizes.

The planning of tests in terms of obtaining valid estimates with limited sample data, and creating a subsystem for adequate planning of the number of tests is considered in [16]. The dependences of the number of test repetitions for different values of accuracy and validity of estimates were obtained.

Improving the validity of determining the certified value of reference materials by accurately determining the quantiles of the expanded uncertainty distribution law is discussed in [17].

Reliability and validity as measurement fundamentals and methods for determining the reliability and validity of any given measurement process are given in [18].

Monitoring the validity result by participation in interlaboratory comparisons is discussed in [19]. It has been shown that participation in interlaboratory comparisons organized by an international provider has provided sufficient evidence of the reliability of measurement results in the field of laser performance measurements.

Thus, assurance in laboratories can be achieved through a variety of methods. Laboratory personnel must choose methods of assurance based on their own experience and knowledge. The chapter proposes to systematize the methods of validity monitoring and develop a unified scheme, the use of which will allow laboratory personnel to choose the optimal number of measures that will properly ensure the validity of the results obtained.

2 Approaches to the Organization and Assurance of Reliability in Laboratories

The process of ensuring the validity, like all other processes in Quality Management Systems, has responsible executors and is managed by the PDCA principle [8]. At the planning stage, monitoring measures are selected depending on the Quality Policy and the medium-term and long-term quality objectives. In practice, this means analyzing which quality management concepts reflect the laboratory's objectives. The section provides cases ranging from initial to full-fledged concepts. Such a wide range of possible concepts is provided to make this work useful not only for quality advanced laboratories but also for those who are just starting to implement the principles of the quality system as a declaration of compliance with certain quality requirements.

The section proposes to classify measures to ensure the validity of results depending on the current quality concepts: Quality Control, Quality Assurance, Quality Management, TQM, and Business Excellence (quality concepts classification is according to [3]). In the same way, the top management and quality team will carry out planning in a way that meets current objectives and does not impose unnecessary financial pressure on the laboratory. Actions to monitor the validity should be proportionate to the financial capabilities and the current adopted quality concept. In laboratories, the problem of moving to each subsequent stage of the quality concept is associated with the effective and efficient implementation of certain requirements, such as:

(1) control of the final products, processes, and services;
(2) continuous control of the process of preparation (production) of products, processes, and services;
(3) Preventing the detection of inconsistencies in work (including non-conforming products, processes, and services) at the preliminary stages, the implementation of the 6σ concept;
(4) setting of practices to manage an organization as a system based on four main components: quality planning, quality assurance, quality control, and quality improvement. Establish the organizational structure, roles and responsibilities, planning, operation, policies, rules, beliefs, objectives, and processes [8];
(5) the continuous improvement of customer-oriented quality as both requiring active management and involving the entire company [4];
(6) application of social tasks with the growing public interest, the common requirement for standards in quality, and the market pressure to deliver high-quality results [7, 9].

To achieve sustainable success and competitiveness of testing services in the market, laboratories are expanding the scope of testing services. The expansion of test methods should be combined with a set of techniques specific to the requirements (4)–(6). These requirements include such areas as a risk-based approach, validity monitoring, improvement, and analysis by management (Fig. 1).

The balance between expanding test methods and strengthening quality can be achieved by choosing the optimal number of validity monitoring measures that, on the one hand, will allow you to meet quality requirements while maintaining the required competitive level.

The basis of testing laboratories' activities is the target customer. This can be an internal customer if the laboratory is owned by the company and operates to test product quality indicators. An external customer is an independent consumer who orders a testing service based on the price-quality ratio. The Laboratory is interested in expanding the demand for the tests it conducts by continuously improving the quality and relative stability of its pricing policy.

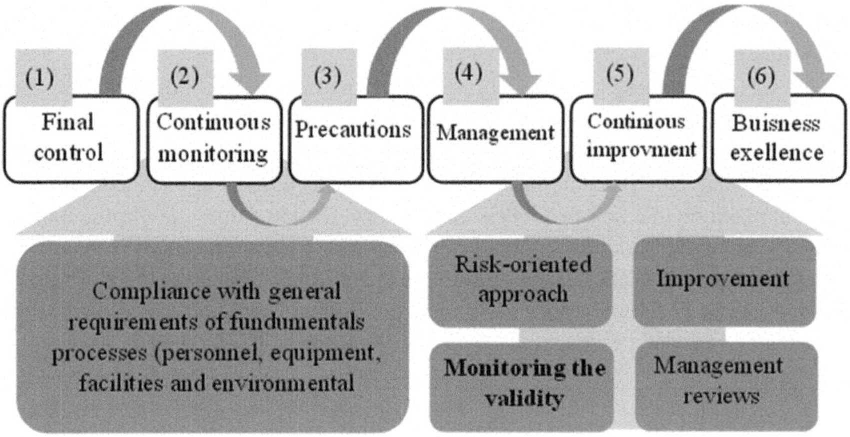

Fig. 1 Classification of validity assurance measures of results depends on the current quality concepts

To this end, inconsistencies from all processes in the laboratory are recorded and monitoring activities are planned to prevent potential inconsistencies. Such actions subsequently lead to the consistency of quality indicators with customer expectations.

The balance between the customer's quality expectations and the laboratory's quality statement and quality concepts can be disrupted. The reason for this is the lack of unified regulated methods for achieving validity, relying solely on the experience and competence of the personnel responsible for monitoring validity. The essence of solving the quality problem is to combine the factors influencing the completeness of validity monitoring as a single system of targeted, continuously performed validity monitoring actions at all stages of testing and in all processes. The system model of validity assurance based on the process approach and the quality concepts adopted in the laboratory is shown in Fig. 2.

The core of the model developed by the authors is accuracy (trueness and prescription) and validity. The completeness of their definition is dictated by the inconsistencies that arise during the management of processes following the requirements [2] and external factors. External factors include: rapidly changing customer requirements, customer satisfaction; business requirements; competition; process requirements outlined in test method standards. All of this is consistent with the available resources, and as a result, we have a Procedure for Ensuring the validity of results.

The main elements of the Procedure for Ensuring the validity of results that form its basis are (Fig. 3):

- ISO/IEC 17025 as the basis for all requirements;
- ISO 10012 in terms of measurement management;
- EN 45501 in regards to intermediate checks on measuring equipment;
- ISO/IEC 17043 and ISO 13528 in the part of proficiency testing by interlaboratory comparison, including statistical methods;

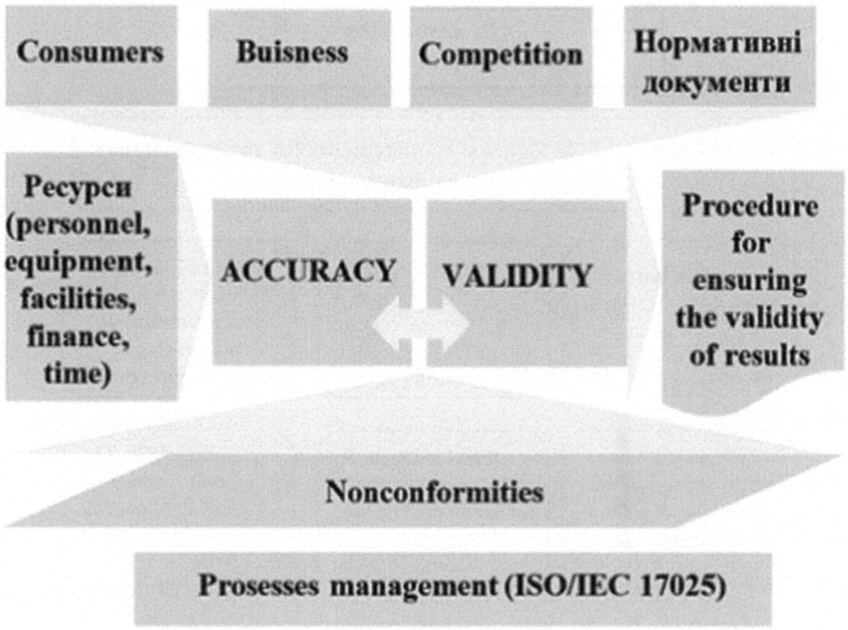

Fig. 2 The system model of validity assurance

- ISO 5725 as a tool for statistical data processing in almost all processes (equipment and personnel management, method validation);
- ISO 7870-2 control charts;
- ISO 13053-2 quantitative methods in process improvement, describes the tools and techniques at each phase of the DMAIC (define, measure, analysis, improve, and control) approach;
- ISO 10018 and ISO 10015 in the part of staff credibility (people engagement and Guidelines for competence);

Group of standards on uncertainty evaluation (ISO/IEC Guide 98-3:2008 Uncertainty of measurement – Part 3: Guide to the expression of uncertainty in measurement (GUM:1995), ISO/IEC Guide 98-3:2008/Supplement 1, 2, ISO 21748 Guidance for the use of repeatability, reproducibility and trueness estimates in measurement uncertainty evaluation, ISO/IEC Guide 98–6:2021 Uncertainty of measurement – Part 6: Developing and using measurement models ISO/TS 21749:2005 Measurement uncertainty for metrological applications – Repeated measurements and nested experiments, ISO/TS 23471:2022 Experimental designs for evaluation of uncertainty – Use of factorial designs for determining uncertainty function) and others (Fig. 3).

Given the wide range of standards and other guidance documents used to develop direct measures for monitoring reliability, it is necessary to develop a unified scheme that takes into account all the identified reliability opportunities.

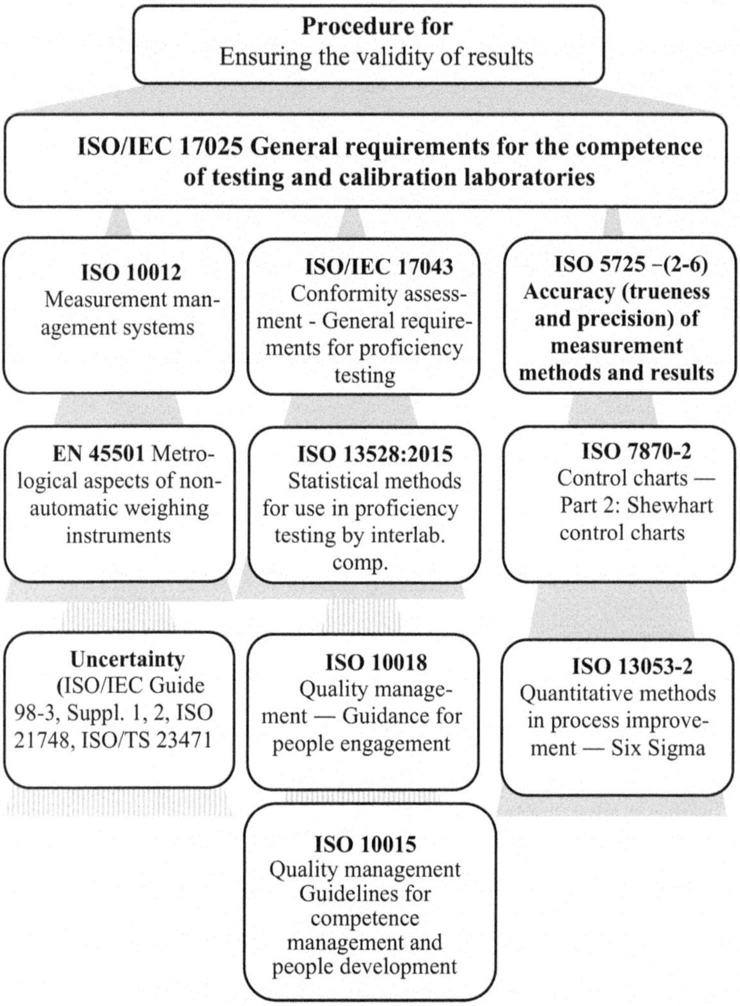

Fig. 3 The system model for validity assurance in laboratories

3 Analyzing Laboratory Processes for Validity Assurance

This paper draws on a wide range of studies aimed at monitoring validity. In particular, the Ishikawa diagram, proposed in [11], showing ways of monitoring the validity of measurement for weighing instruments reflects a practical approach to implementing monitoring of the validity of measurement in accordance with the requirements of [2]. Paper [18] discusses in detail the methods for determining the reliability and validity of any measurement process.

The authors propose a somewhat broader approach, which involves the development of a single concept for monitoring the validity of the adopted quality concept,

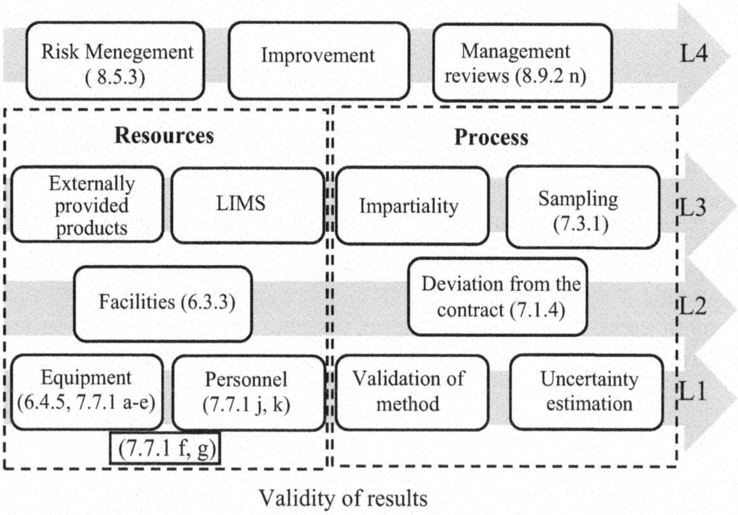

Fig. 4 Processes for assurance of validity

regardless of the type of measurement, based on all processes in the laboratory, not just the measurement process.

The analysis of the main processes in the laboratory allowed us to identify those that directly or indirectly, qualitatively or quantitatively affect the validity (Fig. 4). The clauses of the standard [2] with the requirement to ensure the integrity (if such a requirement is described in the standards as direct) are indicated in parentheses.

Processes related to validity assurance are conditionally divided into levels (L1–L4) depending on their impact. This breakdown will be useful for laboratories that do not need to undergo a formal accreditation procedure. For example, laboratories are owned by a product manufacturer. If a company cares about quality, it creates its own regulations, guidelines, and specifications including for testing laboratories. At the same time, the implementation of validity assurance measures is usually phased, starting with the most important processes. The authors have identified the L1 level processes as basic – the L4 level processes are advanced, for laboratories that can demonstrate the laboratory's validity capability maturity.

It takes a holistic view of the results' validity ensuring to include all of these processes. Let's take a closer look at the processes from Fig. 4 in terms of methods of validity ensuring.

L1 basic processes include personnel, equipment, method validation, and uncertainty evaluation. It should be noted at the outset that each of these validity assurance processes can also be divided into basic and advanced.

Qualitative methods of validity ensuring include those based on methods of qualitative expert evaluation, determined by the principle of realized/not realized. Quantitative methods are those whose implementation involves statistical methods and criteria with a certain confidence P-level.

3.1 Equipment Management

The equipment management process has the following methods to ensure the validity:

1. Calibration of the equipment with obtaining the expanded uncertainty at the operating points with a given confidence level. The operating points are selected depending on the test method.
2. Use of reference materials or quality control materials; it takes place in tests of the composition and properties of a substance.
   For example, gas analyzers and gas alarms use a gas mixture with certain concentrations of components (in the case of gas alarms, the concentration must be above a threshold value). For water analyzers, laboratories usually prepare quality control materials with the specified characteristics; for ph meters, buffer solutions are used.
   For optical measurements, the reference materials are optical filters – glasses with specific optical characteristics.
   For biological measurements (quality indicators of grain, seeds, and food products), the use of reference materials remains the main means of quality control of both equipment and personnel. More details on the use of reference materials for agricultural products are described in [20].
3. Use of alternative instrumentation that has been calibrated to provide traceable results.
   Examples include:
   - use of calibrated thermometers to control the temperature regime in thermostats, and laboratory oven;
   - use of thermocouples as alternative equipment for mercury (alcohol) thermometers;
   - interchange of spectrophotometer and gas (liquid) chromatograph;
   - interchange of equipment that implements express methods with chromatographic methods.
4. Functional checks of measuring and testing equipment. This method is one of the ways to implement the requirement of Claus 6.4.4 [2] and is a purely qualitative method of reliability monitoring. The method consists of the fact that the personnel makes sure that the device responds to switching on (if it is powered by the mains), provides a change in indicators when external influences change, and has no visible damage and other defects that will interfere with its functioning. The list of activities for Functional checks of measuring and testing equipment is planned based on the equipment manual and documented in internal procedures or work instructions.
5. Use of check or working standards with control charts. This method is based on the use of statistical methods for building control charts. The most common is the use of Shewhart control charts (ISO 7870–2). Working standards for such purposes are usually prepared internally by the laboratory according to the quality indicators in the scope of the laboratory.

6. Intermediate checks on measuring equipment. the most common methods are application:
   - calibrated weights for intermediate checks of scales;
   - optical filters for checking photometers and spectrophotometers;
   - solutions with a specified electrolytic conductivity for intermediate checks of conductometers;
   - buffer solutions for pH meters.

For Intermediate checks, inexpensive countermeasure materials are usually used if they are capable of changing characteristics.

3.2 Personnel Management

The human resources management process is the second fundamental mechanism for ensuring the validity, which is based on in-laboratory comparisons: repeated tests under different conditions using statistical criteria.

Most methods of testing biological objects are highly dependent on the competence of the personnel. These methods include: analyzing seed germination, determining the sowing qualities of seed potatoes, analyzing spectrophotometry results, and drawing conclusions based on them. The stage of sample preparation, where it is provided, also depends entirely on the competence of the personnel performing it.

Thus, the validity of the test results directly depends on the competence of the personnel. Personnel performs repeated tests under different conditions depending on the goal set, and the data obtained is processed using statistical criteria, which are used to make decisions about the competence of personnel. Tests conducted by competent personnel are considered to be valid tests.

The requirements of the standard [2] are:

(1) the laboratory shall ensure that the personnel have the competence to perform laboratory activities;
(2) the laboratory shall perform the monitoring competence of personnel;
(3) the laboratory shall develop internal guiding documents (procedures, regulations, work instructions) describing actions related to competence.

Such guidelines should include not only measures to test theoretical knowledge, but also methods to assess practical skills. The latter is achieved both by observing the performance of tests and by applying statistical criteria.

The standard [2] does not establish certain statistical criteria for testing competence. Typically, these are criteria for assessing homogeneity concerning mathematical expectations, and variances of two or more samples. The problem of ensuring the validity lies in the correct choice of statistical criteria. It is shown in [14, 15] that the use of generally accepted criteria leads to significant errors of the second kind when deciding on the competence of personnel.

Errors of the second kind arise due to insufficient amount of data under study, violation of the law of distribution of measurement results prescribed by the criterion, or incorrect choice of criteria.

In [14, 15], the power of the criteria for assessing sample homogeneity depending on the sample size is presented. In [21], the criterion commonly used in En laboratories was studied and it was shown that its use leads to a violation of validity. As a solution, a modified criterion is proposed and investigated that provides the necessary validity.

In [22] methodological aspects of assessing the competence of laboratory personnel in determining the quality of seed potatoes are considered. A method for assessing staff competence for discrete distribution laws has been developed, and sample sizes have been recommended to ensure the validity of personnel competence assessment. The validity assurance methods such as replicate tests or calibrations using the same or different methods and retesting or recalibration of retained items can be used in both personnel and equipment management. If repeated tests are performed by one specialist on different equipment, we check the equipment. If the tests are carried out on the same equipment by equal specialists, we check the specialists.

3.3 Validation of Methods

Method validation is the third pillar of ensuring the validity of results. The requirements for method validation are described in [23, 24]. It is shown in [25] that method validation is an integral part of ensuring the quality of test results. There are two main approaches to method validation; the interlaboratory comparison approach and the single-laboratory approach. The methods of the intra-laboratory approach are similar to those used in the processes of equipment and personnel management, and the methods of interlaboratory comparison are based on [26]. The requirement for method validation (verification) is mandatory in [2] Claus 7.2. The completeness of the definition of validation characteristics depends on the competence of the personnel and deserves separate detailed consideration. In general, it can be said that validation (verification) of a method is proof of its valid use.

3.4 Evaluating the Uncertainty of a Test Result

Evaluation of the uncertainty of the test result is the final stage that combines quantitative methods of validity evaluation (Fig. 5).

Quantitative methods of validity monitoring are characterized by the confidence probability *P*, which is one of the most common quantitative characteristics associated with validity. Confidence probability gives an idea of the controllability of the process and allows you to assess risks. Expanded Uncertainty is a characteristic

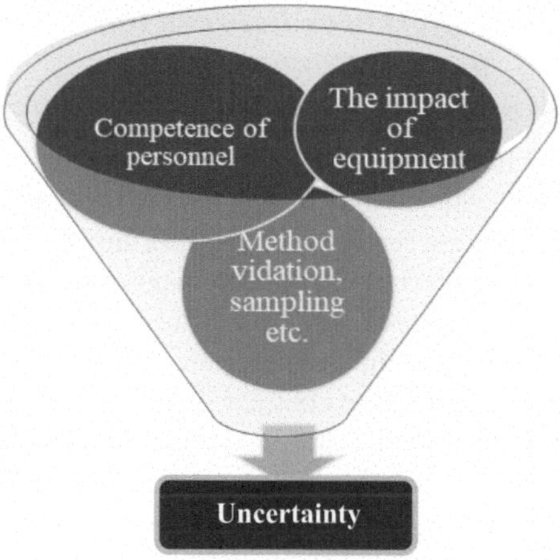

Fig. 5 General representation of the components of measurement uncertainty

that contains information about both the accuracy and validity of the test result. The customer of testing services draws a conclusion about the quality of testing in the laboratory based on it. Improving the accuracy of the extended uncertainty evaluation is an urgent issue that has been addressed in many works, including [17, 27–29].

3.5 Determination of Requirements for Facilities and Their Control

Determining the requirements for facilities and their control is the next level L2 process.

The laboratory shall monitor, control and record environmental conditions in accordance with relevant specifications, methods, or procedures or where they influence the validity of the results [2], Clause 6.6.3. In fact, only a small number of tests significantly depend on environmental conditions. For most laboratory tests, wide temperature and humidity limits are set, so the authors classified this characteristic as L2.

Experimental determination of the exact limits of environmental conditions, depending on the influence on the measurement result, is a task that is set during method validation/verification. This is not a mandatory validation characteristic, so the authors refer to the determination of the exact limits of environmental conditions as a sublevel advanced.

3.6 Differences Between the Request or Tender and the Contract

The issue of possible differences between the request or tender and the contract is an organizational one, but cannot be ruled out in the context of ensuring the validity. If the tests are carried out for conformity assessment, especially if they are certification tests, no deviations from the approved methods are permitted. If the tests are of a research nature or are carried out on samples for the customer's own needs, deviations requested by the customer are possible, provided that the possible loss of validity is specified in the contract. It is mandatory to provide quantitative data on validity if such an impact has been assessed during method validation. This is also not a mandatory validation characteristic; the authors refer to it as an advanced sublevel.

3.7 Externally Provided Products

Externally provided products have an indirect but important impact on assurance. For example, [2] does not single out this process as a tool for ensuring reliability, but indicates that the laboratory shall ensure that only suitable externally provided products that affect laboratory activities are used.

This includes reagents, test systems, pipette tips, filter paper, and other materials that the laboratory purchases to support the testing process.

Typically, laboratories are limited to conformity documents (manufacturer's declarations, quality certificates) provided by the vendor. However, practice shows that it is necessary to carry out internal checks of products before their use, with documentation of the personnel responsible for making the decision. For example, before using a reagent in testing, the customer is required to test it on samples for internal laboratory control similar to the test objects.

If the filter paper is used for the germination of seeds, it is necessary to check whether the paper does not introduce additional contamination. For this purpose, a biological test should be performed on a sample of filter paper from the batch received. Uninfected seeds that are not susceptible to the disease are used. The proof of the filter paper suitability assessment is the test report of the sample for intra-laboratory control with the conclusion that no diseases were detected in the sample under investigation.

Pipette tips can affect validity due to manufacturing defects. This is especially true for tests that involve adding reagents in microdoses. If the entire batch is defective, the laboratory does not use it. If individual tips in a batch are defective, the elimination of such drops, and thus the validity of the results, depends on the skill of the personnel. The skill of the pipette operator is checked, for example, during intermediate pipette checks by repeated testing.

When it comes to verifying the services provided, for example, calibration services, and equipment adjustment, laboratories usually do not order repeat work, as this imposes an unnecessary financial burden. Laboratories are guided by the reputation of companies and their own experience of collaboration.

3.8 Control of Data and Information Management

Laboratory Information Management System (LIMS) is used for the collection, processing, recording, reporting, storage, or retrieval of data, minimizing the influence of the human factor. Automation of the subprocesses, shown in Fig. 6, allows to increase the validity of the results by reducing the likelihood of technical errors from data transfer from one document to another, typos, etc.

The problem of LIMS is the correct choice of quality indicators that would simultaneously meet the internal requirements of the laboratory and the standard [2]. The lack of clear standardization during LIMS implementation leads to the fact that each organization chooses favorable characteristics and quality indicators, and metrics for performance evaluation. The laboratory interprets the obtained values of the selected metrics as maximum values, and grades the evaluation scale for each characteristic, taking into account its interpretation of the metric values and possible degrees of compliance. As a result, it obtains the maximum values of each characteristic, and, accordingly, the maximum value of software quality.

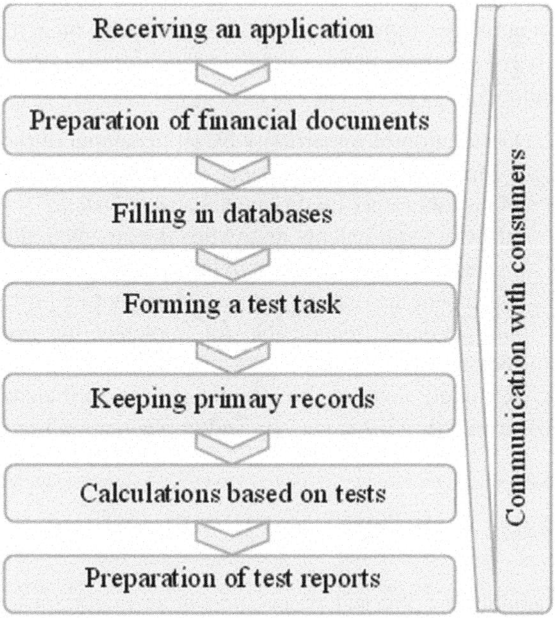

Fig. 6 Laboratory subprocesses that require LIMS implementation

This approach leads to formal satisfaction of software quality as a result:

- incomplete coverage of standardization objects by standards;
- the developer choosing standards that are favorable to him and adapting these standards to his needs.

There is also a lack of comprehensive methodologies that would allow assessing not only the impact of each individual characteristic on software quality but also the interaction of characteristics.

The analysis conducted in [30] revealed the presence of a significant number of LIMS-specific characteristics that are not present in software quality assessment standards. It has been shown that the presence of specific characteristics of LIMS does not allow to realize the functional completeness of the requirements of the standard [2] and requires the development of a comprehensive system of LIMS quality indicators.

3.9 Ensuring the Impartiality

Impartiality in the activities of the testing laboratory is an integral part of the requirement [2] present in all standards and Guides in conformity assessment. This process is classified by the authors as L3 since there is no direct dependence or impact on validity in [2] or ISO 15189. An in-depth analysis of the main requirements for impartiality conducted in [31] showed that the main requirement is the need to identify risks to impartiality, eliminate and minimize such risks. In terms of ensuring the validity, this can be realized as follows.

1. Declaration of impartiality by all personnel of the laboratory, either internal or external, who could influence the laboratory activities, in the appropriate forms of the laboratory quality management system.
2. Identify requirements that, when documented, do not fully eliminate threats to impartiality. Identify risks to impartiality
3. Document the requirements, implementation procedures, record-keeping forms, and personnel responsible for implementing and monitoring the impartiality measures.
4. Routinely analyze the effectiveness and efficiency of preventive actions with impartiality risks, identify and implement improvement measures.

3.10 Sampling

Sampling or physical preparation is an important test step that provides, among other things, information on the degree of homogeneity of the material under test concerning various factors. If the overall validity is underestimated, for example by ignoring the validity of sampling, this leads to erroneous decisions and associated risks. This is especially important during certification tests, when, by the certification scheme, the decision on the conformity of a batch of products is made based on obtained sample as a result of sampling.

Ensuring the validity of sampling usually has two components – qualitative and quantitative. The qualitative component is assessed by the expert method through authorization, competence monitoring, and ongoing training of samplers.

This is realized in the following subprocesses:

- supervision of the sampling process;
- theoretical surveys on the peculiarities of sampling scheme depending on the object of selection;
- commission decision-making on authorization;
- analyzing reports on sampling.

The quantitative component involves comparing the test results of two samples taken from the same batch by different samplers. It is also possible to test an arbitration sample if it was taken by a different sampler.

A quantitative estimate of sampling confidence can also be obtained as a component of the total uncertainty due to sampling.

Guide [32] has methods and approaches for measurement uncertainty arising from sampling. The estimation of sampling uncertainty for different objects has been repeatedly considered in the literature [33].

Standard [2] states that the sampling method shall address the factors to be controlled to ensure the validity of subsequent testing or calibration results. Paper [34] shows that among the factors to be controlled is the amount of data under study.

In the theory of analysis of variance, the number of observations in groups is not important for determining the influence of factors. However, since we are dealing with sample data of a limited size, it is necessary to assess the validity of the obtained estimates of the influence of factors.

The criterion for the validity of the estimates of the influence of the studied features is the ratio of their deviation to the deviation of unaccounted factors (F). If $F > F_{\alpha}$, then the estimates obtained are significant and the effect of sampling on the uncertainty of the analysis should be taken into account. In the last equation, $F_{\alpha}(\nu_1; \nu_2)$ is the limiting value of the Fisher distribution for degrees of freedom ν_1 and ν_2, which depend on the sample size under study and the level of significance α. To obtain a valid estimate of the uncertainty component of sampling, the paper presents a methodology. It is shown that the validity of the estimate decreases with limited data.

3.11 Risk Management, Improvement, and Management Reviews

The authors propose to refer the processes of Risk management, improvement, and management reviews to the fourth level in terms of ensuring the validity. These processes are basic for confirming the competence of laboratories, but their direct impact on ensuring the validity is determined by the motivation and focus of senior management on quality improvement.

The extent to which senior management can influence the reliability of the results is determined by the planned objectives of the laboratory. Thus, the objectives of the laboratory or suggestions for improvement based on the results of the analysis by senior management may include the development of measures to improve the validity by the L1-L2 processes Fig. 4.

These objectives are related to the following:

- acquisition of high-precision equipment for internal checks – additional financial costs, planning of those responsible for conducting such checks – involvement of additional human resources;
- additional measures to monitor personnel competence: to ensure the required validity, the number of repeated tests is usually increased, and processing of the results requires a specialist with an understanding of statistical methods, which requires additional human resources. This is because the personnel are distracted from performing scheduled tests;
- planning activities for in-depth method validation: planning test experiments under different conditions (including changes in environmental conditions, and deviations from contracts). Analyzing, interpreting, and documenting the results of validation requires additional human resources;
- detailed assessment of all components of the total uncertainty – requires additional human resources and a specialist in writing methods;
- testing of externally provided products – involvement of additional human resources and finances for consumables;
- automation of all processes with the help of LIMS additional financial costs of purchasing LIMS;
- experimental determination of the uncertainty of sampling – involving additional human resources.

Following the requirements [2] for the competence of laboratories and the mechanisms for ensuring the validity described in this section, the general algorithm for improving validity in testing laboratories is as follows:

(1) compile a list of risks, and assess their potential impact on the reliability of test results;
(2) record nonconformities, analyze their scope and causes;
(3) during the general analysis by the management, identify processes that require additional assurance measures;
(4) analyze available resources for planning improvements;

(5) set quality objectives in terms of assurance for the next reporting period that reflect the current quality policy.

References

1. Klymenko, L.P., Pizintsali, L.V., Aleksandrovska, N.I., Yevdokymov, V.D.: Metrology, standardization and quality management, Mykolaiv, Ukraine (2011)
2. ISO/IEC 17025:2017: General requirements and for the competence of testing and calibration laboratories
3. Weckenmann, A., Akkasoglu, G., Werner, T.: Quality management – history and trends. TQM J. **27**(3), 281–293 (2015). https://doi.org/10.1108/TQM-11-2013-0125
4. A roadmap to quality. An e-learning manual for implementing total quality management, vol. 1. UNIDO, Vienna (2017). https://www.unido.org/sites/default/files/2009-04/A_roadmap_to_quality_volume_1_0.pdf
5. Edgeman, R.L., Bigio, D., Ferleman, T.: Six sigma and business excellence: strategic and tactical examination of IT service level management at the office of the chief technology officer of Washington, DC. Qual. Reliab. Engng. Int. **21**, 257–273 (2005). https://doi.org/10.1002/qre.635
6. Dr. Robin Mann Musli Mohammad Ma Theresa A. Agustin: Understanding business excellence. An awareness guidebook for SMEs. Asain productivity Organization (2021). https://www.apo-tokyo.org/wp-content/uploads/2021/05/Understanding-Business-Excellence.pdf
7. Transition Support: Business excellence criteria (2018). https://transition-support.com/business_excellence_criteria.html
8. ISO/IEC 9001:2015: Quality management systems – requirements
9. Juran, J.M., De Feo, J.A.: Juran's Quality Handbook: The Complete Guide to Performance Excellence, 6th edn. United States (2010). ISBN: 9780071629737
10. Vâlcu, A., Călin A.: Ensuring the validity of results by intermediate checks in the field of mass measurements. In: XXII World Congress of the International Measurement Confederation (IMEKO 2018). IOP Publishing (2018). https://doi.org/10.1088/1742-6596/1065/4/042033
11. Vâlcu, A.: Monitoring the validity of reported results in weighing operations according to ISO/IEC 17025:2017. OIML Bull. **62**, 5–11 (2021)
12. Oliveira, E., Alvarenga, A., Costa-Felix, R.: Ensuring the validity of results of hydrophone calibration by a mutual comparison calibration method. J. Phys.: Conf. Ser. **1826**(1–8), 012010 (2021). https://doi.org/10.1088/1742-6596/1826/1/012010
13. Kolbasin, A.: Ensuring the validity of the results of calibrations and tests by statistical methods. Ukr. Metrol. J. **4**, 8–14 (2021). https://doi.org/10.24027/2306-7039.4.2021.250375
14. Eremenko, V., Mokiichuk, V., Samoilichenko, O.: A study of the power of the Cochran criterion with a limited number of observations. Inf. Process. Syst. **6**(64), 35–38 (2007)
15. Eremenko, V., Mokiichuk, V., Samoilichenko, O.: Determination of statistical estimates of the power of homogeneity criteria on samples of small volume. Inf. Process. Syst. **4**, 30–33 (2008)
16. Korniienko, I., Korniienko, S., Dmytriiev, V., Pavlenko, A., Kamak, D.: Determination of quality parameters for assessing the stochastic characteristics of the tested sample of weapons and military equipment. Inf. Process. Syst. **4**(163), 56–65 (2020). https://doi.org/10.30748/soi.2020.163.06
17. Eremenko, V., Mokiichuk, V., Samoilichenko, O.: Investigation of Johnson's transformation for the problems of improving the accuracy of metrological characteristics of standard samples. Inf. Process. Syst. **4**(71), 85 (2010)
18. Friedman, C.P., Wyatt, J.C., Ash, J.S.: Measurement fundamentals: reliability and validity. In: Evaluation Methods in Biomedical and Health Informatics. Health Informatics. Springer, Cham (2022). https://doi.org/10.1007/978-3-030-86453-8_7

19. Kolobrodov, V., Sokol, V., Polikarpov, O., Sokol, B.: Monitoring the work of the test laboratory by participating in the proficiency testing program in the field of the laser performance measurements. Ukr. Metrol. J. **2**, 3–8 (2022). https://doi.org/10.24027/2306-7039.2.2022.263548
20. Samoilichenko, O., Priadko, O., Мокійчук B.M.: Certification of reference materials for pharmaceutical products and evaluation of their metrological characteristics, 169p. NULES of Ukraine, Kyiv (2022)
21. Eremenko, V., Mokiichuk, V., Pashchenko, N., Samoilichenko, O., Priadko, O.: Analysis of the impact of personnel competency on uncertainty during calibration. East.-Eur. J. Enterprise Technol. **3**(3–117), 35–42 (2022). https://doi.org/10.15587/1729-4061.2022.259779
22. Samoilichenko, O., Priadko, O., Mokiichuk, V., Slyva, Y.: Methodological aspects of assessing the competence of laboratory personnel determining the quality of seed potatoes. Anim. Sci. Food Technol. **13**(2), 63–71 (2022). https://doi.org/10.31548/animal.13(2).2022.63-70
23. Eurachem Guide: The Fitness for Purpose of Analytical Methods. A Laboratory Guide to Method Validation and Related Topics, Second edn.
24. ICH harmonised tripartite guideline: Validation of analytical procedures: text and methodology Q2(R1)
25. Samoilichenko, O.: Validation of analytical methods as an integral part of quality assurance of test results. Electrotech. Comput. Syst. **6**(2012), 228–234 (2012)
26. ISO 13528:2015: Statistical methods for use in proficiency testing by interlaboratory comparison
27. Zakharov, I., Botsiura, O.: Advanced methods for measurement uncertainty evaluation. In: 2022 XXXII International Scientific Symposium Metrology and Metrology Assurance (MMA) Sozopol, Bulgaria, pp. 1–4 (2022).https://doi.org/10.1109/MMA55579.2022.9993256
28. Witkovsky, V.: Implementation of the characteristic functions approach to measurement uncertainty evaluation. Ukr. Metrol. J. **1**, 38–43 (2022). https://doi.org/10.24027/2306-7039.1.2022.258818
29. Samoilichenko, O., Eremenko, V., Mokiichuk, V.: Peculiarities of estimating the extended uncertainty on the basis of approximating Johnson distributions. In: VIII International Scientific and Technical Conference "Metrology and Measurement Technology. Metrology-2012, 82–24 (2012)
30. Mokiichyk, V., Pashchenko, N., Samoilichenko, O.: Software quality evaluation indicators of the laboratory information systems. Inf. Process. Syst. **2**(153), 165–171 (2018)
31. Samoilichenko, O., Priadko, O., Mokiichuk, V., Pashchenko, N., Bal-Prylypko, L., Slyva, Y., Tkachuk, V., Silonova, N., Nikolaienko, M., Rozbytska, T.: The impartiality of conformity assessment bodies is an integral component of the food safety management system. Potravinarstvo Slovak J. Food Sci. **16**, 765–776 (2022). https://doi.org/10.5219/1784
32. EURACHEM/CITAC Guide: Measurement Uncertainty Arising from Sampling A Guide to Methods and Approaches, Second edn. (2019)
33. Bisson, K.M., Kiko, R., Siegel, D.A., Guidi, L., Picheral, M., Boss, E., Cael, B.B.: Sampling uncertainties of particle size distributions and derived fluxes. Limnol Oceanogr. Methods **20**, 754–767 (2022). https://doi.org/10.1002/lom3.10524
34. Svensmark, B.: Extensions to the theory of sampling 2. The Sampling Uncertainty (SU), and SU as alternative to variographic analysis. In: Proceedings of WCSB10: TOS Forum Issue 11, pp. 41–46 (2022). https://doi.org/10.1255/tosf.133

Methodology for Controlling Greenhouse Microclimate Parameters and Yield Forecast Using Neural Network Technologies

Mariia Morozova

Abstract Developments in the combination of technologies for data analysis, sensors, and self-driving vehicles are very popular nowadays. They use network solutions, control systems, platforms, and applications. A significant area of application of neural networks is product quality control connected with environmental factors. Artificial intelligence technologies allow real-time monitoring of microclimate parameters. This makes it possible to subsequently influence the general state of the grown products and report any problems detected in real-time. Systems using artificial intelligence technologies can operate 24/7. Therefore, the creation of a methodology using neural network technologies is intended to control the parameters of the greenhouse microclimate, which should have a positive influence on the quality of the harvest and increase yields. After analyzing the accumulated information on mathematical models of greenhouses, a new model was created that is applicable for calculating coolant consumption, steam consumption, and carbon dioxide emissions. The optimal greenhouse microclimate conditions were also analyzed and determined: temperature, humidity, and carbon dioxide concentration, according to which it is possible to predict the yield. A structural schema of the greenhouse control and monitoring system is proposed. The necessary and sufficient components of the system for maintaining the microclimate in the greenhouse for growing oyster mushrooms were identified and selected, such as temperature and humidity sensor, carbon dioxide sensor, circulation pump, ultrasonic humidifier, and ventilator. Thus, a neural network was developed to predict yields and control microclimate parameters such as temperature, humidity, and carbon dioxide content. Based on the mathematical model, a program was designed to train the neural network. For an accurate forecast, a neural network was developed that is based on a multilayer perceptron with three hidden layers.

M. Morozova (✉)
National Technical University of Ukraine "Igor Sikorsky Kyiv Polytechnic Institute", Kyiv, Ukraine
e-mail: m.morozova@kpi.ua

V. Eremenko and A. Zaporozhets (eds.), *Advanced Information-Measuring Technologies and Systems I*, Studies in Systems, Decision and Control 439,
https://doi.org/10.1007/978-3-031-40718-5_7

Keywords Greenhouse · Microclimate · Neural network · Perceptron · Yield forecast · Coolant consumption · Steam consumption · Carbon dioxide emission · Harvest · Oyster mushrooms · Monitoring · Control · Temperature · Humidity

1 Justification of the Task's Relevance. Review and Analysis of Microclimate Parameters

Every year, greenhouse farms and enterprises are paying more and more attention to high-quality microclimate maintenance. Properly selected microclimate maintenance technology is one of the most important components that help to increase yields. Efficient use of energy resources is an additional opportunity to significantly reduce the cost of production. A modern automated microclimate control system should maintain the set mode and make the most of the capabilities of the executive systems.

The introduction of automated microclimate control systems can be considered the next step. Currently, specialists who build new greenhouses are also designing water treatment systems and other systems for monitoring production technology to preserve the environment. And these applications take the ways of growing living organisms to a new level.

Growing conditions for the cultivated organisms in the greenhouse should correspond to the optimal conditions for photosynthesis in the plant cell, and for hybrids should correspond to the process of biosynthesis.

The quality and quantity of the crop depend on many factors. No experienced specialist can analyze all the factors and make the right decision. That is why modern technologies are very relevant in this area.

Growing vegetables, mushrooms, and berries is a profitable area of greenhouse farming. Mushrooms are in good demand among consumers. Therefore, the relevance of growing them is growing, but at the same time, the amount of information and research in this area is insufficient.

The mushroom industry is now in the stage of moving to a highly commercialized cultivation method called "mushroom factory" [1, 2]. A mushroom factory (similar to a plant factory [3]) provides the preferred conditions for growing edible mushrooms in a controlled environment that is controlled and monitored by using highly engineered technology. The mushroom factory promotes highly efficient mushroom cultivation by offering a standardized and year-round production regime, minimizing the limitations associated with the natural climate. For example, in China, the total yield of mushroom organisms in 2018 was 3.28 million tons, which is only 8.6% of the total mushroom harvest. This percentage is projected to reach 20–30% by 2030 [4].

Growing mushrooms in greenhouses is one of the most relevant and popular areas in modern greenhouse farming. It is important to choose the right functional elements of the greenhouse, equipment, select mushroom varieties, adjust the temperature and humidity conditions, ventilation, and proper care.

Growing mushrooms in greenhouses is particularly cost-effective because it does not require additional costs for the purchase of fertilizers, agrochemicals and other additives, unlike plants. However, it is necessary to monitor the microclimate parameters.

Compared to traditional family mushroom farming, mushroom cultivation in a monitored environment is a highly complex biosystem [5]. More research has been conducted focusing on the food perspective of mushroom growing, studying sustainable substrate materials, new chemical additives, mushroom morphologies, and life cycle estimation [6, 7].

However, not much research has been done to solve the issues related to the cultivation of mushrooms. Important environmental parameters, including temperature, humidity, light, and gas concentration (CO_2), significantly affect the growth and development of mushrooms. That is why the basic rule in modern fungus farming is to ensure that individual environmental parameters are maintained at an optimal level at each growth stage to achieve the highest yield [1, 8, 9].

Mushrooms can be grown in a greenhouse all over the year. To make a profit, the greenhouse must be equipped with a stove, and electric, or gas heating. Mushrooms grow well in glass and film greenhouses. However, the basic condition (minimum lighting) must be met. So, the building must be protected from the sun.

For mushrooms, you need to prepare a separate area and cover the light with a darkened film or agrofibre. This will create the most comfortable conditions for the development of mycelium. Special attention should be paid to the humidity level, which should be high enough. To do this, it is necessary to constantly spray the substrate and walls of the greenhouse with water. Small particles of sawdust can accumulate water and then return it. Therefore, sawdust should be placed on the floor to increase humidity.

The profitability of a greenhouse also depends on the weather conditions in the region. Natural factors determine the costs of maintaining the right temperature and humidity in the greenhouse.

Mushroom cultivation in a greenhouse is most often carried out when the greenhouse is not occupied by vegetable crops. The most popular mushroom for growing is the oyster mushroom. The advantages of oyster mushrooms include high yields and a short reproductive cycle. Champignons are considered more capricious. The soil for them is prepared using a complex technological process.

Each type of mushroom needs its own microclimate. For example, oyster mushrooms can be grown in basements, as long as the temperature and humidity are maintained at a certain level. Champignon mushrooms can be grown in a conventional film greenhouse. The variety of mushrooms also matters. In any case, the lighting, temperature, humidity level, and type of fertilizer should be optimal for a high yield.

Oyster mushrooms are best grown in a greenhouse that usually grows vegetables from April to September. When the greenhouse is empty, you can place substrate blocks with mushrooms in it. Alternatively, you can start cultivating this type of mushroom at any time of the year if only cucumbers are grown in the greenhouse. After all, these crops are very similar in terms of climatic parameters. When growing oyster mushrooms, you need to constantly maintain the humidity level in the greenhouse at 80–95%. To do this, you can use traditional water spraying.

The main factor in lighting is that direct sunlight should not fall directly on the mushrooms and that the air should not be overheated. The normal temperature regime for oyster mushrooms is to maintain the temperature to +25 °C. Land under such conditions, fruit bodies will grow. The fleshiness and enlargement of oyster mushrooms directly depend on how they are cared for. The lower the temperature in the greenhouse, the better the quality and size of the oyster mushrooms.

So the approximate turnover of oyster mushroom culture in a greenhouse is 1.5–2 months. And during this period, the harvest will be obtained from 3 to 4 waves of fruiting, amounting to 45% of the weight of the finished substrate.

First, the straw is processed. It is cooled to a temperature of +28… +30 °C, filled into prepared containers (bags or cassettes), and the sowing mycelium is added at the rate of 3–5% of the total substrate weight. Then the mycelium must be constantly inspected for humidity content.

After planting the mycelium, the containers are placed in a chamber where the mycelium will germinate. The temperature should be maintained at +22… +24 °C and the humidity should be 60…65%. Ventilation must also be provided. Light negatively affects the mycelium and slows its growth, so the room should be dark. Mycelial overgrowth of the container depends on the type of substrate. On average, this process takes 10–20 days.

The temperature should be +2… +4 °C when the mycelium intertwines with the substrate. Humidity should be maintained at 85…90% and ventilation should be provided for 2–3 greenhouse volumes per hour. Lighting at this stage is necessary and should be up to 10 h per day.

After 7–10 days, the surface of the mycelium is covered with small mounds, the so-called primordia. When the rudiments of mushrooms appear, the air exchange is increased to 8–10% of the greenhouse volume. Humidity and temperature should be maintained at a level corresponding to the type of mushroom.

During the growth of mushrooms in the greenhouse, the temperature should be + 16 °C… +22 °C. If the temperature drops below +10 °C, the mycelium will no longer grow. Or if the temperature rises above +27 °C, the mycelium will die. When the mycelium grows on the surface of the substrate, the temperature should be lowered to +15 °C. This is the optimal temperature for mushroom growth. If everything is done correctly, the harvest will be ready in 14–17 days.

Since fungal cells lack chlorophyll and chloroplasts, they are not characterized by photosynthesis. Mushrooms are heterotrophic. That is, they absorb ready nutrients, the oversupply of which is converted into a stored nutrient (glycogen).

Today, a greenhouse process control system can include:

- A system for monitoring microclimate parameters. The system consists of a set of sensors that records temperature, humidity, CO_2, and light level (light intensity, total solar radiation).
- Substrate parameter control system. The system includes sensors that allow to determine the electrical conductivity, pH, and humidity of the substrate.
- Electronic substrate weighing system. The system provides information on moisture loss over a certain period.
- System for monitoring drainage parameters. The system analyzes the pH parameters of the drainage solution and records the time of drainage and the time of its completion.

By installing special temperature and humidity sensors and controllers to take readings from the sensors, the entire humidification system can be set up to operate automatically 24/7.

For normal growth and development, mashrooms require high air humidity, as mycelium and fruiting bodies consist mainly of water. Relative humidity of 85–95% is considered optimal for the growth and development of mushrooms. At this relative humidity, the fruiting bodies are formed and have a good presentation. The humidity of the substrate should be at the level of 64…68%, then the oyster mushroom grows well. To maintain the appropriate level of humidity, you can use traditional water spraying or use systems for automatic steam supply so that condensation does not accumulate on the mushrooms.

Oyster mushrooms are characterized by a high growth rate. The inoculated blocks are stored indoors at a temperature of +22… +24 °C and humidity of 85…90%. After 12–15 days, the blocks are covered with 90–100% of white mycelium. At this point, the temperature should be gradually reduced to +14… +16 °C, and the humidity should be maintained at 85…90%. The process of mushroom formation and growth begins [5–9].

Research on the internal environment of a mushroom greenhouse can be divided into three main categories, (1) development of instrumentation that controls environmental conditions and performs the process of environmental control; (2) computer modeling to simulate indoor air flows and analyzing the distributions of certain important environmental factors; (3) improving facility design, including advanced ventilation and equipment control.

The researchers designed a wireless remote monitoring and data transmission system for a king oyster mushroom (Pleurotus eryngii) factory [10].

They integrated multi-sensor fusion technology with GPRS (General Packet Radio Service) wireless communication and image transmission to collect real-time data about the indoor environment (temperature, relative humidity, and CO_2 concentration).

Another study [11] was later published, in which Zhao and Zhu developed a single-chip wireless platform for simultaneous monitoring of environmental factors on eight mushroom farms. The researchers aimed to build an accurate system for remote sensing, avoiding the problems of complicated wiring, high expense, and instability of data transmission.

Researchers implemented modern industrial control systems (ICS) to make the environmental control system at the mushroom farm more efficient through automation and computer technology, together with practical experience in the greenhouse [12]. As the development of Internet communications, such as the Internet of Things (IoT), is on the increase, more and more new technologies are being implemented in agricultural production [13, 14].

In addition, IoT is seen as one more information and industrial wave following the introduction of personal computers, the Internet, and mobile networks. This technology provides quicker and more stable data transmission without the problems associated with wired technologies [15, 16].

An android device was also constructed by Marzuki and Ying to monitor all environmental factors on a mushroom farm in real-time using an IoT interface named "ThingSpeak" [17]. This control system consisted of three different regimes with specific settings of optimal conditions, following the phase of mushroom cultivation. Analogously, an IoT-based monitoring and control system was introduced for shiitake cultivation by a team of Malaysian investigators. They implemented a wireless sensor network (WSN) and mobile computing technology to maximize the production effectiveness of local mushroom farms (Fig. 1). With the help of the control system, the yield of shiitake was improved by 192.6% [16, 18] over the usual cultivation methods.

In Fig. 1 T is the temperature; H is the Humidity; CO_2 is the Carbon Dioxide; GA represents the growth area; IA represents the incubation area.

Data on environmental parameters such as temperature, humidity, and CO_2 are usually monitored and recorded in real-time.

New technologies such as artificial intelligence (AI), IoT-based system, big data analysis, and computer vision are expected to be applied to control mushroom factories in the nearest future [19–25].

A study performed by Grant described the airflow in tunnels for growing mushrooms in Ireland [26]. The author designed the duct system to ensure that conditioned air is delivered properly, at the appropriate speed, and as uniformly as possible.

The publication [27] presents an advanced IoT-based system of climate control for oyster mushroom cultivation, which was presented at the 10th Conference on Systems Engineering and Technology (ICSET) [28]. In the authors' previous IoT-based climate control system [27], fixed threshold values were used to automatically adjust the oyster mushroom growing environment in the mushroom house. There were two limitations. The first limitation was: the inability to adjust the mycelium environment to accommodate weather changes during the wet and dry seasons. Malaysia has an equatorial climate with hot and humid weather throughout the year and two monsoon seasons from late May to September and October to March, respectively. Therefore, this work [27] proposed to overcome the limitation of using fixed thresholds to automatically control the climate of a mushroom greenhouse using a fuzzy logic approach and considered the hardware specification of the system to improve the stability of the system due to power supply issues. The project demonstrates the implementation of an IoT climate control system in a mushroom greenhouse placed in Bandar Putra (Malaysia). The publication [29] describes how

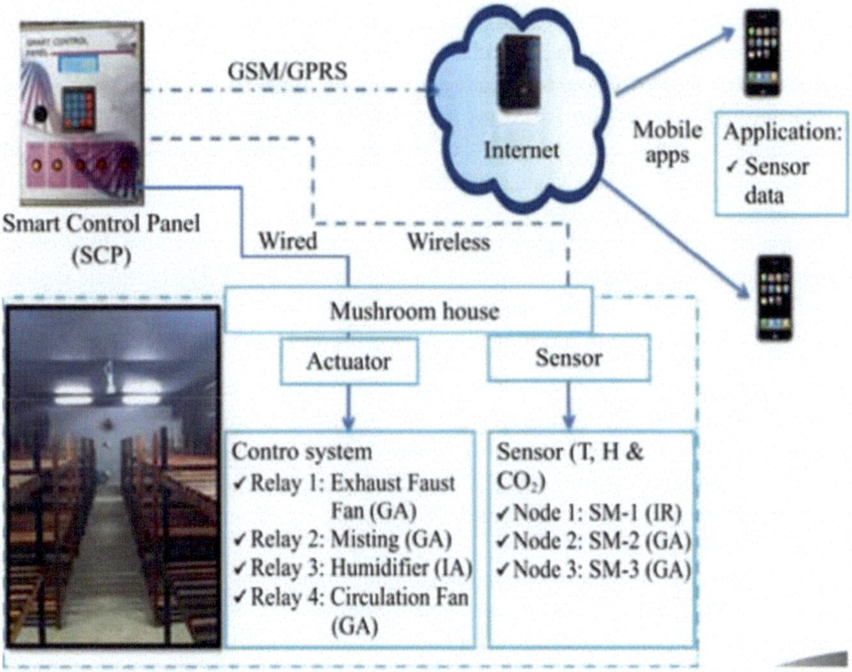

Fig. 1 Schematic structure of the mushroom factory monitoring and control system using IoT [1, 16]

the implementation of fuzzy logic for the IoT controller was successful in providing better environmental management for mushroom cultivation and also reduced water consumption, resulting in increased yields of their harvest.

To solve the problems in the well-known smart greenhouse EPAM (Effective Programming for America), a team specializing in creating user interfaces was formed within its own Innovation Lab. The team's work is related to projects based on open-source software. EPAM wins an award from Oracle for creating a "smart greenhouse" [30].

Developers consider that smart greenhouses have more potential for improvement than conventional greenhouses. But unlike Ukrainian developments, foreign developments are too expensive for many small businesses.

Humidity levels in both air and soil are inversely proportional to temperature. As the temperature rises, the humidity decreases. High air and soil humidity are observed in the colder months of the year.

Specialists continue to develop new mechanisms to automate the maintenance of the microclimate. This is especially true for ventilation, which helps to normalize the temperature in the greenhouse. Each type of automatic temperature controller has its pros and cons:

- Electric regulators are easy to adjust, quite sensitive, but unreliable due to possible power outages. They use a fan and a thermostat that turns it on.
- Bimetallic regulators are cheap and autonomous in operation, but low-powered and unreliable. They are based on the difference in the coefficients of thermal expansion of the connected plates of two metals. When the plate is heated, it bends and opens the hole, and when it cools down, the plate straightens and closes it.
- Hydraulic regulators are highly reliable and autonomous in operation, but expensive. They use the property of liquids to expand when heated. The hermetic system is filled with liquid freon, water, oil, or a liquid that expands when heated. A corrugated brass tube or a retractable stem is used to open the hole.

The concentration of carbon dioxide has a significant impact on the formation of plants and mushrooms. An incorrect O_2/CO_2 ratio can damage the harvest. For example, a high CO_2 concentration is considered to be one of the main factors that stop mycelium growth in the mushroom substrate and the appearance of mold. If the concentration of CO_2 exceeds 0.08% during oyster mushroom growth, the mushrooms grow deformed.

The appropriate microclimate is most conveniently provided by automated systems. Using a network of sensors and a control unit, it is possible to create comfortable conditions for organisms and ensure the required CO_2 concentration. A correctly configured ventilation system provides air supply and air recirculation in the building.

2 Development of a Mathematical Model of the Greenhouse

The next step is to search for a mathematical model of a greenhouse for growing mushrooms. To maintain the microclimate in the greenhouse, it is necessary to create an intelligent system for analyzing and adjusting the parameters of temperature, humidity, and carbon dioxide. Then it is possible to forecast greenhouse yields based on these parameters. It should be noted that mushrooms are aerobic microorganisms, so carbon dioxide is expected to be released during their formation.

To date, there are many works devoted to greenhouse microclimate models. These models take photosynthesis as the foundation of the vegetation process.

A characteristic feature of mushroom greenhouses is the vegetation process that occurs with the production of heat, water, and carbon dioxide into the greenhouse air. This feature appears because mushrooms are aerobic microorganisms, which involve the emission of carbon dioxide, in contrast to the absorption of CO_2 by plants during photosynthesis. In this regard, there is a need to modify the typical greenhouse microclimate models, where photosynthesis is the basis of the vegetation process.

The microclimate model proposed in [31] was taken as the basis for the required microclimate of a mushroom greenhouse.

The equation of the thermal balance of energy that affects the change in air temperature inside the greenhouse:

$$\rho \cdot V \cdot C \cdot \frac{dT(t)}{dt} = G_{hrhs} - \left(\sum G_{hlesb} + G_{hlhfa}\right), \quad (1)$$

where ρ—air density; V—air volume; C—specific heat capacity of air; T(t)—air temperature inside the greenhouse; G_{hrhs}—heat receipts from the heating system; G_{hlesb}—heat loss through the enclosing structures of the building; G_{hlhfa}—heat loss for heating fresh air.

The equation of mass balance of water in the greenhouse atmosphere:

$$\rho \cdot V \cdot C \cdot \frac{dX(t)}{dt} = L_{fac} \cdot_{ahfa} - L_{oa} \cdot_{ahoa} + L_n, \quad (2)$$

where ρ—air density; V—air volume; X(t)—absolute humidity in the greenhouse atmosphere; L_{fac}—fresh air consumption; X_{ahfa}—absolute humidity of fresh air; L_{oac}—the outgoing air consumption; X_{ahoa}—absolute humidity of the outgoing air; L_n—steam consumption.

The equation of mass balance of carbon dioxide in the greenhouse atmosphere is determined from the balance of carbon dioxide masses:

$$\rho \cdot V \cdot C \cdot \frac{d_2(t)}{dt} = L.C_C - L_{oa} \cdot C_{CO_2out} + V_{CO_2}, \quad (3)$$

where ρ—air density; V—air volume; $C_{CO_2}(t)$—absolute content of CO_2 in the greenhouse atmosphere; L_c—fresh air consumption; C_c—absolute content of CO_2 in the atmosphere; L_{oac}—the outgoing air consumption; C_{CO_2out}—the absolute content of the outgoing CO_2; V_{CO_2}—the emission of CO_2 into the greenhouse air during the oxidation process.

Thus, given the equations of the greenhouse microclimate model, it is possible to write down the system for the state variables of the greenhouse parameters—for temperature, absolute humidity, absolute content CO_2 [32]:

$$T(t) = \frac{G_{\Gamma} \cdot C_{tepl} \cdot (t_1 - t_2) + t_z \cdot (k \cdot F + L_c \cdot C) \cdot e^{\left(\frac{(k \cdot F + L_c \cdot C) \cdot t}{\rho \cdot V \cdot C}\right)} + T_0}{(k \cdot F + L_c \cdot C) \cdot e^{\left(\frac{(k \cdot F + L_c \cdot C) \cdot t}{\rho \cdot V \cdot C}\right)}}, \quad (4)$$

$$X(t) = \frac{(L_c \cdot X_c + L_n) \cdot e^{\left(\frac{L_c}{\rho \cdot V} \cdot t\right)} + X_0}{e^{\left(\frac{L_c}{\rho \cdot V} \cdot t\right)}}, \quad (5)$$

$$C_{CO_2}(t) = \frac{(L_c \cdot C_c + V_{CO_2}) \cdot e^{\left(\frac{L_c}{\rho \cdot V} \cdot t\right)} + C_{CO_{20}}}{e^{\left(\frac{L_c}{\rho \cdot V} \cdot t\right)}}, \quad (6)$$

where T_0, X_0, $C_{CO_{20}}$ are initial values.

From formulas (4), (5), (6), it is possible to distinguish the constants:

$$c_{sum} = k \cdot F + L_c \cdot C; \tag{7}$$

$$x_{sum} = \frac{L_c}{\rho \cdot V}. \tag{8}$$

The calculated data for the implementation of the mathematical model of the greenhouse microclimate were taken as follows:

- Greenhouse parameters:

  $k = 6$ J/(m^2 °C)—heat transfer coefficient of the enclosing structure;
  $F = 100$ m^2—the area of the enclosures;
  $V = 400$ m^3—the volume of air in the greenhouse;

- Constants for temperature regulation in the greenhouse:

  $C_{tepl} = 4200$ J/°C kg—specific heat capacity of the coolant (water);
  $t_1 = 95$ °C—temperature at the inlet of the heat exchanger;
  $t_2 = 70$ °C—temperature at the outlet of the heat exchanger;
  $t_Z = 10$ °C—outdoor air temperature;
  $L_c = 1.1$ m^3/s—air flow rate for ventilation;
  $C = 1005$ J/°C kg—specific heat capacity of air;
  $\rho = 1.225$ kg/m^3—air density;
  $G_T = 0.074$ kg/s—maximum coolant consumption;
  $T_{min} = 10$ °C—minimum temperature value;
  $T_{opt} = 17$ °C—optimal temperature value;
  $T_{max} = 30$ °C—maximum temperature value;

- Constants for regulating humidity in the greenhouse:

  $X_c = 0.0041$ kg/m^3—absolute humidity of fresh air;
  $X_0 = 0.013$ kg/m^3—initial air humidity;
  $L_n = 0.014$ kg/s—maximum steam consumption;
  $X_{min} = 0.006$ kg/m^3—minimum humidity of fresh air;
  $X_{opt} = 0.012$ kg/m^3—optimum humidity of fresh air;
  $X_{max} = 0.029$ kg/m^3—maximum humidity of fresh air;

- Constants for regulating carbon dioxide in the greenhouse:

  $C_c = 0.38$ g/m^3—absolute content of carbon dioxide in the air;
  $C_0 = 0.4$ g/m^3—initial carbon dioxide content in the air (it is $C_{CO_{20}}$ in (6));
  $V_{CO_2} = 0.1625$ m^3/h—the maximum value of carbon dioxide emission from the compost;
  $C_{min} = 0.6$ g/m^3—the minimum value of carbon dioxide;
  $C_{opt} = 0.825$ g/m^3—the optimal value of carbon dioxide content;
  $C_{max} = 1.2$ g/m^3—the maximum value of carbon dioxide content.

Now let's transform the formulas of the microclimate model proposed in [32] to find the coolant consumption, steam consumption, and the value of carbon dioxide emission.

$$G_T(t) = \frac{\frac{T_0 \cdot c_{sum} \cdot e^{\left(\frac{c_{sum} \cdot t}{\rho \cdot V \cdot C}\right)} - t_z}{e^{\left(\frac{c_{sum} \cdot t}{\rho \cdot V \cdot C}\right)}} - t_z \cdot c_{sum}}{C_{tepl} \cdot (t_1 - t_2)}, \tag{9}$$

$$L_n(t) = \left(\frac{X_0 \cdot L_c \cdot e^{sum \cdot t} - X_0}{L_c} \cdot e^{sum \cdot t}\right) - L_c \cdot X_c, \tag{10}$$

$$V_{CO_2}(t) = \left(\frac{C_0 \cdot L_c \cdot e^{sum \cdot t} - C_0}{L_c} \cdot e^{sum \cdot t}\right) - L_c \cdot C_c. \tag{11}$$

According to these formulas (9)–(11), it is possible to calculate the expenses for the correct control of the devices while maintaining the optimal microclimate in the greenhouse.

The yield of mushrooms in a greenhouse depends on the microclimate. The most important are temperature, humidity, and carbon dioxide levels. And each parameter depends on each other. If at least one parameter deviates from the norm, increases or decreases, it will change the yield. So, it is necessary to control the parameters together and constantly adjust the heating, humidification, and air circulation. Because cultivated organisms are very sensitive to microclimate deviations from optimal values, and this affects the yield. Yield U was divided into sub-bands, with limit values from 0 to 1: "bad" yield—[0; 0.25), yield "below average"– [0.25; 0.5), "average" yield—[0.5; 0.75), yield "above average"—[0.75; 1), "high" yield—[1] (Fig. 2). Of course, it is possible to make another gradation, more thorough and with more sub-bands, but it is considered sufficient to control greenhouse conditions (Table 1).

For maximum yield, the normalized value is equal to one. This can be interpreted as follows: with a fencing area of 100 m^2 and the number of mushroom blocks

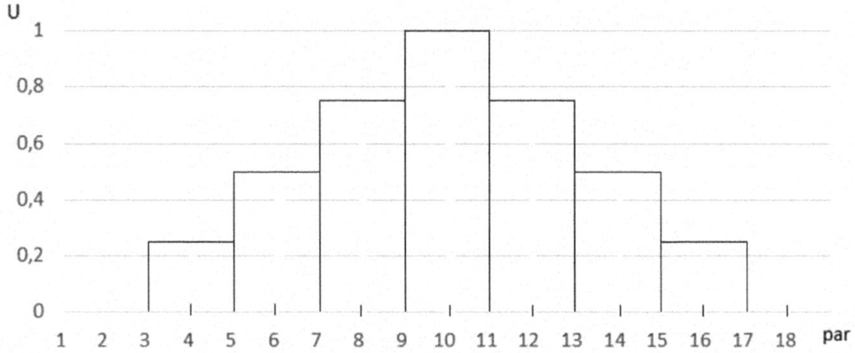

Fig. 2 Dependence of the yield value on the parameters of the microclimate

Table 1 Parameters (par) of the microclimate and yield (U) depending on changes in temperature (T), humidity (X), carbon dioxide (CO_2)

Par	T, °C	X, kg/m^3	X, %	CO_2, g/m^3	U
1	0	0	74	0	0
2	11	0.007	75	0.6	0
3	12	0.0074	76	0.64	0.25
4	13	0.0078	77	0.68	0.25
5	14	0.008	78	0.7	0.5
6	14.5	0.0088	79	0.72	0.5
7	15	0.0094	80	0.74	0.75
8	15.5	0.01	81	0.75	0.75
9	16.5	0.011	83	0.8	1
10	17	0.012	85	0.82	1
11	17.5	0.013	87	0.85	0.75
12	19	0.0145	89	0.87	0.75
13	20	0.016	90	0.9	0.5
14	22.5	0.018	92	0.92	0.5
15	23	0.019	94	0.95	0.25
16	25	0.022	95	0.96	0.25
17	27	0.024	98	0.98	0
18	30	0.025	98	1	0

of 500, with a minimum ripening time of 2 months for the oyster mushroom crop, it is possible to obtain a maximum yield of 1500 kg. That is each mushroom block yields up to 3 kg of harvest.

3 Development of the Structure of the Greenhouse Monitoring and Control System

First, the substrate is prepared, then the bags are placed in space for mycelium germination. It is in this area that the microclimate parameters must be controlled to create optimal conditions for the cultivation of healthy mushrooms.

The structural scheme of the greenhouse monitoring and control system is shown in Fig. 3. Data from temperature, humidity, and carbon dioxide sensors are transmitted to the controller. Next, they are processed and analyzed by software (a designed neural network). The neural network produces recommendations for controlling the actuators: circulation spout, ultrasonic humidifier, and ventilators.

Following the structure scheme, it is necessary to determine the components. The HDC1080 sensor from Texas Instruments is suitable for temperature and humidity

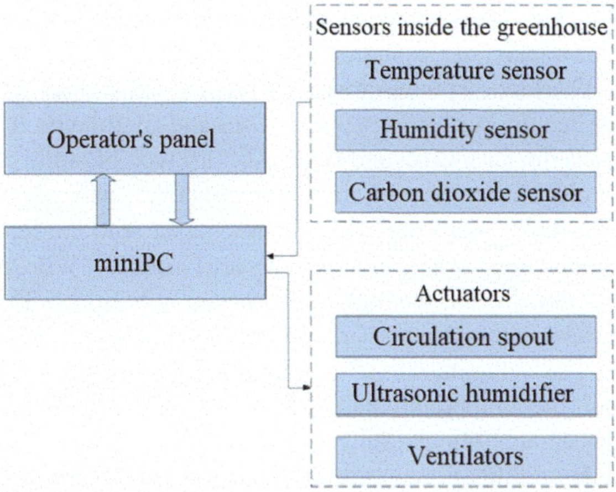

Fig. 3 Structural scheme of the greenhouse monitoring and control system

control. The HDC1080 is an integrated humidity sensor with an embedded temperature sensor and heating element. Due to its factory calibration, the HDC1080 has an error of ±2% and an effective temperature range of −40 °C… +125 °C. The humidity measurement can be performed with a resolution (discreteness) of 8/11/14 bits, and the temperature measurement can be performed with discreteness of 11 or 14 bits [33]. The I2C interface is used to interact with the management controller. To minimize power requirements, the HDC1080 has two working modes: measurement mode and sleep mode.

The amount of sensors is chosen according to the results presented in [31]. The authors note that to determine the temperature at any point in the greenhouse, it is necessary and sufficient to obtain the temperature value from the area located in the geometric center of the greenhouse. It is supposed that humidity is distributed equally throughout the volume of the greenhouse.

The calculation of the relative humidity value into an absolute value is made using the formula:

$$X_a = \frac{6.11 \cdot e^{\frac{17.67 \cdot T}{T + 243.5}} \cdot X_r \cdot 2.17}{273.15 + T}, \quad (12)$$

where X_a—absolute humidity, g/m^3; X_r—relative humidity; T—temperature.

Economically, it is more profitable to heat the area by warming water, to use the water type of heating. The heating system can consist of a heating boiler or furnace, pipes, radiators, a chimney, and a circulation pump.

The circulation pump is used to transfer the water heated in the boiler to the pipes. The water in the system is usually circulated forcibly under the pressure created by the circulation pump. In low-cost greenhouses, water heating is often used in the process of natural water circulation due to the pressure difference in the system. Therefore, the water heating type can be either with or without a pump. This is determined by financial resources and the requirement to maintain the temperature at a given level in the greenhouse. Since in this system, it is necessary to control the heat transfer, a circulation pump is needed.

Circulation pumps are of two types—wet and dry. The wet type of pump is used in the heating system of private houses and apartments. The most popular pump brands are Wilo, Grundfos, and DAB. The dry type of pump consists of a pumping part and an electric motor with an electric impeller, which is located on the motor shaft in a hermetic enclosure. There is no contact with the liquid in these pumps. The most popular and reliable brands today are Grundfos, Wilo, and Ebara. For regulation, it is suggested to use the high-performance Wilo Stratos D pump with electronic control. It is a wet rotor pump with low operating expenses and can be installed in pipes [34].

The maintenance of the humidity level needs to be considered in combination with the temperature and airflow rate. The more substrate load per square meter of space, the more carefully the microclimate system needs to be designed.

Humidification options in mushroom spaces: Watering the floor; Additional and temporary methods of getting rid of dryness; Low and medium pressure injectors; Disc dehumidifiers; Ultrasonic dehumidifiers. To maintain sufficient humidity in the oyster mushroom greenhouse, it is recommended to use the ultrasonic humidifier TM "Vdoh-Nova" [35].

In a mushroom greenhouse, it is also necessary to control the level of carbon dioxide. Increasing the concentration of CO_2 in the areas where mushrooms are grown over the specified norm causes a degradation in product quality and a corresponding decrease in yield. To control the concentration of CO_2 in the greenhouse, it is suggested to select the VENTS CO_2 carbon dioxide sensor [36]. This sensor detects the level of carbon dioxide concentration in the building and gives out a signal that further controls the work of the ventilator. Other types of CO_2 sensors (concentration converter) AG-06e can also be used [37].

The air from the oyster mushroom growing area and the air from the outside are mixed in a separation chamber. It is then heated by a heat extractor, passed through a ventilator, humidified, and then distributed throughout the chamber by air ducts with air injectors.

The airflow rate is generated and supported by a properly selected ventilator and a recirculation device. The fresh air is introduced and the air is removed from the greenhouse by using an extractor ventilator. This is known as supply and extract ventilation.

Ventilators of the YWF2E300S\B type for greenhouses are equipped with an asynchronous motor with rolling bearings and an external rotor. This construction ensures a long service life of the ventilator and the possibility of installation in any position. The capacity of the ventilators is 2000 m^3/hour. It is proposed to control the elements of microclimate maintenance with the help of Raspberry Pi.

4 Neural Network Development for Prediction and Control of Microclimate Parameters

Traditionally, neural networks have been considered to be very complicated. To use them, it is necessary to know mathematics very carefully, program in C++, and comprehend the nuances of parallel and high-performance calculations. However, the situation has recently changed, and a large number of ready-to-use libraries for training neural networks are now available. With the help of these libraries, it is quite possible to apply neural networks to solve practical applications.

The high-level programming language Python and the libraries Numpy, Keras, and TensorFlow were used to realize the project.

NumPy is an extension of Python that provides support for large multi-dimensional arrays and matrices along with a large library of high-level mathematical functions for operations with these arrays.

The Keras library is a high-level interface for creating neural networks. Keras is written in Python and runs over such lower-level implementations as TensorFlow, CNTK, and Theano.

A neuron is a computational unit that receives information, performs simple computations on it, and then transmits it to others. Neurons are categorized into three main types: input (X1...Xn), hidden, and output (Y1...Yn). If a neural network consists of a large number of neurons, the term layer is introduced. Accordingly, there is an input layer that receives information, there are n hidden layers (usually no larger than 3) that analyze the information, and there is an output layer that outputs the result. Each neuron has two main parameters: input data and output data. In the case of an input neuron, it is input = output. The input field contains the summary information of all neurons from the previous layer; after that, the information is normalized using the activation function f(x) and gets into the output folder.

For a neural network to be efficient at a given task, it needs to be trained. There are two main types of training: supervised (with a teacher) and non-supervised (without a teacher).

1. Supervised learning is a process of learning with a teacher, in which samples of training examples are provided to the network. Each sample at the network input is then processed within the neural network structure. The output signal of the network is calculated. It is compared with the corresponding value of the target vector, which is the desired output of the network. Then the training error is calculated. The training error is an indicator of the accuracy of the model's training set. The error can be used as a condition for stopping training.

After that, the weights of the connections within the network are changed depending on the chosen algorithm. The vectors of the training set are fed sequentially, and errors are calculated. The weights are adjusted for each vector until the error over the entire training set reaches an acceptable value.

2. Non-supervised learning is a learning process in which the training set consists only of input vectors. The training algorithm adjusts the weights of the neural

network so that the output vectors are in agreement. Sufficiently close input vectors should produce identical outputs. Similar vectors are grouped into classes. An input vector from a certain class will produce a certain output vector. But it is impossible to predict the output before training. Therefore, the network outputs must be transformed into some understandable form. Usually, it is not difficult to identify the relationship between the input and the output established by the neural network.

Iterative algorithms can be used to solve the multidimensional optimization problem:

1. Local optimization algorithms with the computation of first-order partial derivatives.
2. Local optimization algorithms with the computation of partial derivatives of the first and second degrees.
3. Stochastic optimization algorithms.
4. Global optimization algorithms.

The neural network training algorithm is iterative. Its steps are called epochs or cycles. For a neural network to work effectively, many epochs of training are required.

One of the most important aspects of neural networks is the activation function, which introduces nonlinearity into the network.

There are many activation functions. The main ones are Linear; Threshold; Sigmoid; Hyperbolic tangent; Rectified linear unit (ReLU). The range of values and processing speed are the main differences between the functions.

The question arises: which activation function is preferable to use?

Considering that neural networks are designed to solve complex tasks, the activation function used should be sufficiently reliable and satisfy the following requirements:

- The activation function must be differentiated. This is necessary for the backpropagation algorithm;
- It should be simple and speedy to process;
- The output must be centralized relative to zero and for backpropagation.

The choice of the neural network structure is the most difficult task. It includes choosing the number of layers, the number of neurons, and the number of connections for each neuron. There are different strategies for finding the optimal neural network structure, such as gradual extension, building a complex neural network with further simplification, and alternate extension and simplification. The number of layers is selected as follows. If a particular task does not require complex calculations, you can limit yourself to a single-layer perceptron.

Three-layer networks are used to provide a representation of complex, multiconnected areas. The more layers, the more functions the network can implement. But gradient methods converge worse and it is more difficult to train a neural network.

The number of neurons in the hidden layer is selected mainly by visual inspection and by optimization according to an external criterion.

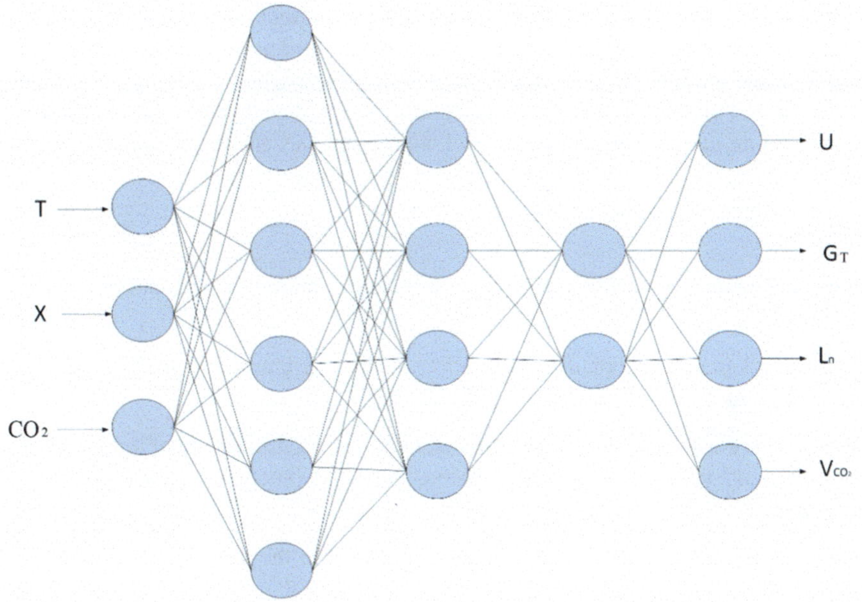

Fig. 4 The structure of the neural network

The neural network model was created based on a perceptron with three hidden layers (Fig. 4).

The neural network receives values from sensors: temperature (T), humidity (X), and carbon dioxide (CO_2). The number of neurons in the first hidden layer is 128; in the second layer the number of neurons is 64, and in the third layer the number of neurons is 16. With three hidden layers, the neural network is more flexible. And the predicted yield value is more highly probable. At the output of the neural network, there is a yield value for the three input parameters. In addition, there is the coolant consumption, steam consumption, and the value of carbon dioxide emissions. These output parameters are necessary for further regulation of the microclimate parameters using appropriate technical equipment to obtain the highest possible yield.

When configuring the model, the ReLU activation function was assigned to the first, second, and third hidden layers, as it is a good approximator.

Other activation functions were set for the output parameters:

- Sigmoid function was used for the yield output. Since the yield is expressed as a normalized value from 0 to 1, like a probability. Therefore, the use of the sigmoid function allows determining accurate limits when predicting the yield;
- A linear function was used to obtain the output of coolant consumption, steam consumption, and the value of carbon dioxide emission (G_T, L_n, V_{CO2})).

The mean square loss function (MSE) is applied to obtain training results for microclimate parameters. The linear loss function (MAE) is used to estimate the quality of training.

The weighting coefficients for the neural network outputs were set as follows: for yield, the weighting coefficient is 0.4; for coolant consumption (G_T), steam consumption (L_n), and the value of carbon dioxide emission (V_{CO2}), the weighting coefficients are 0.2 each. This choice is targeted since finding the predicted yield value is a priority.

According to the mathematical model, a program was created to train the neural network.

The stages of implementation of the neural network training method were identified:

Stage 1. Create ranges for measuring temperature, humidity, and carbon dioxide emissions.

Stage 2. Calculate the coolant consumption, steam consumption, and the volume of carbon dioxide emissions using formulas (9)–(11).

Stage 3. According to the determined ranges for measurement of temperature, humidity, and volume of carbon emission, the yield function for each parameter is created.

Stage 4. Separating the values that were determined in the three stages into ranges for training.

Stage 5. Generating a data array for the training set:

The value of the air temperature in the greenhouse (°C).

1. The value of air humidity (%).
2. Carbon dioxide content in the air (g/m^3).
3. Yield indicator (from 0 to 1).
4. The value of coolant consumption.
5. The value of the steam consumption.
6. The value of carbon dioxide emissions.

Stage 6. Mixing the array elements for better training of the neural network.

Stage 7. Saving the training set to a file for further training of the neural network.

Stage 8. Training using the gradient descent method.

To train the projected neural network, seven parameters are submitted to its input, namely:

- readings from three sensors (readings of temperature, humidity, CO_2);
- the position of the regulators to be set (G_T, L_n, V_{CO_2}), calculated by the formulas (9)–(11);
- predicted yield (ideal), according to the established gradations (Fig. 2).

Experimental data calculated using the formulas from the mathematical model are represented as a file. It consists of comma-separated values, each line corresponding to one sample.

The algorithm of the neural network is shown in Fig. 5.

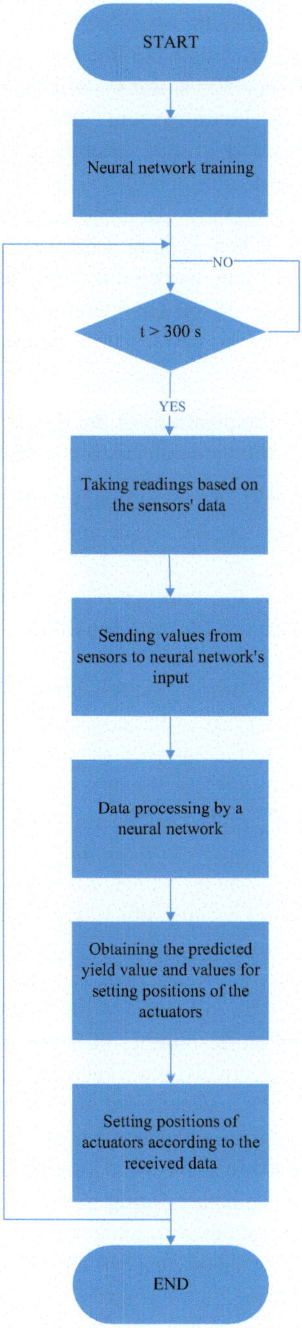

Fig. 5 The algorithm of the neural network

After training, the neural network can forecast the current yield and control the microclimate parameters according to the recommended positions of the outcomes. Every 300 s, the system queries the sensors, receives data from them on temperature, humidity, and carbon dioxide emissions, and sends these values to the neural network's input. After the neural network processes the values, it obtains the predicted yield value and the values for setting the positions of the coolant consumption, steam consumption, and carbon dioxide emission controllers.

When making a decision (at the stage of functioning of the previously trained neural network), only three parameters are input to the neural network: values from three sensors (temperature, humidity, CO_2).

Based on these three input parameters, the neural network generates recommendations:

- the positions of the regulators to be set and controlled (G_T, L_n, V_{CO_2});
- the value of the forecasted yield.

When experimenting with the system, it is necessary to change the program parameters for the maximum values of temperature, humidity, and carbon dioxide. So, $G_{T_{max}} = 0.21$ kg/s, $L_{n_{max}} = 0.021$ kg/s, $V_{CO_2} = 1.9625$ m^3/h.

If the number of samples in the range of the measured value (temperature, humidity, carbon dioxide) is cap = 32, then with the maximum values, the yield value is 0.94.

A similar result is obtained with other values of the parameter (cap).

At cap = 48 and cap = 64, the yield is 0.94.

If the number of samples in the range of the measured value (temperature, humidity, carbon dioxide) is cap = 64, but the time after which the system wakes up and "pings" the sensors are changed, this does not have any effect on the forecasted yield value. For example, changing the value from 300 to 600 s. In other words, the time can be set according to the necessities of the greenhouse. The optimal value is 5 min because this time is enough to support optimal microclimate conditions. After 5 min, the system will restart and will correct the temperature, humidity, and carbon dioxide values using the control and measurement equipment.

The influence of the number of neurons on the forecast was also investigated. When the number of neurons in the hidden layers was changed to 100, 50, and 12, respectively, the forecasted yield became worse.

The influence of an excessive number of neurons is also investigated. When the number of neurons in the hidden layers was changed to 256, 128, and 32, respectively, the forecasted yield value equals 0.99.

Thus, according to the selected components and for optimal functioning of the neural network, the required number of neurons in the hidden layers would be 256, 128, and 64, respectively. The number of samples in the range of the measured value is equal to 64 to achieve optimal microclimate parameters.

Under optimum conditions for the cultivation of mushrooms, the maximum yield is achieved. The ideal yield has been obtained based on the recommendations above. Comparative diagrams of real yield (i.e., forecasted by the neural network) and ideal yield are shown in Fig. 6.

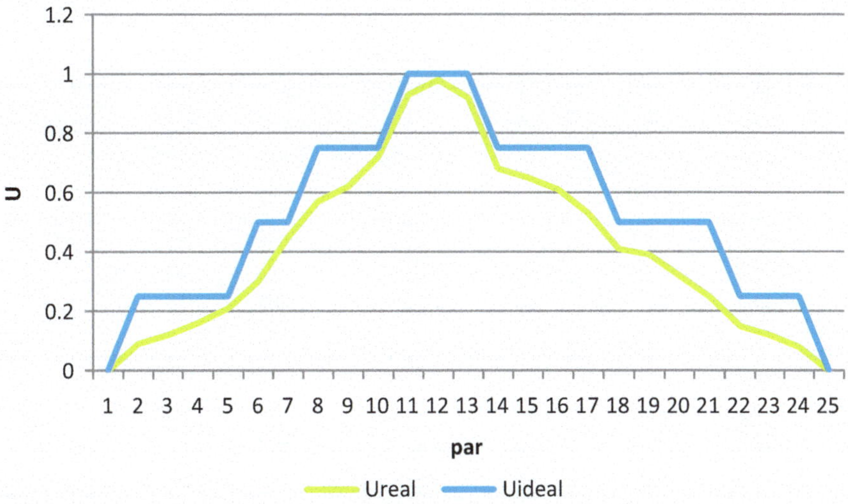

Fig. 6 The influence of microclimate parameters on yield

Three parameters were combined in the neural network (using the multiplication function "×"), according to temperature, humidity, and carbon dioxide volume:

$$U = U_T \times U_x \times U_{CO_2}, \tag{13}$$

where U_T is yield depending on temperature; U_X—yield depending on humidity; U_{CO2}—yield depending on carbon dioxide volume.

Let's introduce the parameter par, which represents the temperature, humidity, and carbon dioxide volume at present.

The results are summarized in Table 2. In Table 2 parameter Δ is the deviation between Ureal and Uideal.

Figure 6 shows that when the microclimate parameters were combined (using the multiplication function), the boundaries of optimal conditions for maintaining the microclimate became more constrained. This is due to the necessity of controlling all parameters for effective mushroom cultivation, since with a gradual change in the parameters to the worse values, the forecasted mushroom yield will be reduced.

A more detailed analysis for the yield values near the value of 1 is shown in Fig. 7 and is given in Table 3 (parameter Δ is the deviation between Ureal and Uideal).

It can be seen that the yield value obtained from the neural network is close to the ideal yield value.

The study of the change in yield according to the temperature parameter is shown in Fig. 8. In this case, the other parameters are invariable. That is, the humidity is 83%, and the carbon dioxide volume is 0.825 g/m^3.

Table 2 The result of the neural network

Par	T, °C	X, kg/m^3	X, %	CO_2, g/m^3	Ureal	Uideal	Δ
1	11	0.007	74	0.6	0	0	0
2	12	0.008	75	0.61	0.09	0.25	0.16
3	12.5	0.0084	76	0.625	0.12	0.25	0.13
4	13	0.0088	77	0.65	0.16	0.25	0.09
5	13.5	0.0092	78	0.675	0.21	0.25	0.04
6	14	0.0096	79	0.7	0.3	0.5	0.2
7	14.5	0.01	80	0.725	0.45	0.5	0.05
8	15	0.0104	81	0.75	0.57	0.75	0.18
9	15.5	0.0108	81.1	0.775	0.62	0.75	0.13
10	16	0.0112	82	0.8	0.72	0.75	0.03
11	16.5	0.0116	82.5	0.81	0.93	1	0.07
12	17	0.012	83	0.825	0.98	1	0.02
13	17.5	0.0127	85	0.84	0.92	1	0.08
14	18	0.0134	87	0.855	0.68	0.75	0.07
15	18.5	0.0141	89	0.87	0.65	0.75	0.1
16	19	0.0148	90	0.885	0.61	0.75	0.14
17	19.5	0.0155	92	0.9	0.53	0.75	0.22
18	20	0.0162	93	0.915	0.41	0.5	0.09
19	20.5	0.0169	94	0.93	0.39	0.5	0.11
20	21	0.0176	95	0.945	0.32	0.5	0.18
21	21.5	0.0183	96	0.96	0.25	0.5	0.25
22	22	0.019	98	0.975	0.15	0.25	0.1
23	23	0.02	98	0.975	0.12	0.25	0.13
24	25	0.021	98	0.975	0.08	0.25	0.17
25	27	0.025	98	1.00	0	0	0

The research of yield changes according to the humidity parameter is shown in Fig. 9. In this case, the other parameters are unchanged. That is, the temperature is +17 °C, and the carbon dioxide volume is equal to 0.825 g/m^3.

The research of yield changes following the carbon dioxide parameter, which is shown in Fig. 10. The other parameters are unchanged. That is, the temperature is equal to +17 °C, and the humidity is equal to 83%.

Thus, when two parameters are maintained at optimal values and the third parameter is varied, the value of the real yield is approximated to the ideal yield.

The research of yield changes according to the carbon dioxide and humidity parameters is presented in Table 4. At the same time, the temperature parameter is unchanged, and equal to 17 °C.

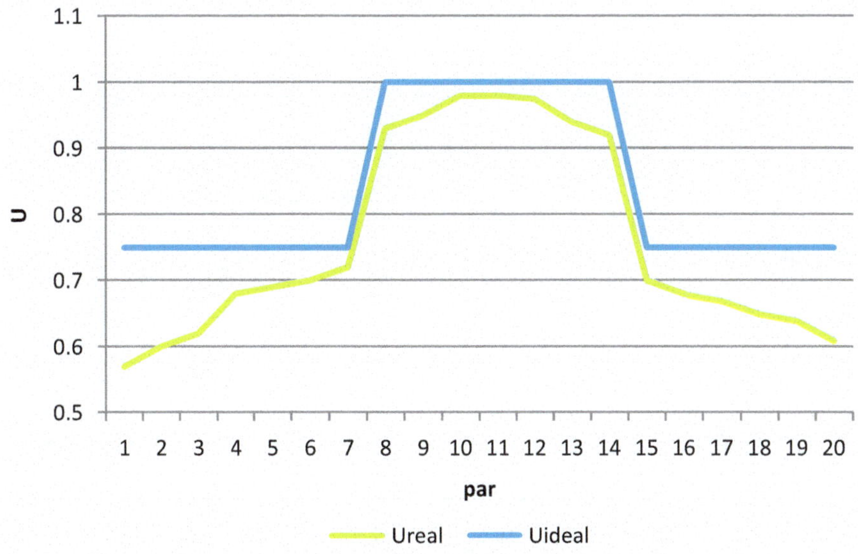

Fig. 7 The influence of microclimate parameters on yield in a narrowed range

Table 3 The result of the neural network

Par	T, °C	X, kg/m³	X, %	CO_2, g/m³	Ureal	Uideal	Δ
1	15	0.0104	81	0.75	0.57	0.75	0.18
2	15.25	0.0106	81	0.76	0.6	0.75	0.15
3	15.5	0.0108	81.1	0.775	0.62	0.75	0.13
4	15.75	0.011	81.6	0.8	0.68	0.75	0.07
5	16	0.0112	82	0.8	0.69	0.75	0.06
6	16.15	0.0113	82.2	0.802	0.7	0.75	0.05
7	16.3	0.0115	82.3	0.808	0.72	0.75	0.03
8	16.5	0.0116	82.5	0.81	0.93	1	0.07
9	16.7	0.0117	82.7	0.815	0.95	1	0.05
10	16.9	0.0118	82.9	0.82	0.98	1	0.02
11	17	0.012	83	0.825	0.98	1	0.02
12	17.15	0.0122	83.8	0.83	0.975	1	0.025
13	17.3	0.0124	84.4	0.835	0.94	1	0.06
14	17.5	0.0127	85	0.84	0.92	1	0.08
15	17.7	0.013	86	0.845	0.7	0.75	0.05
16	18	0.0134	87	0.855	0.68	0.75	0.07
17	18.3	0.0138	88	0.865	0.67	0.75	0.08
18	18.5	0.0141	89	0.87	0.65	0.75	0.1
19	18.7	0.0144	90	0.875	0.64	0.75	0.11
20	19	0.0148	90.5	0.885	0.61	0.75	0.14

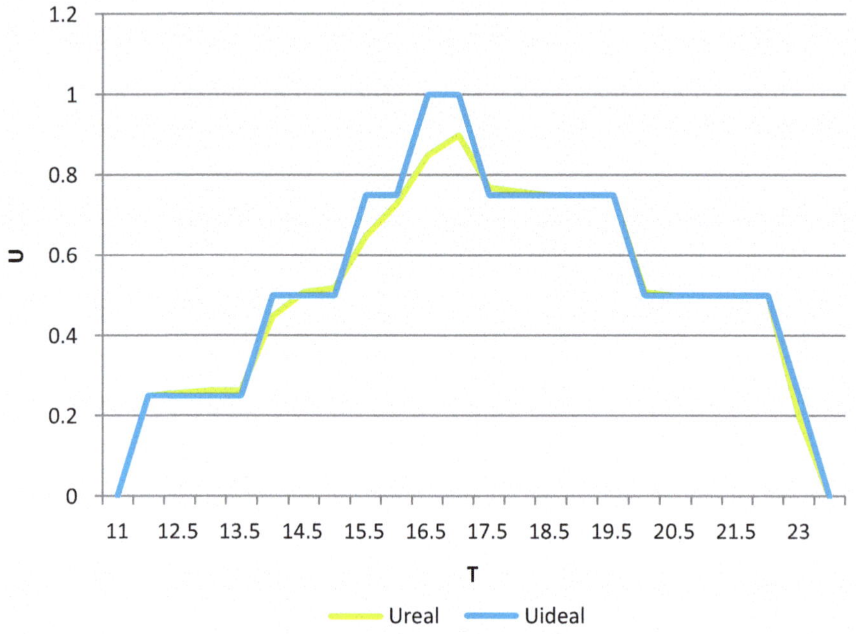

Fig. 8 Dependence of yield (U) on temperature (T, °C)

The research of yield changes according to the carbon dioxide and temperature parameters is presented in Table 5. At the same time, the humidity parameter remains unchanged, and equal to 83%.

The research of yield changes according to the carbon dioxide concentration and temperature is presented in Table 6. In this case, the parameter carbon dioxide concentration is unchanged, and equal to 0.825 g/m^3.

The tables demonstrate that with one constant microclimate parameter, the yield changes because the other two parameters are not constant. Therefore, all three parameters affect the yield.

According to the temperature change, the coolant consumption rate changes. As the temperature in the greenhouse increases, consumption becomes lower (Fig. 11).

According to the change in humidity, the steam consumption indicator changes. As the humidity in the greenhouse increases, the steam consumption becomes lower (Fig. 12).

As the carbon dioxide concentration in the greenhouse increases, the CO_2 emission from the compost (V_{CO_2}) decreases (Fig. 13).

The CO_2 sensors measure the value of this parameter. It is not recommended to highlight this CO_2 parameter. This is indicated by the parameter V_{CO2}. The CO_2 emission from the compost is slowed down by the work of the supply and exhaust

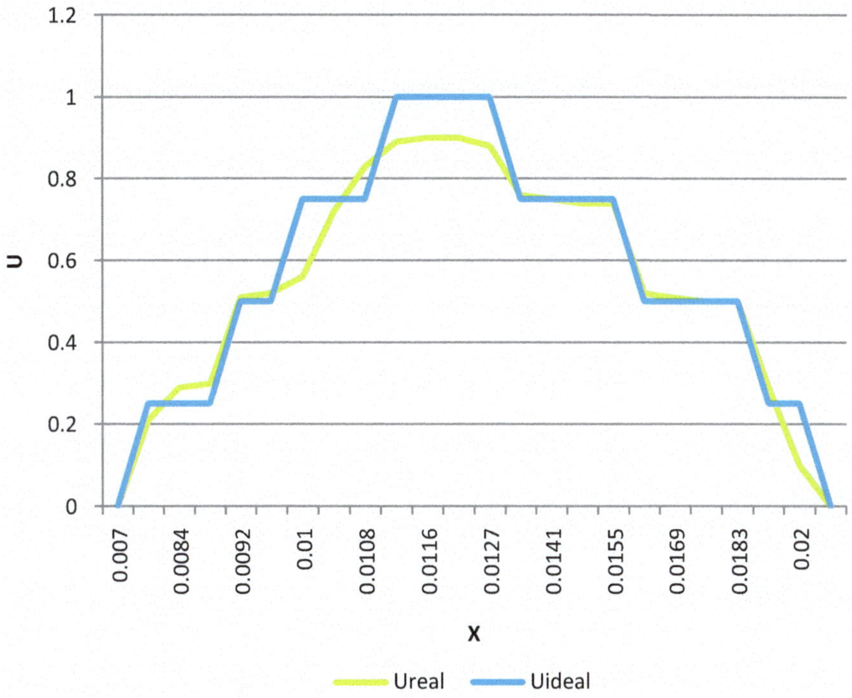

Fig. 9 Dependence of yield (U) on humidity (X, kg/m^3)

ventilation systems. As can be seen from the table data and figures, when critical high CO_2 values are exceeded, which are not desirable for oyster mushroom cultivation, the parameter of CO_2 emission from the compost decreases. That is, the neural network signals the necessity to remove it. Therefore, there are supply and exhaust ventilation systems that include ventilators to reduce CO_2 values.

To estimate the quality of neural network training, a linear loss function was used. A loss function is a function that describes an event or a value of one or more parameters; a real number that intuitively represents some "expenses" associated with that event. The optimization task attempts to minimize the loss function.

The loss function provides a numerical estimate of the value by which the prognosis deviates from the actual values. The linear loss function (MAE) is defined by the formula:

$$L(y, y') = |y' - y|, \tag{13}$$

where y is the desired output; y′ is a real way out.

As can be seen, the error of the neural network decreases during training on 50 epochs (Fig. 14).

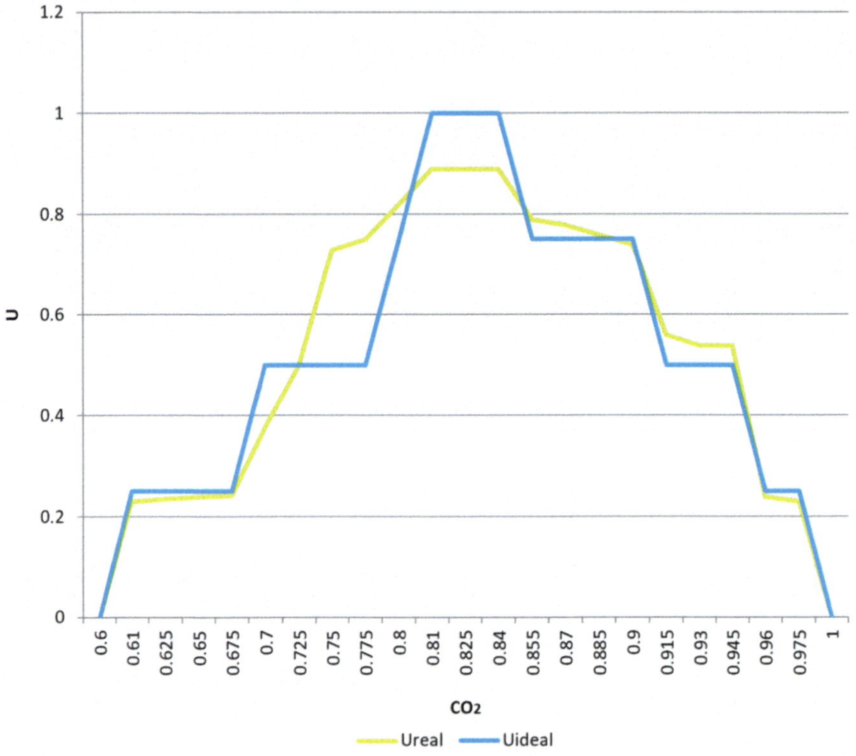

Fig. 10 Dependence of yield (U) on carbon dioxide volume (CO_2 concentration, g/m^3)

When the program is running, it is necessary to first enter the temperature, humidity, and carbon dioxide value in the air. The neural network generates a forecasted yield value in the range from 0 to 1 for the given conditions. Recommendations for maintaining optimal conditions in the greenhouse are also displayed: coolant consumption, steam consumption, and carbon dioxide emissions.

It was estimated that the highest yield would correspond to 1500 kg, with a greenhouse enclosure area of 100 m^2, 500 mushroom blocks, and a minimum oyster mushroom harvest maturation time of 2 months.

Based on the data from the temperature, humidity, and carbon dioxide sensors, the neural network provides recommendations for correcting these parameters (using actuators) and the forecast value for the current yield. This reduces energy consumption, speeds up and optimizes the process of analyzing input and output data, and thus increases the probability of obtaining a high yield in the short term. The data calculated by the formulas in the mathematical model are received and represented as a file with samples of values. The software has been designed that uses machine learning, soft computing (artificial neural networks), and mathematical modeling methods.

Table 4 Research of yield changes following the parameters of carbon dioxide concentration and humidity

X, kg/m^3	CO_2, g/m^3	U
0.008	0.6	0.069
0.0084	0.625	0.11
0.0088	0.65	0.12
0.0092	0.675	0.12
0.0096	0.7	0.17
0.01	0.725	0.27
0.0104	0.75	0.49
0.0108	0.775	0.65
0.0112	0.8	0.82
0.0116	0.81	0.895
0.012	0.825	0.9
0.0127	0.84	0.88
0.0134	0.855	0.59
0.0141	0.87	0.58
0.0148	0.885	0.57
0.0155	0.9	0.57
0.0162	0.915	0.26
0.0169	0.93	0.26
0.0176	0.945	0.25
0.0183	0.96	0.12
0.019	0.975	0.11

Table 5 Research of yield changes according to carbon dioxide concentration and temperature

T, °C	CO_2, g/m^3	U
12	0.6	0.064
12.5	0.625	0.06
13	0.65	0.062
13.5	0.675	0.065
14	0.7	0.14
14.5	0.725	0.25
15	0.75	0.38
15.5	0.775	0.48
16	0.8	0.61
16.5	0.81	0.85
17	0.825	0.9

(continued)

Table 5 (continued)

T, °C	CO_2, g/m^3	U
17.5	0.84	0.76
18	0.855	0.59
18.5	0.87	0.58
19	0.885	0.58
19.5	0.9	0.57
20	0.915	0.25
20.5	0.93	0.26
21	0.945	0.25
21.5	0.96	0.13
22	0.975	0.13

Table 6 Research of yield changes according to carbon dioxide concentration and temperature

T, °C	X, kg/m^3	U
12	0.008	0.093
12.5	0.0084	0.13
13	0.0088	0.13
13.5	0.0092	0.13
14	0.0096	0.22
14.5	0.01	0.27
15	0.0104	0.36
15.5	0.0108	0.55
16	0.0112	0.72
16.5	0.0116	0.86
17	0.012	0.9
17.5	0.0127	0.75
18	0.0134	0.75
18.5	0.0141	0.58
19	0.0148	0.57
19.5	0.0155	0.56
20	0.0162	0.26
20.5	0.0169	0.25
21	0.0176	0.25
21.5	0.0183	0.25
22	0.019	0.21

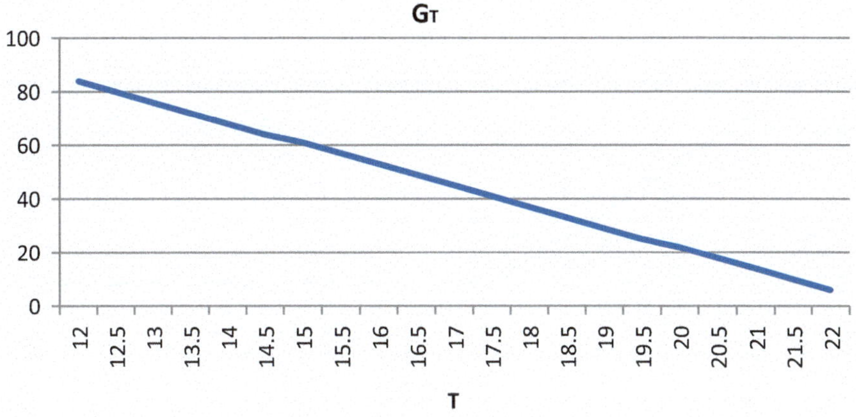

Fig. 11 Dependence of coolant consumption G_T (%) on temperature T (°C)

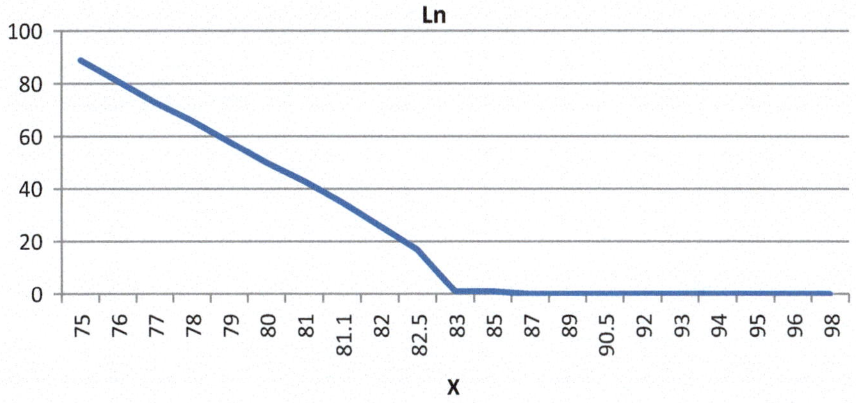

Fig. 12 Dependence of steam consumption L_n (%) on humidity X (%)

Almost all of the existing computer programs are designed to automate the process of controlling microclimate parameters and only allow to control microclimate parameters and send command signals to the actuators. The newly developed methods and control algorithms solve the task of forecasting yield and ensure adaptation to changes in influencing factors. This increases the probability of high yields. And this provides greater potential for efficient energy management of systems that include the appropriate software. The corresponding level of the results is documented in publications and conferences [38–46].

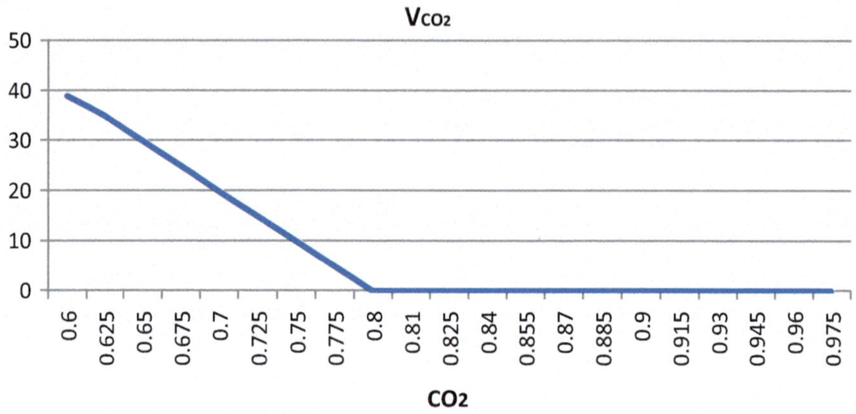

Fig. 13 Dependence of carbon dioxide emission (V_{CO_2}, %) on the value of carbon dioxide concentration (CO_2, g/m^3)

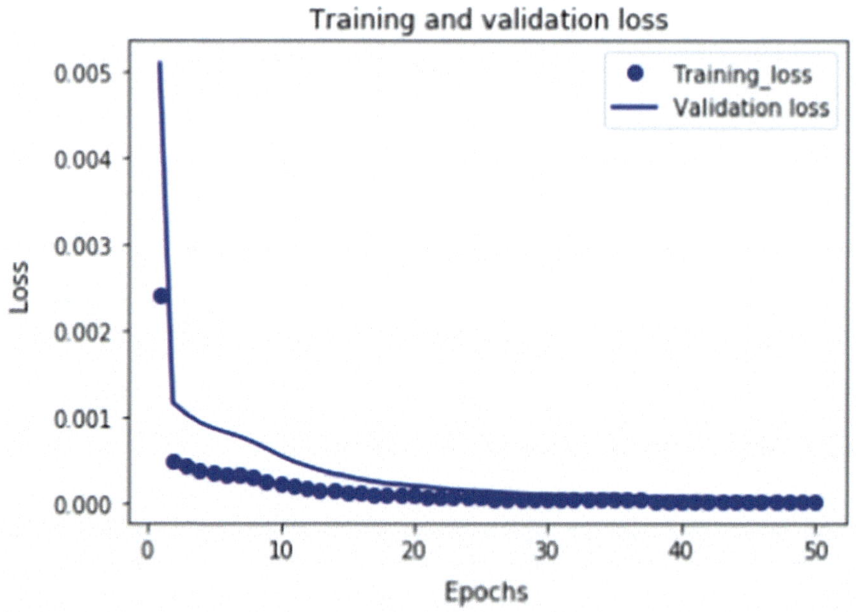

Fig. 14 Dependence of neural network error on the number of training epochs

References

1. Chen, L., Qian, L., Zhang, X., Li, J.Z., Zhang, Z.J., Chen, X.M.: Research progress on the indoor environment of mushroom factory. Int. J. Agric. & Biol Eng **15**(1), 25–32 (2022). https://doi.org/10.25165/j.ijabe.20221501.6872
2. Royse, D.J.: A global perspective on the high five: Agaricus, Pleurotus, Lentinula, Auricularia & Flammulina. In: International Conference on Mushroom Biology and Mushroom Products, pp. 2010–2015 (2014)
3. Wu, C.R., Cheng, R.F., Fang, H., Yang, Q.C., Zhang, C.: Simulation and optimization of air tube ventilation in plant factory based on CFD. J. China Agric. Univ. **26**(1), 78–87 (2021)
4. Li, C.T., Tan, Q., Bian, Y.B., Xie, B.G., Liu, Z.Q., Li, Y.: The status and prospection of edible mushroom industry in China. J. Fungal Res. **17**(1), 1–10 (2019)
5. Sakamoto, Y.: Influences of environmental factors on fruiting body induction, development and maturation in mushroom-forming fungi. Fungal Biol. Rev. **32**(4), 236–248 (2018)
6. Yang, D., Liang, J., Wang, Y., Sun, F., Tao, H., Xu, Q., et al.: Tea waste: an effective and economic substrate for oyster mushroom cultivation. J. Sci. Food Agric. **96**(2), 680–684 (2016)
7. Grimm, D., Wösten, H.A.B.: Mushroom cultivation in the circular economy. Appl. Microbiol. Biotechnol. **102**(18), 7795–7803 (2018)
8. Takakura, T.: Research exploring greenhouse environment control over the last 50 years. Int. J. Agric. Biol. Eng. **12**(5), 1–7 (2019)
9. Jang, K.-Y., Jhune, C.-S., Park, J.-S., Cho, S.-M., Weon, H.-Y., Cheong, J.-C., et al.: Characterization of fruitbody morphology on various environmental conditions in Pleurotus ostreatus. Mycobiology **31**(3), 145 (2003). https://doi.org/10.4489/MYCO.2003.31.3.145
10. Xie, C., Zhu, X., Ning, X., Xing, K.: Design of remote monitoring system for eryngii greenhouse environment in Southern Ningxia. In: IEEE International Symposium on Industrial Electronics, pp. 1–5. IEEE, Taipei (2013). https://doi.org/10.1109/ISIE.2013.6563599
11. Zhao, L., Zhu, X.J.: The development of remote monitoring system for cultivation environment of Pleurotus eryngii. In: 2015 IEEE International Conference on Information and Automation (ICIA), pp. 2643–2648. IEEE, Lijiang (2015). https://doi.org/10.1109/ICInfA.2015.7279731
12. Kochhar, A., Kumar, N.: Wireless sensor networks for greenhouses: an end-to-end review. Comput. Electron. Agric. **163**, 104877 (2019). https://doi.org/10.1016/j.compag.2019.104877
13. Zervopoulos, A., Tsipis, A., Alvanou, A.G., Bezas, K., Papamichail, A., Vergis, S., et al.: Wireless sensor network synchronization for precision agriculture applications. Agriculture **10**(3), 89 (2020). https://doi.org/10.3390/agriculture10030089
14. Lopez-Iturri, P., Celaya-Echarri, M., Azpilicueta, L., Aguirre, E., Astrain, J.J., Villadangos, J., et al.: Integration of autonomous wireless sensor networks in academic school gardens. Proceedings **18**(11), 108 (2018). https://doi.org/10.3390/ecsa-4-04899
15. Ganchev, I., Ji, Z., O'Droma, M.: Designing a low-cost data transfer unit for use in IoT applications. In: 2016 8th International Congress on Ultra Modern Telecommunications and Control Systems and Workshops (ICUMT), pp. 85–88. IEEE, Lisbon (2016). https://doi.org/10.1109/ICUMT.2016.7765337
16. Kassim, M.R.M., Mat, I., Yusoff, I.M.: Applications of internet of things in mushroom farm management. In: 2019 13th International Conference on Sensing Technology (ICST), pp. 1–6. IEEE, Sydney (2019). https://doi.org/10.1109/ICST46873.2019.9047702
17. Marzuki, A., Ying, S.Y.: Environmental monitoring and controlling system for mushroom farm with online interface. Int. J. Comput. Sci. Inf. Technol. **9**(4), 17–28 (2017)
18. Kassim, M.R.M., Harun, A.N., Yusoff, I.M., Mat, I., Kuen, C.P., Rahmad, N.: Applications of wireless sensor networks in Shiitake Mushroom cultivation. In: 2017 Eleventh International Conference on Sensing Technology (ICST), pp. 1–6 (2017). https://doi.org/10.1109/ICSensT.2017.8304516
19. Hu, C., Zhu, X., Zhou, Y.: The use of NoSQL in product traceability system construction. In: 2018 5th International Conference on Information Science and Control Engineering (ICISE), pp. 574–577. IEEE, Zhengzhou (2018). https://doi.org/10.1109/ICISCE.2018.00124

20. Tongcham, P., Supa, P., Pornwongthong, P., Prasitmeeboon, P.: Mushroom spawn quality classification with machine learning. Comput. Electron. Agric. **179**, 105865 (2020). https://doi.org/10.1016/j.compag.2020.105865
21. Cruz-Del Amen, J.D., Villaverde, J.F.: Fuzzy logic-based controlled environment for the production of oyster mushroom. In: 2019 IEEE 11th International Conference on Humanoid, Nanotechnology, Information Technology, Communication and Control, Environment, and Management (HNICEM), pp. 1–5. IEEE Publishing (2019). https://doi.org/10.1109/HNICEM48295.2019.9072902
22. Hendrawan, Y., Anta, D.K., Ahmad, A.M., Sutan, S.M.: Development of fuzzy control systems in portable cultivation chambers to improve the quality of oyster mushrooms. IOP Conf. Ser.: Mater. Sci. Eng. **546**(3), Article 032013 (2019). https://doi.org/10.1088/1757-899X/546/3/032013
23. Kassim, M.R.M., Mat, I., Yusoff, I.M.: Applications of internet of things in mushroom farm management. In: 2019 13th International Conference on Sensing Technology (ICST), pp. 1–6. IEEE Publishing (2019). https://doi.org/10.1109/ICST46873.2019.9047702
24. Najmurrokhman, A., Kusnandar, Daelami, A., Nurlina, E., Komarudin, U., Ridhatama, H.: Development of temperature and humidity control system in Internet-of-Things based oyster mushroom cultivation. In: 2020 3rd International Seminar on Research of Information Technology and Intelligent Systems (ISRITI), pp. 551–555. IEEE Publishing (2020). https://doi.org/10.1109/ISRITI51436.2020.9315426
25. Wahab, H.A., Manap, M.Z.I.A., Ismail, A.E., Ong, P., Ismon, M., Zainulabidin, M.H., Noor, F.M., Mohamad, Z.: Investigation of temperature and humidity control system for mushroom house. Int. J. Integr. Eng. **11**(6), 27–37 (2019). https://doi.org/10.30880/ijie.2019.11.06.004
26. Grant J.: Modelling airflow in Irish mushroom growing tunnels. In: II IFAC/ISHS Workshop: Mathematical & Control Applications in Agriculture & Horticulture (Acta Hortic), vol. 406, pp. 399–408 (1996). https://doi.org/10.17660/ActaHortic.1996.406.40
27. Ariffin, M., Azizi, M., Ramli, I., Zainol, Z., Amin, M., Nazrul, M., Ismail, M., Adnan, R., Ahmad, N., Husain, N., Jamil, N.: Enhanced IoT-based climate control for oyster mushroom cultivation using fuzzy logic approach and NodeMCU microcontroller. Pertanika J. Sci. Technol. **29**(4) (2021). https://doi.org/10.47836/pjst.29.4.34
28. Ariffin, M.A.M., Ramli, M.I., Amin, M.N.M., Ismail, M., Zainol, Z., Ahmad, N.D., Jamil, N.: Automatic climate control for mushroom cultivation using IoT approach. In: 2020 IEEE 10th International Conference on System Engineering and Technology (ICSET). IEEE Publishing (2020), pp. 123–128. https://doi.org/10.1109/ICSET51301.2020.9265383
29. Alpay, Ö., Erdem, E.: The control of greenhouses based on fuzzy logic using wireless sensor networks. Int. J. Comput. Intell. Syst. **12**(1), 190–203 (2018). https://doi.org/10.2991/ijcis.2018.125905641
30. Developing a Smart Greenhouse: EPAM Wins Oracle Prize. EPAM System Inc. (2019). https://www.epam.com/our-work/case-studies/developing-a-smart-greenhouse-epam-wins-oracle-prize. Accessed 02 Mar 2023
31. Peshko, M.S.: The revealed mathematical model of the microclimate of the mushroom greenhouse. Young Sci. **9**(32), 42–48 (2011)
32. Koshkin, D.L.: Simulation model of the system for controlling the microclimate parameters in mushroom greenhouse, TSATU **2**(15), 230–236 (2015)
33. HDC1080 low power, high accuracy digital humidity sensor with temperature sensor. Texas Instruments Incorporated (2016). http://dl.z-thz.com/eBook/zomega_ebook_pdf_1206_sr.pdf. Accessed 02 Mar 2023
34. Wilo-Stratos-D. Wilo (2017). https://wilo.com/il/en/Solutions-Finder/Selection-and-configuration/Heating-replacement-guide/en/wilo/wilo-stratos-d. Accessed 02 Mar 2023
35. Vdoh-Nova humidification system: TM "Vdoh-Nova" (2017). https://www.truba.ua/f/vdohnova/photo-6703. Accessed 02 Mar 2023
36. VENTS CO2 sensor: Vents company (2017). https://vents.ua/series/co2. Accessed 02 Mar 2023
37. CO2 Transmitter: Automated environmental control systems in composting and mushroom growing. TERA Company (Ukraine) (2018). https://mushroom-climate.com/en/co2-transmitter/. Accessed 02 Mar 2023

38. Morozova, M.: Evolutionary computation for engineering design tasks. In: XVI International Young Scientists' Conference on Applied Physics, Kyiv, Ukraine, June 2016, pp. 165–166
39. Morozova, M., Gusar, V.: Neural network technologies for solving ecological problems of the agro-industrial complex. Equip. Technol. AIC **4**(103), 19–22 (2018)
40. Morozova, M.: Possible combinations of genetic algorithms and neural networks. In: XVIII International Young Scientists' Conference on Applied Physics, Kyiv, Ukraine, May 2018, pp. 135–136
41. Morozova, M., Vasylieva, Y.: Using neural network for doing forecast and correcting the temperature error. In: XIV International Scientific Conference "Electronics and Applied Physics", Kyiv, Ukraine, October 2018, pp. 144–145
42. Morozova, M., Morozov, O.: Optimizing algorithm, based on reinforcement learning and neural network. In: XIV International Scientific Conference "Electronics and Applied Physics", Kyiv, Ukraine, October 2018, pp. 140–141
43. Vasylieva, Y.D., Morozova, M.M.: Neural network for controlling greenhouse microclimate parameters. In: XII All-Ukrainian Scientific and Practical Conference of Students, Postgraduates and Young Scientists "Look into the Future of Instrumentation Engineering", Kyiv, Ukraine, May 2019, pp. 429–432
44. Morozova, M.M.: Application of neural network technologies in design of intelligent systems for controlling microclimate parameters. In: Forty-First Practical and Cognitive Internet Conference "Scientific Thought of Present and Future", Dnipro, Ukraine, February 2021, pp. 6–10
45. Morozova, M.M.: Structure of neural network for yield forecasting and control of microclimate parameters. In: Forty-Third Practical and Cognitive Internet Conference "Scientific Thought of Present and Future", Dnipro, Ukraine, May 2021, pp. 3–8
46. Morozova, M.M., Morozov, O.Y.: Kompyuterna programa "HARVEST v.1". svidotstvo avtorskogo prava na tvir, № 003823/АП/21 vid 15.06.2021

The manufacturer's authorised representative in the EU is Springer Nature Customer Service Centre GmbH, Europaplatz 3, 69115 Heidelberg, Germany. If you have any concerns regarding our products, please contact ProductSafety@springernature.com

Printed and bound by CPI Group (UK) Ltd, Croydon, CR0 4YY
01/10/2026
02225855-0006